Highways and Byways of Beekeeping

Alan Wade

To Lynne
and to our children
Andrew and Alice

And to enlightened beekeepers
Des Cannon
past Editor of The Australasian Beekeeper
Frank Malfroy and Jenny Douglas
Beelinebeewares
Victor Croker and David Leemhuis
Australian Honeybee
Dave Flanagan
freelance beekeeper
Dannielle Harden and Peter Czeti
and Jenny and Peter Robinson
Canberra Region Beekeepers
Ian Wallis and Frank Derwent
fellow traveller beekeepers

Northern Bee Books

Highways and Byways of Beekeeping
Copyright © Alan Wade
Canberra Region Beekeepers

All rights reserved. No part of this publication may be reproduced, stored in a retrieval system, transmitted in any form or by any means electronic, mechanical, including photocopying, recording or otherwise without prior consent of the copyright holders.

ISBN 978-1-914934-90-2

Front cover photos:
Alan Wade
Black Chinese Honey Bee (*Apis cerana heimifeng*), or honey-feng, Apiary on the road to Xia He in Gansu Province, High Tibetan Plateau, 21 June 2015

Back cover photo:
Alan Wade
Tim Geoghagen

Published 2023 by
Northern Bee Books,
Scout Bottom Farm,
Mytholmroyd
West Yorkshire
HX7 5JS (UK)
Tel: 01422 882751 Fax: 01422 886157

Design and artwork DM Design and Print

Preface

Highways and Byways is about bees and some of the ways they have been best kept, lessons learnt by eternally curious and observant beekeepers. In that sense *Highways and Byways* is not a standard text on how to keep bees.

Having kept bees for rather too many decades, I find I am now and then surprised to observe an element of comb architecture, bee practice, or bee behaviour that I had never encountered before. Often in plain sight, I pondered what else I had missed or learnt from watching bees, from conversing engagingly with other beekeepers and from delving into old journals.

It is always helpful to build on what one already knows. With this in mind I decided to gather together my notes about bees and on beekeeping practice. The result, *Highways and Byways of Beekeeping* is an anthology of those writings. Some of those mind dumps have appeared in the Canberra Region Beekeepers newsletter *Bee Buzz Box* and have made it as far as the pages of *The Australasian Beekeeper*. All of these recollections were intended to inform practical beekeeping practice but, in the serendipitous way one follows bees, I often strayed more into writing about them rather than about how to keep them.

All this led to my discovering facts about bees that I would never have encountered from casual encounters with other beekeepers. It led me down many back lanes of beekeeping and ultimately into keeping bees in ways that I had never envisioned or intended to practice. One foray, keeping honey bees with an additional queen, particularly took my fancy.

Historically running bees with an extra queen has resulted in apiarists reporting spectacular honey crops. However with bees there is no free lunch: bees rarely conduct themselves in any predictable manner. This style of beekeeping is time expensive and fraught with difficulty – bees know best and operate most simply with a single queen. For me tackling two-queen beekeeping has proved wonderfully rewarding. Much of what I know about about the role of the queen bee and what to expect of her comes from that perhaps injudicious foray.

The focus of this book lies elsewhere, the pursuit of good beekeeping practice and discovering more about honey bee biology. In seeking to discover more, I have learnt a good deal about how bees other than the western honey operate and the origins of their social structure. This I have found both interesting and a base upon which to adjudge the merits of many novel apiary routines.

Highways Part I explores beekeeping practices learnt in large measure from the bees themselves. The opening essays examine the pivotal role of the honey bee queen and the nuanced interplay of swarming and colony requeening. *Highways* then ventures into various topics that bear upon the art of practical beekeeping. Principal among these are the long road to the evolution of the modern framed hive from the makeshift skep, the many facets of hive construction and design that impinge upon colony performance and the bee language of chemical communication. Part I concludes with a few notes on keeping track of bees and a nod to two-queen hive operation.

Byways heads down a different path. It delves deeper into the biology of honey bees other than those that can be kept in hives and lists the many pests and diseases that exploit the bounteous resources of the honey bees we keep. Together these are a marker for the nuanced performance of managed bees. *Byways* is capped off with a handful of bee keeper character sketches.

Highways and Byways of Beekeeping champions good beekeeping practice drawing in the contributions of sometimes little known but quite extraordinary beekeepers. It is humbling to also acknowledge that many improvements to beekeeping practice have arisen from a clear understanding of how bees operate unmolested in an old tree hollow.

Alan Wade
Christmas Day 2022

Contents

Introduction	vii
Part I — Highways	1

I   The Honey Bee Queen   4

Long live the queen	7
The heiress to the throne	39

II   The Honey Bee Swarm   58

Out of the box	65
Keeping the lid on the box	84

III   Requeening the Honey Bee Colony   98

Direct requeening	99
Requeening with swarm control	109

IV   Honey Bee Keeping   129

Art thou a skeppist?	130
Bee space and bee hive architecture	149
Going overboard	189
Let us spray: Honey bee pheromone chemistry	213
Keeping bees in good fettle	225
Keeping bee records	237
Rivers of honey	243

Part II Byways   253

I   Bee Back Lanes   255

Social organisation of ants, bees and wasps	256
Global distribution of honey bees	265
The giant honey bees	278
Phoretic honey bee mites	301
Beetle mania	313
Where rust and moth doth corrupt	321
Bee pests and diseases	325

II   Bee Keeper Sketches   335

E.W. Alexander –	
The North American inventor of the two-queen hive	336
George Wells – The British inventor of the doubled hive	339
Stan Hughston - Stringy Hughston's coffin hives	348
Don Peer – The wisdom of lost beekeeping practice	355
Tom Theobald – Where beekeepers fear to tread	365

Acknowlwedgements   369

Bibliography   369

Introduction

While I have kept bees for over four decades, other beekeepers, good friends, have kept them for as long as seventy years. By and large the people from whom I have learnt most from are commercial beekeepers. Far too preoccupied tending their bees and making a living to ever record their insightful management practices, I have found them more than willing to share their inside knowledge. It is to them, and from others of whom I have read, that made compiling this companion to beekeeping practice possible.

One unsung beekeeper, Dave Flanagan, was particularly helpful. These days Dave lives in Wyalong well west of here. Des Cannon – and former editor of *The Australasian Beekeeper* – and Dave led me to the remote Channel Country hamlet of Wanaaring where Stringy Hughston had run a truly crazy setup, 2400 tripled hives – colonies with three queens sharing common coffin supers above queen excluders: Dave had worked on his operation as a young man over thirty years ago. *Highways and Byways of Beekeeping* recounts or alludes to several such back-lane ventures.

Another inspiration for compiling this anthology of beekeeping practice came from reading about pioneer beekeepers, those who threw out the rule book and, though now hardly remembered, kept bees in entirely new ways. Amongst these there were those who had corresponded generously with journal editors and who had shared their inventions or those that were lent upon heavily to reveal their secrets.

One such gentleman was George Wells, the late 19th Century inventor of the doubled hive, another E. W. Alexander, a prodigious inventor and pioneer of multiple queen hives, both of whom more anon.

Then there were also those of the mid 20th Century who left an impressive trail of applied research discovery. Particularly notable were those that came out of the Rothamsted Research Centre in the United Kingdom and a bevvy of researchers in Clayton Farrer, Winston Dunham and John Holzberlein in America. Amongst the UK researchers of high distinction were James Simpson who discovered most of what we know about swarming and the centenarian Colin Butler who conceptualised, discovered and synthesised the principal honey bee queen bee pheromone 9-ODA.

The new findings of most beekeepers are few: my only claim to discovery has been confined to observations of a few atypical behaviours of colonies sporting two and occasionally three queens. Perhaps there is nothing new under the sun, but finding the wisdom of others, not least listening to what the bees are telling us, is truly instructive. For it is the bees, rather more than their keepers, that do most of the work and know best how to go about it.

Highways has several distinctive themes, some key facets of bee biology as it instructs beekeeping practice and an outline of how the modern frame hive arrived and a few key factors that both keep the bees and keeper of bees in good order.

Byways is more discursive and less instructive on the matter of routine management of bee hives. In one sense diving deeper into general biology of honey bees – as *Byways* attempts to do – sets the boundaries to our understanding of honey bees, not least from what we learn from those bees that cannot be kept in the box. In *Byways* Part I we visit bees other than *Apis mellifera* and we touch on the pests and diseases intent on making a meal of honey bee labour. In Part II we glimpse the endeavours of a handful of beekeepers whose endeavours are so instructive.

Part I
Highways

Highways of Beekeeping sets out with the story of the honey bee queen and swarming, two interrelated and key elements of the biology of honey bees. An understanding of how they relate is central to optimising the performance of managed honey bees. Highways also focuses on the origins of the modern hive, its critical design features, the means by which bees communicate and other facets of keeping bees, an understanding of which is inimical to good beekeeping practice.

Over all the many years I have kept bees I have always been faced with one nagging question: 'How is it possible for bees in a hive that I opened only a few short weeks ago to now be in such a different condition?' It takes both a bee and keeper of bees perspective to understand just how dynamic the honey bee colony is and how quickly things, like the weather outside, can change.

The extrinsic factors governing the 'success of honey bees' are determined by the soundness of the nesting cavity and the intertwined climatic, weather and plant flowering conditions external to the nest. The intrinsic factors that govern the performance of honey bees inside the nest are those of nutrition, the prevalence of predators, pests and disease, the calibre of the queen, and the propensity of bees to swarm, themselves all influenced in varying degree by conditions beyond the hollow or hive in which the the colony of bees reside.

Failing to produce a bumper crop – poor conditions and a paucity of nectar bearing flowers aside – boils down to one pervading factor, colonies headed by poor queens. It is as simple as that. For the honey bee colony is an expression of that one bee, the queen, and the many drones with whom she has mated.

At an apiary or wild population level, however, ascertaining the condition of bees cannot be distilled to the makeup and performance of a single honey bee colony. In assessing the success of honey bees surviving in the wild, or the beekeeper succeeding in her or his beekeeping endeavour, one needs also to accommodate the confounding effects of swarming and myriad external factors impinging on the general bee population.

In an apiary the ways bees are kept and managed will largely determine how well the average colony performs. In the wild, the ability of bees to fend off disease and predators, to bud off new colonies and to collect sufficient stores to overwinter will be the key determinants of their success. The latter might be better measured in terms of the number of colonies present in a district from year to year.

I   The Honey Bee Queen

In assessing the role of the honey bee queen we can start by examining the origins of the ancestral single queen hive, moving onto the natural mechanisms of queen replacement attendant to both colony renewal and colony reproduction. *The honey bee queen* concludes with an overview of conventional requeening practice extended to some very sophisticated requeening techniques involving purposeful swarm control and colony building employing a second queen.

II   The Honey Bee Swarm

The honey bee swarm examines the underlying nature of swarming and the biological purpose it serves across the honey bee genus.

III   Requeening the Honey Bee Colony

In *Requeening the honey bee colony* we turn to the gamut of conventional swarm control practices and the limits presented by the natural bee instinct to reproduce.

We then visit extraordinary accounts of advanced swarm control practice, a 1952 swarming roundup forum followed by an incisive 1954 panel that outlined the use of two queens to effect automatic requeening, swarm control and colony buildup.

IV   Honey Bee Keeping

Honey bee keeping starts with an account that reflects the transition from traditional skep hives to frame hives and the enhanced benefit incurred if hives are also well insulated and well designed. We then take a fleeting glance at honey bee chemistry (most bee behaviour is mediated by bee pheromones and olfactory sensing) and dive into a few basics of keeping practice, making sure bees are always well provisioned and having spare queens at hand to replace those that are patently failing. Floral resources and siting of hives aside, keeping your bees as safe as houses is about the most a keeper of honey bees can do to ensure that bees thrive.

I
The Honey Bee Queen

To a large extent, the conditions under which a queen is raised, her genetic makeup, and that of the drones she has mated with, will be the key determinants of how well a honey bee colony performs and how productive her colony will be. For the hollow tree wilding hive, the capacity of a queen to produce both daughter queens and drones and to cast swarms – the ability to contribute to perpetuate honey bee populations – will be the arbiter of her success.

> *The king was in the counting-house*
> *Counting out his money*
> *The queen was in the parlour*
> *Eating bread and honey.*
>
> **Sing a song of Sixpence**
> Mother Goose

For the apiarist, a prolific queen whose progeny are efficient harvesters of pollen and nectar, whose offspring practice hygienic behaviour and who are good comb builders are the key sought after attributes. Queen traits required for honey bee survival and those required for productive beekeeping need not be the same. Honey bee colonies, performing well in an apiary, are not those needed for bees to best defend themselves or perpetuate their line in the woods.

Hives headed by young queens are more productive and swarm less often than do unmanaged or wild hives.

Thomas Seeley pointedly notes that beekeepers have built artificial nesting cavities – we call them hives – to a size formula rather than that which best suits the needs of bees. Most custom made hives are poorly insulated and are often fitted with entrances that do not allow bees to best defend themselves. The pioneering efforts of Seeley and Morse in determining the natural choice of nesting cavity and those of Derek Mitchell in characterising the insulating properties of hive materials are salient.

Until now, we have not been fully cognisant of hive architectures that benefit both bees and beekeepers.

A right royal introduction

The common view of the honey bee queen is that of a monarch attended to by a bevy of slaves, better termed 'workers'. Alongside this queen are a smattering of couch potatoes often designated as 'lazy drones'. The reality is that the queen, workers and drones are all integral to the functioning of the colony, none fully in control of, but each dependent on, each other. We return elsewhere to chemical signalling as the prime method of communication, pheromones that cement their relationships and cohesiveness.

This said, a distinguishing feature of the queen is that she is the mother of all the progeny of the hive. There are of course exceptions: a few bees, mainly drones, drift to join hives not of their natal origin. The queen's other functions are to maintain a measure of hive homeostasis, to ensure that rival queens are not raised, to regulate the number of drones the colony deems it needs and to ensure that she herself is closely attended to and maintains her position as the sole egg layer:

> *You can fool all of the people some of the time,*
> *And some of the people all of the time,*
> *But you can't fool mom.*

Captain Penny's Law

In many respects the queen is, like all her subjects, only part of the colony fabric. Within the limits of her makeup she must produce as many offspring as the colony can raise and must accompany any swarm departing the nest. In large measure these functions are controlled by the colony, not by the queen. The process of collective decision making, outlined in Seeley's *Honeybee Democracy*, are essential to the complex operation of the hive.

Despite the shared decision making expressed by the superorganism colony, the queen is nevertheless the sole source of new generations of bees. The denizens of the hive, all her progeny are themselves only an expression of her genetic makeup and that of the drones she has mated with. The selective pressures the environment places on the colony are channeled through the queen to the next generation, the queens and drones she gives birth to. In that sense the queen is preeminent, but in the everyday working routine of a honey bee colony, she is but an eminent member of that putative royal family.

The honey bee queen tracks the origins of eusociality and the single queen hive, the natural processes of queen succession and the lessons learnt in how best to effect requeening.

Long live the queen[1]

A woodland bee colony has a peak summer population of around 20,000 worker bees and, for much of the year, supports a population of around 200 drones and always one queen. Except for a few drifting bees, all bees in residence are the progeny of this solitary colony queen.

Members of this honey bee community are routinely replaced and the queen is no exception. All bees age, can be weakened by disease or may die accidentally. Planned queen succession, like other hive member replacements, is essential for colony survival.

A tree on the author's property (Figure 1.1) has harboured one and occasionally two honey bee nests intermittently over the past forty years. How long bees may have occupied the same nest cavities is anyones guess but it may have been from the earliest settlement of Canberra around 1824 as the tree is likely two to three hundred years old and probably predating European settlement.

Figure 1.1 Nest entrance to honey bees in a two-to-three hundred year old Red Box (*Eucalyptus polyanthemos*) in the suburb of Kambah in the Australian Capital Territory.

Claims that a beekeeper should never requeen his or her hive and that the same queen has been there for many years are easily refuted. Bees routinely orchestrate queen departure then installing a daughter queen and do this for a variety of reasons. Replacement is normally achieved by swarming – where the old queen departs and a new queen takes up residence or by supersedure where the queen is simply replaced.

For wild colonies annual queen replacement is the likely norm. Sometimes bees will replace their queen more frequently. A colony will change out its queen whenever it performs unsatisfactorily, say due to age or miss-mating or where queen egg-laying capacity drops off precipitously. If, on the other hand, a colony grows quickly it may swarm, or swarm repeatedly, a large proportion of its productive bees departing and taking with them colony stores. The hive will then need a replacement queen, one supplied by a ripe swarm cell, a luxury not extended to the swarm if its queen fails to establish a fledgling colony.

In extreme cases, colonies may 'swarm themselves to death', that is nearly all bees abscond with swarms. In the more normal circumstance, the bee population will be rebuilt and stores made up to enable the parent colony to survive through winter and into the next season. Swarms – when they survive – make up for colony losses. So the number of wild colonies in an undisturbed landscape remains fairly constant, constrained only by the number of nesting sites and the availability of forage. Seeley[2] reports that of forest honey bee colonies in New York State:

> ...most (97%) survive summers, but only 23% of founder colonies and 84% of established colonies survive winters. Established colonies have a mean lifespan of 5–6 years and most (87%) have a queen turnover, probably by swarming, each summer.

Seeley also noted that the colony density in this local study area had returned to pre-*Varroa* levels – that is after initially collapsing – signalling the resilience of bees faced with new diseases and periodic drought. So while a colony may appear to be potentially immortal, in practice any unmanaged colony will succumb on average in half a decade but then be quickly replaced by another swarm.

The role of the honey bee queen

The condition of any regular honey bee colony is that of a single long-lived queen working with a large retinue of bees, her offspring. Together they occupy a hollow cavity of typically of 30-60 litres capacity (Figure 1.2) with a preference for a small hive entrance[3]. The dynamic living component, including bees combs and stores, is a colony of bees often designated the term superorganism[4]. That term can be equally employed to other eusocial insects such as stingless bees[5] as well as to ants and to rather less social vespid wasps: paper wasps, hornets and yellow jackets.

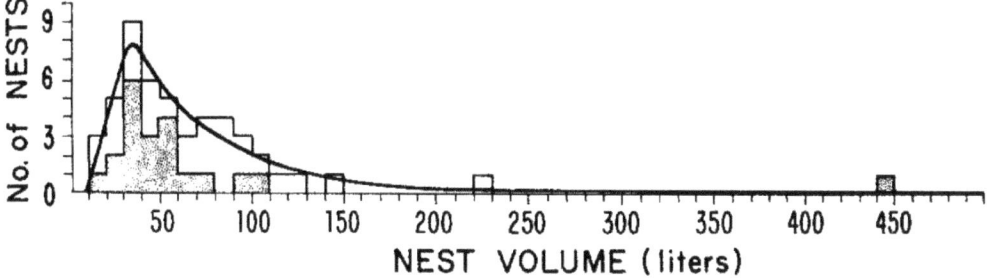

Figure 1.2   Seeley's plot of nest volume selection by wild honey bees.

Such is the specialisation of these most advanced of bees that they have ceded all of their reproductive rights to just the one individual, the queen. This is an over simplification of the complex eusocial relationship the queen has to other members of the colony. Wikipedia defines eusociality[6] as:

> *the highest level of organisation of animal sociality characterised by cooperative brood care, overlapping generations within a colony of adults and division of labour into reproductive and non-reproductive groups.*

While worker bees are essentially sterile, it is interesting to note that ancestral monogamy, essential to the rise of such a specialised role for the queen and for the evolution of eusocial behaviour, has since been modified by polyandry (multiple drone mating of the honey bee queen)[7]. This confers advantages for colony survival where workers altruistically raise workers and new queens less related to themselves. However, we will see that while the scenario of colonies headed by more than one queen have been actively selected against, it is not uncommon to find a second queen during periods of queen transition. The presence of one or more additional daughter queens is ephemeral, the colony universally reverting to the single queen condition.

These evolutionary developments confer survival advantages for honey bees in general but, at the level of the individual colony, it poses serious risk. On the positive side of the ledger large numbers of insects working cooperatively, foraging over areas that may exceed fifty square kilometres, are a formidable force. The capacity of such a large aggregate of insects to defend their colony is similarly well-renowned. This defence is exemplified by simultaneous attacks being launched by multiple colonies of the open nesting giant honey bees in response to disturbance, hence the aptly titled Megapis Syndrome[8].

It follows that to look at a honey bee colony in an apiary as an isolated entity is to ignore the dynamic effect of its interaction with other colonies, predators and competition for food resources. Hepburn[9] argues that foraging range of individual colonies is governed by inter-colony competition for floral resources. In periods of dearth, many tropical honey bees, particularly weak colonies, abscond or migrate while stronger colonies, at least amongst some species, avoid relocation by expanding their foraging range. This results in lowered competition resulting from colonies forced to migrate. Artificial provisioning of colonies with pollen supplements and sugar has been found to help circumvent absconding behaviour in managed *Apis cerana* colonies.

However, taken from the perspective of their being just one female – the queen – supporting a large workforce, any single honey bee colony is extremely vulnerable. Further, since the condition of that queen will in large part determine colony health and vigour, it is important that the worker bee community protect her, maintain her in optimum condition and, in the event of her failing, have a sound and flexible strategy for her replacement. Evolution has come a long way in providing a range of pathways for colony survival and has devised more than one mechanism for queen replacement near perfecting its self queening and survival strategy.

The functional value of the queen

Under favourable spring conditions, a highly fecund European honey bee queen laying up to 1800 eggs a day will result in rapid colony expansion. That, and a program of worker bee replacement, ensures that the maximum amount of energy stores (honey) and body building protein (pollen) are harvested during often remarkably short honey and pollen flows. It also means that the number of bees, both workers and drones, can be down-regulated[10], that is safely retired, under conditions of dearth either during drought or, in temperate regions, coming into winter.

An alternative strategy, adopted by all tropical bees in the *Apis* genus, is one where colonies invest most of their energies in swarming profligately under good conditions rather than banking on storage of honey and pollen or indeed in maintenance of a large workforce[11]. They then abscond under deficit conditions to search for alternative floral sources or seasonally migrate to known reliable stands of flowering plants. This ensures that many small colonies, each with its own queen, are produced in abundance and that some colonies, at least, will survive.

The queen and worker bees signal their ever changing hive condition and defence requirements by means of complex chemical messaging. Honey bees have many glands that serve utilitarian purposes such as wax secretion through to roles in releasing messenger pheromones. They include Dufour's, head salivary, hypopharyngeal, Koschevnikov, mandibular, Nasonov, thoracic, tarsal, salivary, venom and wax glands.

Especially important amongst the pheromones used for communication are a mixture of substances produced by queen mandibular glands. This chemical cocktail, termed queen substance, is the colony glue that controls many important colony activities. Very importantly this mixture limits the construction and development of queen cells and regulates drone comb building and raising of drones. The honey bee queen, unlike most solitary wasps and bees, is entirely unable to fend for herself or provision her offspring so she must communicate her every need and the working force theirs. Neither can found a new colony unaided.

The colony as a unit must work with the queen both to form new nests – through swarming – and, at the level of the individual nest, ensure that any queen and indeed workers and some drones can be reliably replaced when and as needed.

The role of the honey bee drone

Drones have half the normal genetic complement of their female counterparts but the common beekeeper notion of useless drones is anthropomorphic overreach. Their role in honey bee reproduction and in passing on desirable traits is vital: a queenless colony or a colony headed by a drone laying queen may be doomed but their drones may remain functionally useful in being able to pass on their genes.

Because of their haploid state, a half-gene complement, drone sperm are genetically identical. This means that a multiply mated queen will produce cohorts of related super sister workers[12]: each colony will contain discrete groups of super sisters admixed with half sisters, the progeny of other drones. Overall the colony will benefit from having groups of bees having disparate and desirable characteristics such as maintaining colony hygiene.
Since drones mate of the wing, they have evolved both better sight and flying

(pursuit) powers than either workers or queens: their genetic and physical fitness is tied to their being able to successfully compete and mate with a newly emerged queen. However, the social life of the honey bee drone within the colony would appear to be minimal though we may seriously underestimate their value in maintaining the natural condition of bees found in tree hollows.

Drones signal and receive food from worker bees, while their tendency to drift between colonies and to be accepted by other colonies is well recognised[13]. They may play some role in brood nest thermoregulation, they are well known to actively accompany swarms and their response to queen pheromones during mating flights has been subject to intense investigation.

There is a high cost to colonies in investing in the reproductive process. Their propensity to produce many more queens than are needed – both to replace old queens in the parent colony and to found new colonies – as well to as produce an abundance of drones – to support multiple queen matings (polyandry) is astounding. While drone production is widely perceived to significantly reduce the potential of bees to store honey, in the honey bee equation genetically fit drones are figured as important for perpetuating the species.

The colony in transition

In both wild and managed hives it is important that queens and drones are raised under the best conditions possible, a circumstance best replicated under mild spring conditions (Figure 1.3) to ensure their successfully passing on of their genes to the next generation of honey bees. Mature drones (at least three weeks from emergence) are reliably present during the main colony fission stage (swarming season) but must also be present if colonies are to replace their queens at other times, that is for emergency queen replacement and during supersedure. Interestingly the presence of reproductives (both new gynes and drones) is governed by seasonal conditions, especially those associated with the important process of swarming and their being absent during periods of dearth, that is in winter and during drought. Hence simple queen replacement may fail if sufficient drones are not available even if the occasional queen were reared.

a	b

Figure 1.3   Queen larvae:
(a) well attended queen – 5 October 2013; and
(b) well fed swarm cell in a hive at Queanbeyan – 2 October 2017.
Photos: Sarah AsIs Sha'Non and Alan Wade

In rare and exceptional instances a well fed queenright colony will raise and maintain drones (Figure 1.4) year round. We noted that this colony had also raised about fifteen swarm cells at this first early spring inspection signalling its overall prosperous condition.

Figure 1.4 Overwintered and emerging drones in early spring (September 2017), Narrabundah ACT.
Photo: Christine Joannnides

Regular queen changeover is important as it ensures colony vigour but it also serves to maintain gene flow and ultimately to enable bee populations to adapt to local conditions. So equipped the survival of whole wild bee populations – not the individual hive – is more assured. However any colony preparing to replace its queen faces the hazard of unsuccessful daughter queen establishment. Queen failure will be its death knell.

Despite remarkably reliable installation of a new colony queen, the actual workings of queen transition can be difficult to read. My experience in operating two-queen colonies, where there may be as many as ninety five to a hundred thousand bees, has made ascertaining queen condition rather challenging. For the most part however success in managing colonies preparing to swarm remains the single most challenging aspect of keeping bees.

Brood nest signatures

There are several features of overall hive condition that signal that the population of a honey bee colony is out of kilter. These include an abnormal ratio of different stages of brood, hive restlessness occasioned by queenlessness – indicated by hive roar – the presence of active queen cells, the presence of too many bees to tend a large amount of brood and too many bees, as well as too little space, to cure nectar and store honey.

Several rules emerge:

> *Rule 1 – Brood balance ratio rule*
>
> *A good queen, well provisioned and with ample nurse bees to raise brood, will lay to full capacity and at a steady rate. Here the amount of brood at different stages of development will be reflected in the time it takes for a worker bee to develop: eggs 3 days: larva 6 days: pupa 12 days, a simple 1:2:4 ratio*[14].

You are unlikely to ever count brood cells in different stages of development. However a quick scan of brood nest combs showing only eggs and young larvae or only sealed brood will signal that a queen has only just started to lay or that a queen has been lost, is failing or that brood raising has been curtailed.

When, alternatively, there are an overly large number of bees emerging and a surfeit of nurse and house bees, and the colony becomes overcrowded, the queen may lose control over her progeny:

> ***Rule 2** – Colony bee number to brood balance rule*
>
> *When there are surplus bees, those that would result in broodnest overheating or that are no longer employed in processing nectar and storing honey, they will beard outside the hive entrance or hang off combs as replete storage bees signalling that the colony may swarm.*

A healthy expanding colony having an adequate numbers of bees to provision larvae and to maintain the brood nest will be in balance. It will have a large retinue of house and field bees working actively together to raise brood and to steadily build and fill storage combs. Keeping a close tab on fast expanding brood nests in spring will help you judge whether the number of bees relative to the amount of brood is getting out of balance.

The Scottish master beekeeper Bernhard Möbus[15] read these pointers as either an early sign of swarm preparation, or – out of swarming season – a condition where the colony no longer had room to store incoming nectar. His keen observation of brood-bee imbalance being the major factor – apart from queen age – in initiating swarming deserves wider recognition.

Farrar developed some other brood rules[16] especially pertinent to the colony buildup phase (Figure 1.5), that is spring for cold and temperate climate zone beekeeping:

> ***Rule 3** – Queen laying rate rule*
>
> *The number of cells of sealed worker brood divided by twelve equals the egg laying rate of the queen.*

Since sealed brood is relatively easy to observe changes in the amount of sealed brood will signal that the queen's laying rate has proportionally changed. While it is natural for a queen to increase her rate of laying as more bees emerge and can nurture more brood, some tailing off of the amount of brood present will signal that the colony population is approaching its peak.

Farrar also observed that:

> *The most populous colonies produce not only the most honey per colony but the most honey per bee.*

To give substance to this statement his detailed studies demonstrated that the ratio of sealed brood and colony populations decrease 10 to 14 percent for each increase in population of 10,000 bees and that the rate of egg laying by the queen increases with a rise in the population up to 40,000 bees. The import of the workforce in switching its activity from brood rearing to harvesting is tied up with getting the colony to full strength by the commencement of the flow when the proportion of bees tending brood is lowest and when bees are tasked to forage:

Rule 4 – Proportion of bees attending brood rule

The number of cells of sealed brood divided by the number of bees equals the relative productivity per bee.

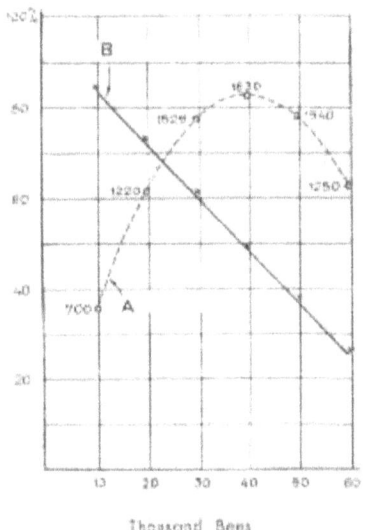

Fig. 1 – Influence of colony populations on daily egg-laying rate and brood rearing. (Cells of sealed brood divided by 12 equals average daily egg-laying rate; cells of sealed brood divided by the number of bees equals the relative brood production per bee.)
A. – daily egg laying;
B. – ratio of sealed brood to bees.

Figure 1.5 Clayton Farrar's plots of spring queen egg laying:
(A) in response to colony size; and
(B) declining colony proportion of colony devoted to brood tending as population expands.

In the early stage of build up around 80% of bees are attendant to raising brood. By the commencement of the main flows 2-3 months later, as little as 30% will be tasked to brood rearing, most bees being involved in harvesting, comb building and nectar curing.

Information on just how productive bees are are shown in Figure 1.6. A careful reading of the graph is needed to understand the yield factors as crop yield depends on many environmental factors and where the bees are kept, but the rules are the same. Put 100% = 100 kg if that is the actual season crop.

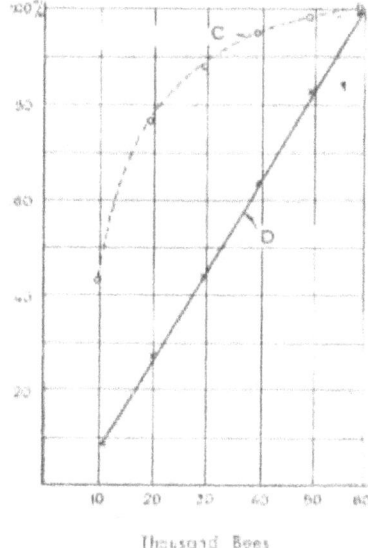

Fig. 2 – Influence of colony populations on colony yields and production per bee. (Colony gain at each population level divided by the yield at 60,000 equals the percentage yield; colony gain divided by its population equals the production factors per bee or relative production per unit number of bees.)
C. – relative production per bee;
D. – colony production.

Figure 1.6 Clayton Farrar's plots signalling positive influence of population size:
(C) on individual bee harvesting potential; and
(D) honey production based on size of population.

Often enough colonies do not take in an early crop simply because there are too few bees to forage and what is taken in is used to raise bees in the buildup phase. A strategy of dividing and requeening colonies late in the season (late summer and early autumn) and reuniting bees in early October (mid spring) when the risk of swarming is largely over is a good strategy for spring blossom and early and, in southeastern Australia, regular Red Box (*Eucalyptus polyanthemos*) flows. Secondly many package bee producers use only medium strength colonies (~ 20,000 bees) because they are primarily focussed on raising bees. By feeding these bees can be regularly shaken and have their numbers build again quickly.

The brood cycle can be manipulated in other ways to advantage. Beekeepers move their Southern Highlands bees to warmer climes and feed them heavily to prepare them for early almond and cherry pollination.

Farrar went on to show how different systems of colony management effect the seasonal population dynamics (Figure 1.7). Honey bee populations respond radically differently at different latitudes and under different climatic conditions. For example bees close to the Arctic must be built to a very tight timetable, but the days are very long and honey flows can be short but very intense. Package bees and swarms follow the trajectory line AC initially losing bees reaching full colony strength in just over 90 days. Healthy overwintered bees with ample stores, starting from a better base line BD, make a peak population of around

50-60 thousand bees in about 50 days (1 1/2 months). And strong enough colonies split and requeened are initially put back – the laying queen is deprived of some nurse bees – but two queens then laying together soon outstrip even the most producing single queen topping out at 90-110 thousand bees timed for the main honey flow – line BEF. In this case, and unless the flow is likely to be extended, the excluder is removed to allow the colony to return to its ancestral single-queen condition.

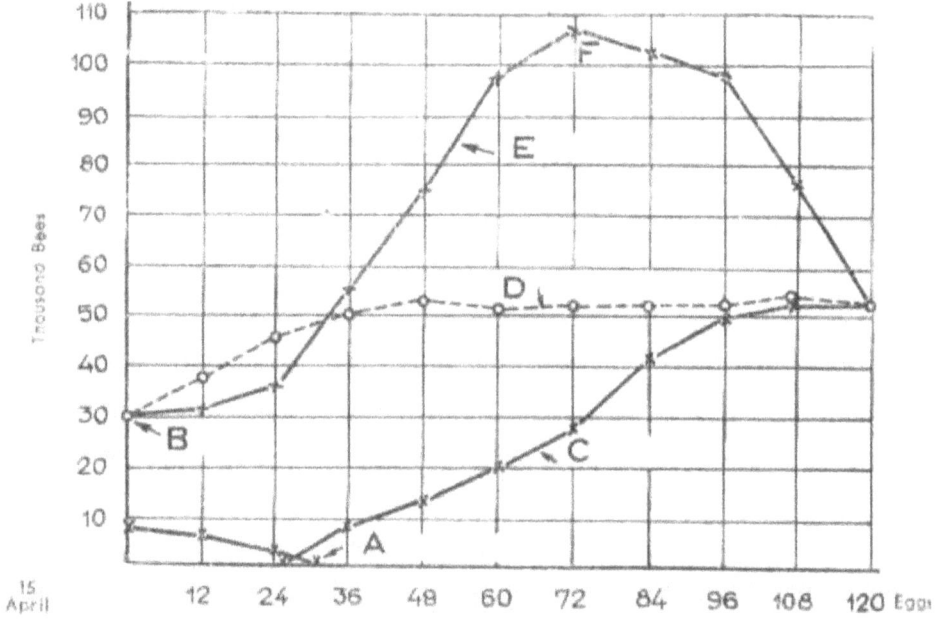

Figure 1.7 Clayton Farrar's plots of population growth projections over time:
(A) from package or swarmed bees;
(B) from established bees; and
(C) from colonies splits headed by two queens and united near the commencement of the honey flow.

Queen bee signatures

A forensic examination of queen cells left behind as a result of queen succession shows that the signatures are different for each method of queen replacement. Whenever a colony replaces its queen, it is simply a matter of finding cells from which queens started, and inspecting them closely.

Emergency queens originate from modified worker cells so are started deep in the worker cell matrix and are located close to where the queen last laid, that is almost anywhere where there were young brood. There may be few to many such cells.

Supersedure queens are always started on the surface of the comb and are usually, but not always, located within the main brood nest. There are rarely more than four or five such cells.

Swarm cells are similarly formed on the comb surface or are built free-form on comb margins. They are characteristically many in number, more than a hundred in colonies of the African Honey Bee (*Apis mellifera scutellata*), and are located mainly at comb margins, but especially along the bottom bar.

Common expressions of colonies transitioning to a new queen are those of interruption to the normal brood raising pattern, or the presence of active queen cell building, exemplified by colonies that are apparently queenless, colonies that are preparing to swarm or where the colony is simply just planning to replace its queen.

The honey bee has evolved to invest maximum resources in maintaining and protecting their long-lived queen. With optimal nutrition, the cultivated varieties of European honey bee queens (as distinct from African and Middle Eastern races of the Western Honey Bee) may not only survive winter and short periods of dearth but, as Butler records, some live for 2-3 years, in rare instances living up to 4-5 years[17].

However, since a very productive queen may lay more than 200,000 eggs in her first year, her egg-laying capacity is always precarious and she may then fail entirely at any time. Honey bees have not only evolved plans to optimise her survival but also to prepare for her orderly replacement.

Supersedure queen replacement

In the least complex of all succession plans, the queen is simply replaced. The worker bees raise a daughter without any antagonism from either the colony or the resident queen. In the Western Honey Bee it is not uncommon to find both mother and daughter queen present[18] after the daughter has successfully mated and settled in as the heiress apparent and then both laying together for an extended period. Indeed Hepburn and Radloff record polygynous African *Apis capensis* colonies[19] comprising as many as two supersedure queens in addition to the old queen and in any case two queens laying together for 1-6 months. They also cite records of a colony with a drone laying supersedure queen that failed to mate coexisting with an old laying queen, colonies headed by two queens of both the same and different races arising from amalgamated swarms or colony invasion, and colonies where the old queen being superseded was relegated to the periphery of the brood nest. In the normal circumstance, however, the net result of supersedure is that the old queen is replaced after a short time, brood rearing is minimally interrupted and the colony is invigorated by the presence of a new and productive queen.

So in supersedure we observe that bees have mastered the art of intergenerational renewal, a gradually failing queen allowing her workers to raise a few daughter queens. The old queen subsequently disappears and an orderly transition is achieved, though she may not survive for long enough to see her daughter installed.

> **Rule 5** – *Supersedure queen cell attachment rule*
>
> *Supersedure queen cell originate inside brood nest and are attached to the comb surface.*

Since supersedure queen cells are attached to the surface of the comb, they will come away freely from the comb face. They are also well formed and, with good nutrition and an area saturated with high calibre drones, they will be as good if not better than can be raised by a queen breeder.

Surprisingly it is the plasticity of the honey bee that allows a managed colony to be operated with two fully functional queens, separated by a queen excluder. Such colonies can form an exceptionally large consolidated brood nest, its stability attributable to the second queen producing additional queen substance.

Butler[20] delineates a very clear picture of supersedure:

> *Queen supersedure, the process by which a colony of honeybees replaces its queen without swarming, is (of) frequent occurrence ... the process of queen supersedure differs radically from that of swarming by which colony reproduction is achieved, and in which one or more new queens are also reared and the old queen leaves the parent nest with a swarm if she is still alive.*

This very eminent researcher was able to conclude that:

> *Queen supersedure is initiated by failure on the part of the queen of a colony to produce sufficient queen substance to satisfy her workers.*

From an ingenious set of experiments, involving tethering of queens and observations on *Nosema* affected colonies, Butler was further able to conclude that:

> *It is, likely, therefore, that a shortage of queen substance is the only immediate cause of queen supersedure, and that disease and injury produce their effects, if any, by reducing the ability of a queen to produce sufficient queen substance.*

Hence, while the physiological state of every queen inevitably declines with age, it is important to recognise that the supersedure impulse is governed entirely by reduction in queen pheromone production. Factors such as age-related reduced egg-laying capacity and injury do not necessarily presage queen supersedure. Indeed the factors initiating supersedure and swarming are complex and there are many recorded instances of intermediate behaviour. Hepburn and Radloff[21] summarise this well in stating:

> *The prevailing explanation for the immediate cause of supersedure is relative insufficiency of the queen pheromones necessary to inhibit queen cell construction which is demonstrably not the case for swarming. While all the genetic factors are not fully understood, it is clear that pollen flows instigate swarming in some African races of Apis mellifera.*

In practice this means that, with supersedure, any newly introduced queen producing limited amounts of queen substance will sometimes be replaced shorty after she has been successfully introduced. This often occurs if she is poorly mated or raised under less than optimal conditions. Meanwhile some old queens whose mandibular glands continue to work perfectly well, and who run out of stored sperm (drone layers), who stop laying altogether or who are injured, aren't necessarily superseded. However from the colony perspective the process of supersedure, once initiated, is usually a reliable succession plan provided the daughter queen is successfully mated.

If a supersedure daughter queen fails to return from a mating flight or mates poorly, the colony will often remain in a condition to raise another queen. This was demonstrated in our Jerrabomberra Wetlands Apiary in the summer of 2016-2017 where attempts by a colony to supersede its previously exceptionally fecund Carniolan queen were deliberately thwarted three times by removal of supersedure cells. In each instance worker bees immediately commenced raising a small new batch of cells just as Doolittle (see discussion below) observed long ago.

Supersedure is important for the survival of the individual colony whereas, with swarming, the focus is on colony increase, that is on reproduction[22]. Nevertheless queen signalling translated to colony queen replacement behaviour is complex as Butler concludes:

> *Unfortunately the interrelationships between the processes of queen supersedure and swarming are still far from clear.*

From the perspective of the apiarist, supersedure may be harnessed to produce well raised queens (Figure 1.8) though, as with any uncontrolled open mating system, the daughter queen may not always perform satisfactorily.

Just as destroying queen cells does not remove the willingness of bees to continue to make swarm preparations, bees will reconstruct supersedure cells (Figure 1.9) if they are removed.

Indeed this was the way queen breeder Gilbert Doolittle[23] made his 1889 pioneering foray into queen raising:

> *After I had this experience with the colony that had two Queens in a hive... I began to watch for a like circumstance to occur, which happened about a year from that time. In the latter case, as soon as I found the cells, they were sealed over, and not knowing just when they would hatch [sic emerge], I at once cut them out and gave them to nuclei. In a few days I looked in the hive again, when I found more cells started, which were again cut off and given to nuclei, just before it was time for them to hatch [emerge]... By this plan I got about sixty as fine Queens as I ever reared, and laid the foundation for my present plan of securing Queens...*

Figure 1.8 Single supersedure cell development mid spring, 20 October 2016.
Photo: Alan Wade

Figure 1.9 Persistent supersedure queen cell development started on comb face amongst sealed worker brood, 2 November 2016.
Photo: Alan Wade

Emergency queen replacement

Now and then a colony loses its queen accidentally. Like the ex Norwegian Blue Parrot[24] she falls off her perch. Natural losses occur when a colony is raided by predators, is overcome by disease, is structurally damaged or is inadvertently or intentionally made queenless by either bees or the beekeeper.

A strong colony has a ready solution: it immediately turns its attention[25] to raising queens by lavishing young worker larvae with royal jelly (Figure 1.10). Note that large diameter queen cells have encroached on neighbouring worker cells and that larvae in the queen cells are noticeably larger than those of neighbouring worker larvae. Careful inspection of these emergency queen cells also signals that they are modified worker cells that the attendant nurse bees have flared and turned downwards. Supersedure cells, on the other hand, are started from scratch, are fewer in number and are attached to the comb surface or, less commonly, to a comb margin.

So beekeeper initiated direct requeening can be seen as a much less seamless transition. Removing the old queen and installing a new caged queen necessitates interruption to the brood cycle. For a caged laying queen there is a period of at least several days before this new queen is accepted and commences laying: in a wild colony that has to raise a queen from a one to four day old larva the colony will take over three weeks to establish a new laying queen. In the worst case scenario the newly installed queen is lost and the colony becomes broodless and, ultimately, hopelessly queenless.

> **Rule 6** – *Emergency queen cell attachment rule*
>
> *Emergency queen cells originate from worker bee larvae so their base originates at the worker comb septum.*

Since emergency queens result from worker bees repurposing newly hatched worker larvae the cell will originate in an area that the old queen last laid eggs. The signatures are either of flared worker cells or of a fully developed queen cell both leaving their telltale origin at the base of a former worker cell.

Figure 1.10 Emergency queen development showing modified worker cells, 21 October 2017.
Photo: Alan Wade

The process of emergency queen replacement is triggered by sudden loss of queen substance. This results in a dramatic change in worker behaviour, colonies audibly roar and emergency queen cells are immediately constructed. It differs from both supersedure and swarming induced queen replacement in that it is unplanned and in that larvae once destined to be workers are reprogrammed to become queen bees. Distinguishing features of emergency queen replacement are not only that it can occur at any time but also that the colony will suffer an extended period of broodlessness and that only a limited number of cells are raised.

Whether a fertilised egg laid by a queen is destined to become a worker or a queen is determined by how well larvae are fed and the quality of that food, not by the genetic makeup of the still young bees. Female bee development is regulated by juvenile hormone (JH): high levels of JH are conducive to queen development, queen versus worker production being a cause célèbre of epigenetic gene expression[26].

If food is rationed to any developing queen larvae, or if a queen is developed from an older four or five day old worker larva, an intercaste queen bee is produced[27]. These gyne progeny have some worker and some queen-like characteristics, are renowned to be quickly superseded and have a limited egg-laying capacity[28].

Swarm queen replacement

A large colony, faced with a shortage of space to process nectar and store honey and pollen or a brood nest congested by young emerging bees surplus to brood raising requirements, is swarm prone. This is especially the case during the late spring ascendant phase of colony development. This brood-bee imbalance results in there being many young bees surfeit to colony needs. Here the process of bees exchanging food, trophallaxis, a process that shares food and that also distributes queen substance and other pheromones, is interrupted.

However conditions and colony size apart, crowding is by no means the only factor triggering swarming. Radloff and Hepburn[29] note, for example, very different swarming behaviours amongst some African honey bees, *Apis mellifera capensis*, *Apis mellifera jemenitica* and *Apis mellifera scuttelata* that harvest large amounts of incoming pollen swarm from demonstrably uncrowded colonies. They report records of extraordinary numbers of queen cells, up to several hundred with *Apis mellifera lamarkii*, many dozens in *Apis mellifera intermissa* and *Apis mellifera sahariensis* and intermediate numbers for *Apis mellifera adansonii* and *Apis mellifera scuttelata*. *Apis mellifera lamarckii* is also remarkable in that swarm cells are developed on the comb face amongst brood.

Disruption of pheromone distribution releases the brake on worker bees raising queens. Queen cells are typically constructed mainly on the periphery of the brood nest (Figure 1.11) well out of range of the influence of workers tending to the queen and receiving queen substance. There they are often found in large numbers. Recognition of the number of queen cells and their placement is perhaps the best way to tell whether swarming or supersedure will ensue. Well-formed supersedure cells can be left as there will be minimal risk of colony division.

> *Rule 7 – Swarm queen cell attachment rule*
>
> *Swarm queen cells originate mainly on the brood nest periphery so are attached to hive frames, the base of holes in the comb matrix or – like supersedure cells – are attached to the comb surface.*

Swarm queen cells are produced in abundance and are built from scratch. And like supersedure queen cells they are not all produced at once: they are produced in sequence to accommodate the scenario where the replacement queen fails to mate or return to the hive. The other characterising feature of swarm cells is that they are located on the periphery of the brood nest, very commonly on or near the bottom bars of brood frames.

Actual swarming should be seen as the end point of series of internal colony changes leading to colony reproduction. The American researcher-beekeeper John Hogg has designated unemployed worker bees as a temporary swarm caste[30], bees that prepare for and orchestrate swarming. Since the first, or primary, swarm leaves the hive with the established colony queen, the partially deserted parent colony will almost always be headed up by a new, initially virgin, daughter queen. In most instances the queen succession process in the parent colony proceeds smoothly: the colony acquires a young and vigorous queen and the colony returns to its single gyne (potential reproductive queen) status.

Figure 1.11   Swarm cells started at comb margin 12 November 2017.
Photo: Alan Wade

In his classic 1924 treatise *Swarm Control*, Demuth[31] describes the swarming process – at least for European races of the Western Honey Bee – in all its manifestations, a topic and gentleman we will return to in *Out of the box*.

Eventually the old queen in the prime swarm (Figure 1.12), having established a new colony in a suitable cavity, will herself be superseded. Given that any new queen may be replaced due to a combination of poor brood and queen pheromone signalling[32], queen substitution must also be of common occurrence amongst any secondary swarm headed by any sub-optimally raised daughter queen. These interpretations of swarming are augmented by the finding that relatively high levels of pollen-based vitellogenin in worker bees preparing to swarm may provide an important impetus for colony division even where colonies are demonstrably not crowded[33].

Like supersedure, the process of swarming provides a reliable means of queen replacement and colony renewal. The exceptions to the swarming-based rules of succession are where the parent colony starves, in response to issuing too many swarms, or where the colony swarms too late in the season. In the latter instance, the swarm will have insufficient time to build combs and stores before winter so it, too, will perish.

What are the consequences of allowing bees to swarm? Firstly much of the productive capacity of the colony is lost. This may be exacerbated by the loss of a high performance queen, as the outcrossed daughter queen will likely be both more disposed to supersedure and swarming. In any managed hive there is ample reason to replace any poor performing colony-raised queen as soon as a new high calibre queen can be found.

A further message is that a colony may have swarmed because it has been two or more years since the colony was last requeened and the colony is now extremely swarm prone if not also of cranky disposition.

Attempting to replace a colony queen at the point of, or just after, swarming is probably unwise as the swarming impulse may result in any newly introduced mated queen leaving with a further issuing swarm. However requeening, combined with swarm control measures, will work well provided every care is exercised to first remove all potential rival queens.

A salutary message is that while swarming is good for augmenting stocks of bees in the wild it may not be for your neighbours wellbeing or for your honey gathering prospects.

Figure 1.12 Classic primary swarm.
Photo: Sarah AsIs Sha'Non

Queenless colony prospects

An old queen, herself depleted of stored sperm, will eventually become a drone layer. It is not uncommon to find colonies in this state at a first spring inspection. This condition is indicated by a normal solid brood pattern with small raised capped drone brood found in worker comb and by the colony being overwhelmed by small drones.

With laying workers and no queen, however, drone brood is more scattered and drone eggs are laid on worker cell walls and sometimes in batches of two or three. In the absence of a functional queen around 7-45% of non-reproducing workers have the potential to become laying workers but this figure climbs 20-70% in colonies preparing to swarm[34]. In a normally functioning queenright hive, however, only a few workers lay fertilised eggs and these are recognised and cannibalised.

Distinguishing the two types of drone-laying colonies is therefore relatively straightforward. Both colony types are characterised by the tell-tale sign of large numbers of small drones and both may retain these drones when normally none would be present. Nearly all workers in such colonies are old and it is almost inevitable that both types of colony will dwindle and perish.

In very exceptional instances laying workers may produce functional female offspring including workers and the occasional replacement queen in a so-called process of thelytokous parthenogensis. This rare trait has been reported in self-requeening of hopelessly queenless Caucasian, *Apis mellifera carnica*, and Italian, *Apis mellifera ligustica*, honey bee colonies. Here workers and virgin queens produce low numbers of viable offspring[35]. This trait is also known from the Tellian Honey Bee, *Apis mellifera intermissa*, and the Syrian Honey Bee, *Apis mellifera syriaca*.

However the South African Cape Honey Bee, cultivated for honey production and for pollination services because of its docility, regularly produces viable queens and workers from worker-bee-only colonies. The end points of Cape Honey Bee queen replacement in both queenright and queenless Cape bee colonies are extremely complex. This complexity arises out of both queens and workers producing viable female worker bees[36]. Queenless Cape Bee colonies may dwindle and die much like those of any other race of *Apis mellifera* but surprisingly often these South African bees raise their own queen from viable worker bee female offspring.

An informative discussion of the risks that the Cape Honey Bee presents to apicultural practice is presented by Barry et al[37]. Hepburn takes a different view[38] arguing that these same traits:

...would also permit rapid modification of honey bee lines for whatever traits that might be desirable, from enhanced propolis collection even to the absolute control over swarming, or elimination of their very high resistance to new queen introductions and enhanced colony defence.

Usurping swarm queen replacement

There are two well-reported instances of natural queen replacement by a queen from another colony. In the first of these late swarms invade colonies of European races of the Western Honey Bee where the resident queen is usurped. Here the colony resources are plundered to establish the invading swarm queen and her offspring.

A more insidious parasitism of honey bee colonies is displayed by the Cape Honey Bee (*Apis mellifera capensis*). Its workers routinely drift to African Honey Bee (*Apis mellifera scutellata*) colonies and take over the role of the queen – as laying workers – steadily replacing the host African bees[39]. In this extraordinary circumstance, where unfertilised workers produce workers parthenogenically, the colony is gradually repopulated by Cape bees[40]. However these laying Cape bee workers, even in aggregate, lay less efficiently than a fertile laying queen, so the colony dwindles and dies.

But like the parasitising Cape Boney Bee, the productive African Honey Bee (*Apis mellifera scuttelata*) caused its own problems. This bee was imported to Sao Paulo, Brazil from South Africa and Tanzania in 1956. The saga of its escaping from the Brazilian research facility is recorded by Mark Winston[41].

Drones of this race, mating with long-domesticated European honey bees, created a new race of so-called killer bees whose menace was amplified by its propensity to swarm and abscond. The geographic range of this new hybrid race has been constrained latitudinally because the bees of European makeup are better adapted to storage of honey and pollen, a necessary condition for honey bee survival in colder climes.

This usurping behaviour should be seen as a whole of population level phenomenon. At the colony level all queen stock in every colony is affected not so just much by queen replacement as by a radical change in their genetic makeup. Even with the Cape Honey Bee (*Apis mellifera capensis*), the genetic signatures of active laying worker reproduction can be seen in the African Honey Bee (*Apis mellifera scutellata*) in a northerly zone beyond the natural distribution of the Cape Honey Bee.

Very presciently, Winston has pointed to the fact that African and European races of the Western Honey Bee had already been artificially translocated over an extended period elsewhere and that honey bee hybridisation of this type had occurred long before this misadventure. Even amongst some African races of the Western Honey Bee, natural hybridisation is well recognised where the focus of interest is the extent to which this occurs and the geographical range of influence this has had on observable traits[42].

Colony signatures

Very often the condition of a hive is most apparent at the colony level, there being no evidence of a colony actively replacing its queen and there being no evidence of brood disease. The absence of stores and potential for starvation (Figure 1.13) is an obvious example of a colony in trouble and is easily diagnosed.

Figure 1.13 Winter colony starvation bottoms-up pattern Jerrabomberra Wetlands Apiary, 22 July 2017.
Photo: Alan Wade

However evidence of a hive performing sub-optimally, and likely queen related, is not always immediately apparent though scattered brood may signal imminent queen failure.

The apparent queenless condition

A colony that has recently swarmed or that has accidentally lost its queen may be free of eggs and young larvae, indeed be even entirely broodless. As a virgin queen (or a queen out on mating flight) is easily missed it is also never certain that the colony is, in fact, queenless.

A colony in transition takes quite some time, sometimes almost a month, to establish a new queen. We have often found a new laying queen during followup inspections a few weeks *after* finding a colony broodless. Had we assumed that the colony were queenless and gone ahead and introduced a queen anyway, she would have been lost.

There is a simple remedy to the presumptive queenless condition. It involves going to a neighbouring healthy hive and removing a frame containing eggs and young larvae. Place this frame in the centre of the brood box, mark the top bar distinctively so it can be easily recognised, and come back a day or two later to see whether any emergency queen cells have been started. If the colony is indeed queenless, it will have been conditioned by the addition of young brood and will more readily accept a caged queen and have started queen cells. Alternatively, if a new queen is already present and searching had failed to find her, no harm will have been done.

I never introduce a new caged queen (or a select queen cell) to a colony unless I am quite certain that it is queenless. One of the great advantages of making up a nucleus colony a day before introducing a new queen is that you can be assured that the nucleus is queenless as again new queen cells will have been started.

The most obvious case of bees being queenless is that of a colony that has long lost its queen. Very often the colony will be weak and there will be many small drones present. Such colonies have laying workers, so called gynecoid queens, and may have small scattered patches of drone brood laid in worker cells. This is a surprisingly easy condition to recognise when you first open hives in early spring. These so called hopelessly queenless colonies are difficult to requeen and the bees are best shaken off frames well away from the hive. The gear is removed or united with a normal functioning hive after a mandatory disease hive check.

There are other situations where a queen may be present but no brood is present. Most notable of these are conditions where nothing attractive to bees is flowering when the queen will stop laying. Most of us will have observed this low-brood condition in late autumn when there is little or no forage and there is little value in the colony raising new forager bees. This said, it is still important to have a good laying queen to supply the colony with young long-lived (diutinous) bees needed to maintain the colony over winter. Small patches of brood will appear periodically even in the depths of winter[43] despite the heavy drain on colony stores to raise them. Bees cluster when the ambient temperature drops to 14 ^{0}C, maintain a broodless cluster temperature at 18-21 ^{0}C but need to raise the temperature to 35 ^{0}C to incubate brood. We have found that taking the time to work out why a colony is broodless is more a matter of common sense than adopting the expedient practice of dropping in a new queen that, in the presence of a queen, is unlikely to survive.

The situation is more complicated when, during the actual period of natural queen changeover, there are a number of incipient queens (queen cells and virgin queens, so-called gynes) present. Requeening a colony while the colony is in this state will likely result in loss of the queen you are trying to introduce. Let's look at these quirky conditions as they may also confound your well intentioned effort to requeen a more normal failing or cantankerous single-queen, queenright hive.

The swarming colony condition

Once swarming preparations are well underway, the colony queen will have stopped laying and there will be no eggs and little or no unsealed larvae. As well, foraging and comb building will have also ceased. So if you are particularly observant, you may find a slimmed down queen ready to fly and bees gorging themselves on honey and pollen making ready to swarm. Further salient features of swarming are that the colony queen never departs until sealed queen cells are present and that the swarm that departs without a queen (or that accidentally loses its queen) will almost invariably return to the hive.

This swarm departure behaviour is, of course, conventional wisdom. Under exceptionally favourable conditions of spring 2021, a New South Wales queen breeder reported swarms departing as soon as queen cells had eggs. Fellow beekeeper Dannielle Harden and I both observed double prime swarms from two-queen hives, a rare phenomenon, under the same seasonal conditions.

In practice determining at what stage queen changeover is at can be difficult to ascertain particularly in very populous colonies. A colony may temporarily abandon swarming and tear down queen cells when weather conditions suddenly deteriorate, though most often it will resume its effort to swarm once conditions improve. Counterintuitively, and under the same poor weather conditions, overcrowding resulting from confinement, can be the trigger to commence swarm preparation.

So far we have recorded the condition of colonies making preparations to swarm. Notable features of any colony that has already swarmed are that a laying queen may not appear for several weeks, that the whole hive will be sparsely populated and that worker bee recruitment will be delayed and limited to that of any remaining emerging brood. In the worst case scenario the colony may be so depleted it may never recover or the remaining virgin queen may fail to mate.

When swarming is over, or after you have taken effective swarm control measures, you can go back to your requeening plan, this time more assured of a successful effort to requeen. The problems of requeening during the normal swarming season – late August through to the end of October in temperate southern hemisphere climes – prompts another question: *'What is the best time to requeen?'*

Most beekeepers prefer to conduct requeening by mid spring before colonies become overly large and difficult to handle. There is, nevertheless a very strong case for routine autumn requeening when colonies very rarely swarm. Instead of allowing really strong colonies to down-regulate their numbers at the end of the season, they can be split and both parts requeened, or if of average strength, just requeened. This will provide a much needed supply of colonies headed by young queens long before queens can be raised or are commercially available in spring.

Another advantage of autumn requeening is that such queens make an earlier start to laying than do older queens. Well provisioned, colonies come away quickly as early as late winter and will rarely swarm during the spring and early summer buildup. This commends late season, say March (early autumn), requeening though you will still need to place a queen order well in advance to be assured that queens can be supplied.

A further fallback position is to consider summer requeening. While large numbers of bees makes finding and removing queens more difficult than in autumn, or indeed in a normally functioning hive in spring, queens tend to be most readily available for purchase in summer.

The queen replacement condition

The supersedure condition is indicated by the presence of up to a handful of queen cells located within the brood nest (Figure 1.14). It is important to distinguish this condition from that of swarming, where large numbers of queen cells are present, and emergency queen replacement, where a few to many queen cells may be present.

Simple queen replacement can occur at any time of the year but is most prevalent when colonies are under breeding stress, that is whenever the queen cannot lay fast enough for number of nurse bees available to raise brood. This is well illustrated by a lesser know facet of supersedure identified by John Hogg[44]:

> *Queen supersedure is never the cause of swarming. But supersedure may occur concurrently with the swarm whenever the queen's failure was caused by being overtaxed while generating the bees for that swarm. Significantly, it is the virgin queen that is then selected to accompany the swarm. The failed queen is retained in the parent hive to be superseded in turn...*
>
> *Apparently, just as the survival instinct of the bees in control won't allow a swarm to leave a hive without a replacement queen in the parent [hive], they won't allow a failing queen to issue with that swarm.*

Figure 1.14 Second batch of supersedure cells in the presence of an old and previously particularly productive Carniolan queen. Jerrabomberra Wetlands Apiary 31 October 2016.

Note eggs in adjacent worker cells, clearly visible here, are absent in colonies preparing to swarm.

When you find supersedure queen cells, simply close the hive as the colony needs a new queen. Removing such cells risks either the old queen dying – when the colony will become queenless – or at best delaying the inevitable process of queen replacement.

In the aftermath of supersedure, a new daughter queen (and occasionally two daughter queens) may be present and actively laying alongside the old queen. We have found two queens – the queen being superseded and her daughter – in a handful of colonies accurately reflecting the observation made by the *British Bee Journal* authors just one hundred years ago[45]:

> It is quite possible that there had been two queens in the hive, a condition that has been prevalent this year, or beekeepers are becoming more observant.

In the normal course of events the old queen disappears, either of natural causes or is removed by the daughter queen. For all practical purposes, queen supersedure will not interrupt apiary operations, though replacing a failing queen should always involve a search of all brood combs for the not unlikely presence of an additional queen.

By understanding the natural processes of queen succession, beekeepers have adopted a range of successful requeening strategies and, in doing so, have been able to largely circumvent problems associated with natural supersedure and swarming. We can now move on to reflect on successful queen introduction practices.

The heiress to the throne[46]

Having examined the origins of the single queen colony, how colonies function and the processes of natural honey bee queen replacement, we can now apply what the bees have shown us when introducing new queens to our bees.

Why requeen?

One of the most daunting tasks facing a beginning beekeeper is knowing how to find and then how to replace a hive queen.

There are many good reasons to requeen bees and there are others to retain queens where they are of good stock and have been performing well. This is because the condition of the queen, as the mother of each and every nest bee, will shape the ongoing condition of the honey bee colony.

Firstly queen age is closely correlated with fecundity so is a good guide for the need to replace her. Young well-raised queens can sustain an egg-laying capacity of around 1800 eggs per day for months on end, well at least under favourable in-hive and environmental conditions. However, the innate capacity of a queen to keep on laying at this rate can be lost after a period of as little as six months in hard-worked migrated commercial colonies. For a sideline beekeeping operation sudden loss of egg-laying capacity should be expected in any colony headed by a queen older than eighteen months of age, a timeframe longer than for a commercial hive. All other things being equal, her productivity and condition will then automatically decline, sometimes precipitously.

Secondly queen ageing increases the propensity of honey bees to swarm. Between 10-40% of unmanaged colonies will swarm in their first year while swarming occurs three times as frequently in two-year old queens compared with one-year old queens[47]. Supersedure also increases the rate of queen replacement but, as we have seen, its frequency is contingent upon factors such as the conditions under which the queen was raised.

Queen breeders have been able to advance selection for strains of bees with a much lower propensity to swarm or to supersede and that have a high-egg laying capability. The take home message is that, if you want to control swarming as well as supersedure and make honey, you should requeen regularly. In practice, especially where nuisance swarming or public safety concerns need to be taken into account, requeening should be conducted at least every 1-2 years. Once bees have swarmed however (Figure 1.15) colony rebuilding and the getting of a honey crop becomes problematic.

Requeening should be conducted somewhat earlier if an individual colony is performing well below par or if it suffers persistent minor disease problems. I have observed chalkbrood outbreaks disappearing a month or so after requeening with a queen producing hygienic bees, most observable first near the brood nest centre but quickly then taking effect in peripheral brood combs.

Overall routine requeening will help keep colonies on track so that one has:

- gentle, tractable bees that are easy to work;
- bees that are less likely to swarm, that is head out the front door;
- populous thriving colonies of bees supported by a queen with high laying capacity;
- stable healthy colonies that are less subject, assuming they are well raised, to supersedure; and
- colonies that have a modicum of disease and parasite resistance.

Figure 1.15 After the bees have bolted (29 January 2013).
Photo: Sarah Asis Sha'Non

Since daughter queen mating is uncontrolled, except where the queen breeder saturates mating zones with good quality drones, there is always a high risk of ending up with an unproductive queen. A high calibre young queen adds to colony vigour – there are many more replacement bees – and doing so is good insurance against cranky, unmanageable and disease prone stock.

Finding the queen

Finding queens in bee hives

Marking queens makes them easier to locate. In a closely managed apiary the absence of a properly marked queen signals that the old queen has been superseded or that the colony has swarmed. Finding a queen comes with practice and is much easier in a tractable colony where minimal smoke is used. Locating a queen in a cranky agitated colony, where bees are unsettled and moving about quickly, can be nigh impossible.

There are circumstances where finding a queen proves difficult for even a very experienced beekeeper. Carniolan and Caucasian queens have a very similar colouration to their workers and a small, poorly raised queen can look very like a worker bee. However if you find eggs and attached to the base of honeycomb cells, you can be quite certain that the queen is present or was over the previous three days, the time it takes eggs to hatch. If you can't find the queen and the bees become so unsettled that you rightly feel there is little chance of doing so do not despair. Little harm is done in waiting for a day or two and again searching the colony thoroughly when the bees are more well settled.

There are special situations where finding and replacing a queen is inherently problematic. A colony preparing to swarm may have a queen that has simply thinned down and stopped laying while a queen undergoing supersedure may also be producing so few eggs that either may be easily missed. However even if you cannot locate her and are unable to locate eggs and young larvae there is still no guarantee that she is not still present. More confusing is the condition of a colony in which there are swarm or supersedure cells, some from which virgin queens may have emerged. Here there may be a number of small, hard to find queens.

Firstly undertake the simple check by introducing a frame of eggs and young brood from a neighbouring hive, after making sure it is free of AFB and AFB, mentioned above. If no emergency queen cells have been started and certainly within forty eight hours there is every chance that a new queen is already present but that she has not yet commenced laying. Search for her starting with the marked frame and once found remove her before introducing the caged queen. This simple reading of a colony will save you the cost and frustration that accompanies loss of that new queen you had hoped would have given the colony a new start.

Finding queens in swarms and in natural cavities

Freshly swarmed bees are often remarkably gentle. They require minimal if any smoke as they are engorged and have no nest or stores to defend. Finding a queen in such a swarm can be quite straightforward if it is located close to or at ground level or if bees can be searched systematically as they enter the prepared swarm capture box. There are many surprises but swarms located on or near to the ground (Figure 1.16) are not uncommon where finding the queen can be easy.

Figure 1.16 Hiving secondary ground-level swarms in South Canberra late October 2016 where queens were easily found.

However any swarm that has resided in an open location for more than a few days and has been unable to locate a new nesting site can be become unsettled and aggressive. Always enquire as to how long bees have been resident when you get that call to collect a swarm and get an update before you head out to make sure the swarm is still present. Often as not you may arrive to discover that the swarm has absconded, it has already built some comb or it is indeed not a swarm at all but bees that have been long resident in a brick wall.

In any case mark the queen if you find her to facilitate her later replacement. If a queen is not present, collected swarms tend to abscond often returning to the parent nest. Once hived most beekeepers delay further trying to locate the swarm queen until the colony has settled into a normal frame gear and has her own brood.

Finding a queen in a colony located in a natural cavity is another matter since the ensuing nest destruction and disturbance makes searching difficult. If, however, the brood nest proves easy to access, it is advisable to check not only for the presence of eggs and brood but also for the presence of brood diseases and infestation by small hive beetle.

Unless you have certain knowledge of the provenance of stock, always requeen as wild stock should never be equated with good performing stock. A possible exception to this rule would be the finding of a marked queen showing that the bees came from a managed colony so may be worth retaining. Apart from wild colonies being far more swarm prone and potentially aggressive, requeening from selected stock will produce more productive and tractable bees.

While many starting beekeepers welcome finding a swarm and even keeping the queen, relying on them to sustain an apiary reflects an age-old skeppist belief that encouraging swarming is good beekeeping practice. Contemporary apiary management focuses on minimising nuisance swarming and consequent loss of the productive capacity of bees.

Avoid the trap of attempting to requeen a swarm immediately after it has been hived. Firstly after swarms are well-known for containing several to many virgin queens. Delaying requeening for about three weeks will enable the bees to settle and sort out any problems they may have with their gyne status and will further enable you to assess whether or not serious brood diseases are present. Plan to requeen colonies established from swarms but avoid doing so immediately.

This all said it is rare for swarms of European races of *Apis mellifera* to settle with a second queen. Hepburn and Radloff[48] indicate that protracted polygyny, colonies with two or more queens, sometimes occurs naturally colonies of *Apis mellifera adansonii*, *Apis mellifera capensis*, *Apis mellifera intermissa* and *Apis mellifera scutellata*. Though the single gyne (one-laying queen) rule applies in most instances, it is advisable to delay requeening directly as there is no immediate guarantee that only one queen is present.

Locating a rogue queens

Any experienced beekeeper will have encountered 'killer bee' colonies that he or she would have preferred not to have had to deal with. There are, however, several approaches for finding a queen in a colony with an exceptionally savage disposition. First, either:

- move the colony well away so that field bees drift to neighbouring colonies; or
- split the colony and move brood boxes sideways to a new stand so that bees drift back to honey supers left on the original stand – an additional lid and base will be needed; or
- insert queen excluders above and between brood nests and then, four or more days later, move the only super containing eggs to a new stand a meter or so away – using a second base and lid – and allow the bees to drift back to queenless units on the old hive stand.

Then, come back after a suitable short interval – an hour or so if bees are freely flying – and search for the queen. The task of finding the queen will be much less daunting and will indeed prove quite manageable.

Alternatively if the bees are reasonably tractable, but repeat searching fails to find the queen, she can be found by using an excluder to sieve the bees. While I do not advocate this as a routine procedure, it can be useful as a last resort.

Firstly move the colony sideways and shake all bees into an empty box placed on top of an excluder and under-supered with yet another empty box on the original hive stand (Figure 1.17). In practice you will need one spare super.

Then, taking great care to keep the top super with the excluder, progressively return brood combs to the lower super after shaking all bees into the empty top box above the excluder. The brood will draw bees down through the excluder while foraging bees will return to the hive entrance. Use enough smoke to force queen retinue bees down through the excluder. You may also find it necessary to use some smoke to prevent the majority of bees crawling out of the top box as this sometimes messy exercise proceeds. The queen will be found alone on the excluder.

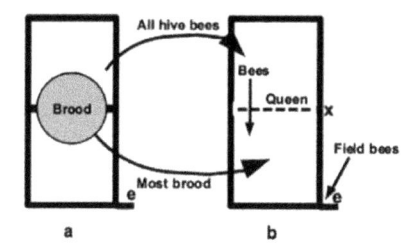

Figure 1.17 Locating queens by sieving bees: e = entrance; x = excluder:
(a) populous colony with hard to find queen; and
(b) bees shaken into top super over an excluder with most brood returned to empty lower super.

The lost queen syndrome

If you lose a queen, none is available, and the colony appears otherwise to be very healthy, there are two simple remedies. One is to get the colony to raise its own queen. The other is to cut your losses and use the gear and bees as best possible elsewhere in the apiary.

If the colony is strong and has not been queenless for long and there is some brood present, use the same procedure described above: simply introduce a frame or two of brood at most stages of development but including some eggs. The colony will usually, but not quite always, raise its own emergency queen cells and the best cell formed can be left. Removing additional emergency queen cells will be important if the queenless unit is very strong as occasionally such colonies will still swarm.

The simpler and more sure alternative for a hive made inadvertently queenless is to paper on surplus boxes with bees, any brood and stores to other queenright colonies. If bees seem healthy but the colony appears to have been queenless for a long time, simply shake bees off all frames and add gear to queenright hives. Any missed queen or cohort of laying workers will be stranded and most bees will drift to other hives.

In all instances take decisive action to avoid spoilage of stores by wax moth and small hive beetle. Take advantage of the fact that bees have a remarkable potential for renewal especially when headed by a young queen. You can always catch a new swarm or divide a colony another time after re-ordering a new queen.

Fostering requeening conditions

Where an old queen or other look-alike queens, so called gynes – queen cells, virgin queens and laying 'gynacoid workers – are present, a colony is functionally queenright. Any attempt to introduce a new queen will result in her being balled by hostile workers or stung by the reigning queen. As Tom Theobald reminds us[49] introduction of a caged queen to a queenright colony will always result in the loss of that new queen.

Many factors enhance requeening success[50], not least feeding bees to make them more queen receptive, in the absence of a honey flow. This will keep them well occupied. It is also a good idea to avoid windy, wet or cold conditions where bees tend to be highly defensive. In an insightful study, John Hogg discovered that the general conditions for acceptance of a single queen were identical to those of simultaneous acceptance of many queens[51]. What John Hogg had discovered was that a colony made queenless will accept any number of queens. However maintaining any additional queen for any extended period of time is of course another matter.

Spring or autumn requeening is ideal since there are many fewer bees to handle and this is the time when the best quality queens are available. Try to work bees on warm sunny days when the majority of the bees will also be out foraging.

The basic principle for successful queen introduction is to create conditions where bees, unrelated to the new queen, can best accept her as their new mother. We need only to replicate the conditions that bees employ to replace their queens naturally: sudden colony queen loss, colony supersedure or colony swarming.

Replicating emergency queen replacement impulse

Bees sense queen loss by a process of chemical messaging: each colony queen has a pheromone signal that the workers recognise and show remarkable fidelity towards. However once a queen is removed her odour and pheromones dissipate and the colony quickly becomes queen receptive. However, instead of allowing the colony to raise its own new queen, you supply one. This sleight of hand short circuits the ten or so days it takes for a colony to raise a virgin queen from already hatched eggs (Table 1.1) and the additional week it takes for her to be mated and to commence laying.

	Queen				Worker				Drone			
Species	Egg	Larva	Pupa	Emergence	Egg	Larva	Pupa	Emergence	Egg	Larva	Pupa	Emergence
Apis mellifera	3	5	8	**16**	3	6	12	**21**	3	7	14	**24**
Apis cerana	3	4-5.5	6-7.5	**13-16**	3	5	11	**19**	3	6	14	**23**
Apis dorsata	3	4.5	7	**14.5**	2.9	4.6	10.9	**18.4**	2.9	4.6	14.3	**21.8**
Apis florea	3	6.8	7.7	**17.5**	3	6.3	11.2	**20.5**	3	6.7	12.8	**22.5**

Table 1.1 Brood development schedule (days) for representative species of honey bee[52]. Note the earliest emergency queen to emerge will develop from a 3-4 day old larva.

Sometimes emergency queen cells will appear after the introduction of a new queen (Figure 1.18). Their appearance is due to gradual diffusion of new queen pheromones and loss of signalling from the old queen: these cells are transitory and disappear by the time the new queen is released and has commenced laying. We are left with a tried and true formula, replicating the emergency queen replacement impulse. To do this we:

- remove the colony queen, split the hive or establish an offset nucleus colony;
- introduce a queen or ripe cell to the queenless unit; and later
- check that the new queen has taken and is laying well.

Figure 1.18
Temporary supersedure queen cells developed shortly after queen introduction. Jerrabomberra Wetlands Apiary, 28 January 2016.

Supersedure queen replacement impulse

There appear to be two main exceptions to the dequeen before you requeen rule. The lesser known of these is where two populations of bees with queens in the same state are united. This is best illustrated by papering together two colonies each with a laying queens separated only by an excluder, a key step in establishing hives with two queens[53]. With some finesse this technique can be adapted to requeening but doing so successfully demands considerable beekeeping acumen.

The alternative is to rely on a new queen superseding the old under the exceptional circumstance where the colony is preparing to do just this. It is of course possible to create conditions conducive to colonies raising queens in the presence of the old hive queen (Figure 1.19) or to recognise, as did Gilbert Doolittle, and to either harvest the cells or allow the colony to proceed and requeen itself.

Surprisingly there are several schemes where a newly introduced queen (or gyne) is introduced in order to replace the existing colony queen. In practice, there is always the high risk that any old, and still well-functioning queen, will remove any new rival queen introduced. How might the direct queen replacement scheme be made to work? Could it circumvent the requirement to locate and to remove the old queen and avoid interruption to the brood cycle occasioned by the dequeen first strategy?

Figure 1.19   Supersedure cells induced by isolating young brood above excluder. Canberra and Region Beekeepers workshop, Jerrabomberra Wetlands Apiary 30 September 2019.

There are two options:

Scheme 1: This involves introduction of a virgin queen or queen cell to a queenright colony sporting an old queen. The notion of using queen cells to requeen is dealt in the sketch *Don Peer – The wisdom of lost beekeeping practice*. The eminent apiarist Dr C. C. Miller outlined a seemingly elegant approach to achieve this outcome way back in 1890[54]:

> *During the past summer I experimented somewhat largely with queen-cells and virgin queens. I wanted to decrease the chances for swarming by means of young queens, for I believe it is pretty generally conceded that a colony, with a young queen of the current year is less likely to swarm than one with an older queen.*
>
> *…I read in Doolittle's excellent book, that it is dangerous to allow a young queen to hatch [sic emerge] in a hive where there was an old queen, as the old queen would be destroyed if the colony did not swarm.*
>
> *…I'll get a young queen into the hive before there is danger of swarming. She'll supersede the old queen, and very likely there will be no swarming that season.*

A closer reading of Miller's narrative of placing queen cells in queenright hives showed that his experiment failed spectacularly: 54 of 55 introduced queen cells were destroyed putting pay to the accepted dequeen first before introducing a new queen wisdom. Aided by a very competent apiarist and sister in law Emma Wilson[55], they found a colony with the classic hinged cap of a newly emerged queen in that 55th colony. On searching they located the old and a newly emerged queen, Miller insisting that both queens be returned to the hive. The old queen prevailed.

Thus began the long search to find conditions more conducive to this short-cut approach to requeening. Outlined elsewhere is the remarkable success of Don Peer in achieving an 80% success rate in 4000 colonies – in a study spanning three years[56] – and the lesser success of others[57]. Many commercial apiarists now employ Peer's technique of introducing queen cells routinely to the top of colonies remote from the brood nest. Their rationale is that a newly emerged queen will replace an old and defenceless queen, the virgin producing no pheromone that the old queen might detect. The question remains: 'Will it work for the average punter?' It would appear conditions needed for it to be successful are strictly limited to those where supersedure would be likely to proceed. It appears that Peer introduced cells to colonies with hard worked queens, his scheme boiling down to:

- introducing a ripe queen cell to the top of queenright hive with an old queen; and
- checking whether the new queen had superseded the old queen and was laying.

Scheme 2: A much more attractive scheme was formulated by two-queen hive aficionados, Clayton Farrar, John Holzberlein, Charles Gilbert and Winston Dunham and others[58] in the mid 1930s to the mid 1950s.

They queened (or queen-celled) into a nucleus colonies formed by a division board (or double screen) located above the parent colony using bees, brood and stores from the parent colony. By mid spring they ended up with expanding colonies, each with a healthy nucleus colony with a young queen in the nucleus hive perched on top of each hive. Bees on steroids! To requeen they simply removed the division between the two colonies using a sheet of newspaper if a solid division board had been employed or simply pulling out the double screen.

In most instances the new upper queen superseded the old queen below mimicking the queen supersedure impulse. Their scheme can be surmised as:

- establishing a new queen (or cell) in a nucleus colony formed from and located above a strong parent hive;
- uniting the parent colony with the nucleus but only after the new queen was established and had emerging brood; and
- allowing the new queen to supersede the old queen, then consolidating the brood.

Swarm requeening impulse

Finally one might consider employing the swarming impulse to requeen, acknowledging that the presence of so many potential replacement queens may confound any requeening effort. However well built swarm cells, though potentially producing more swarm prone queens, can be used to advantage. This is especially the case where colonies are split and where each split – with a ripe swarm cell – will quickly form a new colony averting the need to build queens

from scratch. Without having to find the queen:
- hives preparing to swarm are split ensuring that each has at least one well formed ripe queen cell in each hive; and, if needed
- the splits are supered to minimise the risk of the queenright unit swarming; and
- the splits are united then checked after three to four weeks to ascertain whether the old queen has been superseded.

Having examined what we have learnt from bees to enable us to control the process of hive requeening, we need also to consider the aftermath of requeening and the impact of the stock chosen. Firstly beekeeper initiated requeening results in the complete turnover of the colony stock. This means that all replacement bees are unrelated to those of the just replaced parent colony queen. So apart from requeening altering the genetic makeup of a hive – deemed desirable where the colony is too defensive or disease prone – we need to realise that it takes time for a colony to return to its normal queenright condition. We might also question the calibre of the stock we are attempting to introduce to our bees.

The slow return to the fully queenright condition

Normal colony recovery following requeening Where a new laying queen is successfully introduced, a delicate process, she will normally commence laying within a few days. The conventional wisdom is to delay inspection for uptake for at least a week to avoid the risk of her being rejected. Consequently some queen breeders advise leaving the queen check until around the ten day mark.

Nevertheless a fairly prompt inspection is necessary since a colony sometimes rejects an introduced queen and you may not be able to determine that the colony has not raised one itself. However, in the normal course of events, the previous queen's worker and some of her drone progeny will be entirely replaced within a period of about six-to-eight weeks.

With self-requeening, that initiated by the bees not the beekeeper, the situation could not be more different. It takes sixteen days to raise a queen from a newly laid egg and up to another week to ten days for that queen to mate and commence laying. Under inclement weather conditions the mating process may be delayed for up to a further week. With any further delay, however, requeening may fail entirely and a queen – if she survives – will become a drone layer. With all natural queen replacement, except supersedure, there will always be a period where the colony is functionally queenless.

Emergency queen colony recovery With emergency queen replacement it takes a minimum of ten days for the colony to produce a queen and another week or so for her to be mated and to commence laying (see Table 1.1 above). So with any raised emergency queen, such as that induced by splitting a strong hive, only sealed brood will be present by the time the new queen is laying. Because emergency queens are often raised from older larvae, they often perform sub optimally and are frequently soon superseded.

Where a colony with queen cells is split, swarm queen cells will result in both colonies being quickly headed by well raised queens (see *Direct requeening*).

Supersedure queen colony recovery The transition is much smoother where queen replacement is direct. While the old queen may disappear before a new daughter queen emerges and mates, there will be little or no interruption to the brood cycle. Often enough the old queen will survive until well after the new queen is established and both may continue to lay together for some time.

In the absence of regular hive surveillance, the only evidence that such queen replacement may have occurred will be renewed colony vigour and the finding of an additional queen or only a few cells from which queens have emerged. I mark all queens and, having discounted instances where bees have swarmed, I identify supersedure by the presence of an unmarked queen.

Interestingly if a newly raised supersedure queen fails to return successfully to the hive after mating, a new batch of cells will be started. Thus with supersedure the risk of the colony ever becoming queenless is minimised.

Swarming queen colony recovery With swarming, however, a colony may be effectively queenless for many weeks, so it is quite possible for a surviving virgin queen to be present when all brood has long emerged.

At swarming preparation outset we have noted that the queen will have stopped laying. By this stage a fairly large number of queen cells, in varying stages of development, will have been started. Since the colony queen departs with the prime swarm, the swarm leaves virgin and emerging queens behind. The colony may then settle with a soon to be mated virgin queen or cast further swarms every several days (Figure 1.20). Even where a colony swarms just once, there will be extended delay in the colony returning to its single queen condition.

Figure 1.20 Secondary swarm trajectory at Peter and Jenny Robinson's garden apiary in Lyons in the Australian Capital Territory, 20 October 2020:
(a) initial swarming flight preparation at 13.57.28 with swarm departing hive around 14.11.00;
(b) colony arriving at sedge thicket 20 m away around 14.20.00 and settling over several minutes from 14.20.35; and
(c) six-frame nucleus hive put in place around 14.25.00 with most swarmed bees having entered the hive by 14.30.20.

I found and marked two virgin queens (using different marking colours) in a single colony preparing to swarm at the Jerrabomberra Wetlands Apiary on 20 September 2020. As it turned out the colony, after supering, did not swarm but there were obviously more queens present as I subsequently found that the colony was headed by an unmarked queen.

On a related note we – Danielle Harden and myself – have tracked queen succession in two-queen colonies. Two-queen colonies revert to single-queen colonies of their own accord near the end of the season and likewise do so under dearth conditions. Such reversions were common over the 2018-2019 and 2019-2020 summer drought seasons.

We have been investigating instances in colonies colonies with two queens where the queens were clearly mismatched and where supersedure or swarm cells appeared in one half of the consolidated brood nest. To date, it appears that a two-queen colony self requeening to retain the two-queen condition never occurs. Any new queen raised either does not survive or, if one does, it supersedes both the founding queens.

Despite the reputation of two-queen colonies rarely swarming, two of our two-queen colonies cast very large double swarms simultaneously during the 2021 spring build up period: conditions were particularly propitious and swarming was unusually prevalent. With one two-queen hive the swarms coalesced with two queens sighted though only one survived. In the other instance both swarms were collected, both sporting marked queens that were successfully reestablished in nucleus boxes. Both parent colonies reverted to the single-queen condition despite the presence of numerous queen cells.

This finding is corroborated by Dave Flanagan working with 1200 tripled hives over several years in the early 1990s (see *Stan Hughston – Stringy Hughston's coffin hives*). He noted that queen failure in any one of the three contiguous brood nests resulted in reversion of the tripled hive to a doubled hive (two-queen) condition. Such colonies were routinely restored to the tripled hive status by papering in a new ten frame nucleus colony.

In a surprising twist, we captured a swarm that had departed a doubled hive setup with one of the marked queens: the remaining parent colony queen – the one with the swarm cells was simultaneously superseded. Nevertheless the doubled hive reverted to the single queen condition. Farrar made the same finding for similar two-queen hives[59]. He noted that conditions conducive to initiation of supersedure or swarm cells were referred to the other brood nest:

> *Swarming is less of a problem in two-queen colonies than in strong single-queen colonies, but queen cells started because of a failing queen or crowding in either brood nest will stimulate production of queen cells in the other.*

In another instance I united two single queen colonies, both queens marked, to form a two-queen colony – a standard technique for their establishment. However one of the single colonies already had supersedure cells, that is the queen was already in the process of being replaced.

Instead of sensibly delaying the set up of the two-queen colony (that is until well after the queen had been replaced) I went ahead and united the two colonies anyway. My clear expectation was that one queen would supersede the other and I would end up with a large single-queen colony. However I successfully generated a two-queen colony containing both a marked queen and an unmarked supersedure queen. Here it seems likely that the failing queen was replaced and that the normal reversion to a single-queen condition rule was broken, albeit under an exceptional circumstance.

These findings point to the more practical problem of attempting to requeen hives that are preparing to replace their queens naturally. It is essential to change the intent of the bees to either swarm or supersede their queen for requeening to be successful. Attempts to requeen colonies under such conditions is fraught.

Stock selection

Stock selection is an eternal quest and one that begs the question of desirable frequency for requeening and what stock to requeen with. While there are some advocates of employing wild stock and some even hold that one should never requeen – practises I do not support – I have found that using commercial stock (or raising a few queens myself and evaluating their performance) is the key to hassle free and productive beekeeping. Though never the silver bullet, use of commercial stock reasonably guarantees colony productivity, gentle temperament and a measure of disease, particularly chalkbrood, control. Home raised queens will always mate with drones of uncertain heritage and cannot be relied upon for the more valuable traits that queen breeders in the main have good control over.

Nevertheless I have found, where swarming is controlled (*Keeping the lid on the box*), that within eighteen months 70% of requeened colonies are headed by an unmarked supersedure queen.

I concluded, as have other beekeepers, that colonies headed by young queens will regularly replace their queens irrespective of management practice. A key difference is that, while both supersedure and swarming result in a colony ending up with a younger and more fecund queen, with supersedure the colony workforce is retained.

Queen breeders expect the queens they raise to start laying at around the 10-12 day mark that is from the time of cell introduction. Noting that it takes one or, at the most, two days for ripe queen cells to emerge, this correlates well with the roughly seven days it takes for a queen to be mated and the additional 3-4 days it takes for her to start laying.

According to Harry Laidlaw Jr.[60], a renowned authority on queen raising, catching queens is typically timed at around fourteen days. Our preferred queen breeder catches and cages his queens at around day seventeen the exact date pending the need to accommodate a new batch of cells and weather conditions.

This brings us to the question of the calibre of the queens available on the market. The vast majority of queens sold to beekeepers are untested, that is they are captured and caged as soon as they are laying. For the most part these queens will perform well and are relatively inexpensive. That said, all reputable queen breeders aim to produce queens that are productive and whose progeny are both gentle and disease resistant. As well they will ensure that raised queens (and supporting drone mother colonies) receive optimal nutrition.

You can, however, purchase tested queens, more expensive select tested queens or even artificially inseminated queens, those that are held for longer and that are evaluated for their performance and genetic traits. However their high cost is not warranted unless you are planning to raise large numbers of queens for sale.

The traditional wisdom is that well-raised queens may live for several years but that they tend to be twice as swarm prone by the time they reach twelve months of age. Further their productivity – depending on how hard they have been worked – will suddenly decline, few Australian commercial queens performing by the time, if they survive, they reach two years of age.

Queensland bee researcher John Rhodes[61] demonstrated that queen acceptance can be improved if queens are held in their mating nucs for some weeks (Table 1.2). Rhodes also found that Australian queens tended to store less sperm than is recommended in Europe and North America – maybe explaining their short productive lives – noting that:

> *The 1999 experiment showed a positive correlation between increased queen survival and increased sperm counts with data suggesting that sperm counts of less than two million sperm per queen was associated with low survival of queens following introduction into honey production colonies...*

The practical reality is that not all mailing cage queens will be accepted. It is also a mistake to believe that any well-raised queen will perform well beyond 12-18 months of age making it a good idea to requeen at least every second year.

Day kept in mating nucs	Survival rate (%) at 2 weeks#	Survival rate (%) at 15 weeks#
7	15-28 [22 ± 9]	10-25 [18 ± 11]
14-17	48-70 [61 ± 12]	15-70 [51 ± 31]
21-24	80-93 [86 ± 7]	63-88 [74 ± 13]
28	85-95 [90 ± 7]	60-87 [74 ± 19]
31-35	88-95 [91 ± 4]	73-88 [80 ± 8]

Range [average and standard deviation]. Values are the average survival rates of several cohorts of 20 queens.

Table 1.2 Survival rate of mailing cage queens introduced to dequeened colonies illustrating dependence on time queens are held in mating nucs.

Overall I have found that at least nine out of ten queens, those I have bought from reputable queen breeders, are accepted and that they are laying well at first inspection. I attribute this seemingly high acceptance rate to the fact that I normally introduce caged queens to nucleus colonies or splits rather than to dequeened full strength colonies. Indeed the introduction of queens to nuclei rather than to full strength colonies – though I have no clear evidence for it – may enable queens to mature further in lieu of their being held longer in queen breeder mating nuclei.

From here I use these nucs, once well-established, in a flexible requeening program. This practice obviates the need to synchronise full strength hive requeening with arrival of caged queens in the post. The two-step process gives me full control of requeening at an apiary level rather than at the level of the individual hive and provides insurance against the occasional queen introduction loss.

II
The Honey Bee Swarm[62]

The productivity of a honey bee hive is linked the calibre of the queen and the drones she has mated with. It is also linked to the potential of that colony to avoid swarming. This is, of course, the apiarist's perspective. His or her job is to prevent swarming and to manipulate bees in such a way that they produce a large honey crop or become exceptionally efficient pollinators.

However from the vantage point of honey bees in a tree hollow, their potential to build quickly and to swarm will largely determine the survival of wild honey bee populations.

Bees have not evolved to produce a honey greatly surplus to their needs. A healthy managed colony has an annual maintenance budget of around 120-160 kg of honey and 30 kg of pollen. Bees instinctively choose a small hollow and store just enough to survive periods of dearth, winter in colder climes, and to reproduce. This has led me to conclude:

> *What do bees do best?*
> *They swarm reproductively to form new colonies.*

That beekeepers find swarming a very perplexing issue arises out of the innate desire of honey bees to produce offspring colonies: the raison d'être of bee and beekeeper are intrinsically a polar opposite. So keeping bees under the one lid was never going to be easy: survival is the key mission of the honey bee.

Natural swarming

Despite our ability to house bees in man made structures, bees are emphatically creatures of the wild. With remarkable ingenuity we have learnt to control some, but by no means all, of the inner workings of the honey bee nest. But beyond locating or moving bees to favourable sites and climes where plants bloom in profusion, we have little control over their foraging behaviour and limited ability to prevent their absconding.

In another context, we might ask why honey bee colonies sometimes fail. They can be entirely lost to swarming, that is where bees swarm themselves to death.

Particularly notorious are some strains of the Carniolan bee (*Apis mellifera carnica*). This was something I once witnessed when called out a number of times to collect swarms from the same parent hive of this ostensibly gentle strain of bee. The parent hive became a lost cause.

Wild and unmanaged colonies can and do botch their supersedure and swarming effort, occasionally failing to requeen themselves. Of course colony loss can be also be attributed to starvation, predation and disease rather than to queen misadventure. So while a colony is theoretically immortal, Thomas Seeley has demonstrated that, on average, wild colonies survive for roughly half a decade.

Swarming is the means by which bees makes good their losses. And beekeepers divide colonies, that is they artificially swarm them, to make increase. But bees have always had to also contend with short term survival, either by migration (as occurs amongst many tropical species and honey bee races) or, in the case of species that have adapted to colder climes, through hoarding of stores.

Instinctive storage

All wild honey bees have finessed honey storage to the point where they will survive any short period of dearth. Several species, the northern European race of the Western Honey Bee (*Apis mellifera mellifera*), the Black Chinese Honey Bee (*Apis cerana heimifeng*) and the Malaysian Honey Bee (*Apis nuluensis*) are particularly notable for having evolved to store much larger larders to survive extended periods of extreme cold. There absconding or migrating to flowering pastures is not an option. The Giant Mountain Bee (*Apis laboriosa*) of the lower Himalaya – also a high country species that forages as high as 4100 m – migrates to aestivate, that is survive as a dormant broodless winter cluster, at a much lower altitude.

Given any choice, European honey bee races will choose a nesting cavity – a wall cavity, a compost bin or a tree hollow – close enough in size to that of a single standard full depth Langstroth honey super (~40 L). But the honey bee, any honey bee, will swarm as often as possible and will store only just enough honey to tide it over any long winter or dry season. Swarming is an evolved survival strategy, not an invitation to provide bears, mice, ants and human beings with a free lunch.

At the other extreme some African races of the Western Honey Bee will swarm despite an absence of honey stores and in response to a pollen flow. All but one of the giant honey bees are famed for exhausting stores and migrating vast distances – like migratory beekeepers – to follow flowering plants. Dwarf honey bees similarly abscond – swarm non reproductively – in response to dearth, disturbance and predation as do many tropical races of the Asian Honey Bee (*Apis cerana*) and amongst species closely related to it. A swarming pattern to fit all contingencies and honey bee species simply does not exist.

Imposed swarming

Swarm control measures that work best are those that are in sympathy with bee instincts. Most notable of those is splitting of colonies during the spring buildup. Colony division artificially swarms a hive. As already noted, the hackneyed practice of removing swarm cells, while immediately putting a stop to a hive swarming, does nothing to suppress that swarming instinct. The bees promptly build a new swarm cell batch. A better strategy is to divide such colonies, ideally onto closely juxtaposed hive stands that allows their being reunited after the swarming season has ended.

Beekeeper attitude towards swarm control has evolved in leaps and bounds but over a timeframe of centuries. Traditional honey hunters relied on swarms to replace colonies lost to harvesting and to natural attrition. Over exploitation and destruction of every nest would have put them out of business. So harvesting practices were curtailed. Hunters learnt to protect bee trees and to rely on swarms repopulating nesting hollows.

Eva Crane traces many traditional beekeeping practices. Bees kept in containers – pots, log hives, skeps – would either attract swarms or be provisioned with swarms, mainly those emerging from the many hives that such beekeepers kept. Swarm control may not have even been envisioned: swarms were the single means of making increase. The argument went: 'The more swarms the more honey'. With the wisdom of hindsight we find that supporting that practice was a recipe for diminishing returns.

Then, with the advent of frame hive, came a golden age of beekeeping. Swarm control measures came quickly, notably that propounded by George Demaree (1832–1915). Well practiced, his technique and the splitting of hives greatly improved honey yields.

However none of these practices changed the reliance of wild colonies on bees swarming to replace their number. The way we conduct beekeeping – spreading mites, reduced forage associated with changing land use, large scale migration of bee hives and use of chemicals – has certainly diminished the success of swarm control in boosting hive productivity.

In setting the scene for swarm management, I mused on an allied topic, warming to swarming, I had previously given some attention to. This topic questioned the beekeeper construct of practising swarm control.

Warming to swarming

In a short note to local beekeepers half a decade ago, I pondered the issues most often floated about the need or otherwise to effect swarm control, can I stop or prevent swarming?

While many natural beekeepers advocate allowing bees to swarm, and there is perhaps some merit in allowing them to do so, there is a very a very practical reason for not allowing them to buzz off. When early swarms take off you can wave goodbye to the prospect of a good summer honey crop. Another reason to prevent bees swarming is to factor in the nuisance value that they present to *your* neighbours. There is also the reality that *your neighbours* may have to pay good money to have bees removed from walls and chimneys. For an apiary operating in a public space, as our beekeeping club does, preventing swarming is an imperative both from a safety and public relations perspective. But let us put this issue aside and ask ourselves: 'Can we really and truly prevent swarming?'

In making a good start to a beekeeping season, swarm control – disease inspection aside – would appear to be a first priority. The local spring 2016 and 2021 seasons were such a time and bee management turned out to be particularly problematic. The 2016 season started well and swarming was widespread in south eastern Australia though lack of rain meant that conditions quickly deteriorated. Nothing seemed to flower and almost every fixed location apiary ended up with an abysmal honey crop. It was a truely swarmy-swarm season and nothing came of the beekeeper's best efforts.

The season of 2017-2018 was different. It started badly but quickly turned around to be a bumper harvest year. So seasonal conditions beyond our control play a part and we should never assume we can truly control bees. If you are at hand

and know what to do, most swarm seasons can be well managed. The problem is we are all busy and leaving the bees alone for any period longer than a fortnight means that, in the normal course of events, bees will and do swarm.

The south-eastern Australian bee season of 2021-2022 was quite something else. Conditions were exceptional in every sense. Almost every weed and every tree came into flower, and there was a continuous flow from late winter for a full nine months. The apiary of every beekeeper lost bees to swarms, even those who supposedly knew how to manage bees well. As it turned out, it was a knockout year for honey production. The bees had it their way as well: they swarmed promiscuously.

Can I prevent swarming?

Often I hear people say: 'I cut out swarm cells to prevent swarming'. We know that doesn't stop the bees as their instinct is unchanged. As noted, they will simply raise a new batch straight away.

With supersedure, however, queen replacement proceeds without swarming. Telltale signals of supersedure in progress are indicated by few cells in total located inside the brood nest (Figure 1.21). Simply close the hive and allow this helpful self requeening to proceed. Come back a few weeks later to check that the colony has a new unmarked queen and that she is laying well.

Figure 1.21 Supersedure in progress.

One hundred and twenty five years ago George Demaree boldly proclaimed:

> *You can do it without a doubt, by practicing my new system of preventing swarming: and if you have the ingenuity to apply proper management to suit the new condition, your surplus yield will be larger than by any other method heretofore made known to the public.*

Of course the *Demaree Plan* is now very well known and other measures – reversing brood nests, providing space in the brood nest and space to cure and store nectar above the brood nest, regular requeening, and splitting hives or judicious brood spreading – have been added to the list of practical measures to arrest swarming.

Fed up with reading standard beekeeping swarm control texts and reflecting on my own necessarily limited observations and experience in managing swarming, I decided it were time discover more about what bee researchers had learnt from bees.

Sophisticated swarm control

Two approaches to swarm control, requiring an intricate knowledge honey bee nest dynamics, were published in the May 1952 and the April 1954 editions of the *American Bee Journal*. These watershed documents were the contributions of genius beekeepers in the likes of Mykola Haydak and Clayton Farrar.

The first of these entitled *Swarming Roundup* focussed mainly on very nuanced versions of the Demaree Plan that also facilitated requeening. Their schemes achieved the redirection of bees with swarming intent to building a new brood nest and subsequent replacement of the old queen with a daughter queen that provided renewed colony cohesion.

The second tranche approach, replicating the May 1952 schemes, targeted the appearance of second queen to bolster the queen pheromone complement of the colony, suppressing the natural swarming instinct. The approach did not focus on the remarkably successful approach to running two-queen hives, but borrowed heavily from the approach to establish such hives. The end result was that they ghosted the earlier Demaree schemes, also with the short term objective of replacing the existing colony queen.

The approaches proved adept in avoiding the time consuming practices of finding and removing old queens, the classic requirement for successful requeening. In an approach taken by many commercial beekeepers, a new queen (or queen cell) is introduced to the top of a hive beyond the influence of the existing colony queen. Well managed the new queen becomes established and subsequently supersedes the old queen.

In normal backyard practice this approach is not reliable as the new queen may fail. However, by ensuring that the new queen is well established before supersedure is allowed to go ahead, swarm control and queen replacement – with minimal intervention – can be achieved together. This topic will be perused in finer detail under *Requeening with swarm control*.

With these broad observations, we can now examine the roles of both workers and queens in orchestrating swarming and turn to examining those measures that have been discovered to exercise a modicum of swarm control.

Out of the box

Eva Crane's book *The archaeology of beekeeping* provides an account of Sir George Wheler's 1682 visit to St Cyriacus Monastery located on the Hymettus mountain range in the Athens area of Attica in eastern central Greece. In it Crane relates Wheler's account of manipulation of Greek top-bar hives, the first ever scientific observation of honey bee swarm control (Figure 1.22)[63]:

> *The Hives they keep their Bees in, are made of Willows, or Osiers, fashioned like our common Dust-Baskets, wide at the Top, and narrow at the Bottom, and plaister'd with Clay, or Loam, within and without: They are set the wide end upwards, as you see here, The Tops being covered with broad flat Sticks, are also plaistered with Clay on the Top, and to secure them from the Weather, they cover them with a Tuft of Straw, as we do. Along each of those Sticks, the Bees fasten their Combs, so that a Combe may be taken out whole, without the least bruising, and with the greatest care imaginable. To increase them in Spring-time, that is, in March or April, until the beginning of May, they divide them; first separating the Sticks, on which the Combs and Bees are fastened, from one another with a Knife: so taking out the first Combs and Bees together, on each side, they put them into another Basket, in the same Order as they were taken out until they have equally divided them. After this, when they are both accommodated with Sticks and Plaister, they set the new Basket in the Place of the old one, and the old one in some new Place. And all this they do in the middle of the day, at such time as the greatest part of the Bees are abroad; who at their coming home, without much difficulty, by this means divide themselves equally. This Device hinders them from swarming, and flying away.*

Wheler goes on to provide a detailed scheme for the seasonal management of the bees needed to rebuild the split colonies. He makes special note of the very different needs of bees in the mild Grecian climate to those that would be needed in England:

> *In August they take out their Honey, which they do in the day-time also, while they are abroad, the Bees being thereby, they say, disturbed least. At which time they take out the Combs laden with Honey, as before; that is, beginning at each out-side, and so taking away, until they have left only such a quantity of Combs in the middle, as they judge will be sufficient to maintain the Bees in Winter; sweeping away those bees that are on the Combs they take out, into*

the Basket again, and again covering it with new Sticks and Plaister. All that I doubt concerning the Practise of this here in England, is, that perhaps they gather a less quantity of Honey; and that, should they take like quantity of Honey from the Bees here in England, they would not leave enough to preserve them in Winter. For this hinders not much: For by being less covetous and not taking so much Honey from the poor Bees, the great increase and multiplying of them would soon equalise, and far exceed the little Profit we make by destroying them. This is done without smoak; wherefore the Antients call this Honey '\Αχαπνιξον Unsmoaked Honey. And I believe the Smoak of Sulphur, which we use takes away very much the Fragrancy of the Wax; and sure I am, the Honey can receive neither good Taste, nor good Smell from it.

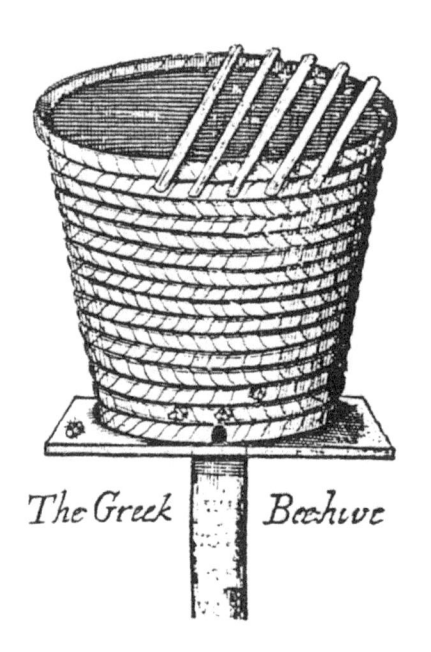

Figure 1.22
Greek top bar hive sketched by Wheler [1651-1724].

In this account we have a record of the colony split, a well recognised swarm control measure. By and large however swarming continued to be encouraged as it was the only practical way of obtaining bees, there were few methods of moving bees other than for short distances and hive management was limited to varying the size of the hive receptacle.

It was to take another 100 years for the Swiss naturalist François Huber[64] [1750-1831] to appear on the scene and to make the first systematic in-hive observations. Letter IX, *On the formation of swarms* describes his 1788-1799 study of the swarming process in exquisite detail:

I can add but a few facts to the information M. de Reamumur has communicated relative to swarms... In the course of spring and summer, the same hive may throw several swarms. The old queen is always at the head of the first colony; the others are conducted by young queens. Such is the fact which I shall now prove; and the peculiarities attending to it shall be related...

The Reverend Johann Dzierżoń[65] [1811-1906], sometimes referred to as the true father of beekeeping[66], became famous for other discoveries including bee parthenogensis, the production of royal jelly and the spacing of frames so they could be easily removed. Amongst his most notable contributions, he pioneered practical principles for transfer of fixed-frame colonies and bees to moveable-frame hives:

> *It would be superfluous to speak of the season when the occupation should take place if many bee-keepers in their eagerness to introduce these hives into their apiaries did not come to grief by transferring stocks from log or straw-hives in autumn—a time quite unsuitable. They not only thereby do themselves harm, but create prejudice against the new hives, because of the unfavourable result is ascribed to the hives rather than to the unsuitable time when the changes are made.*
>
> *New hives should be occupied when the bees are naturally ready for moving. There is no need to wait the full arrival of swarming time, as is the case of other empty hives, because we can provide comb for the bees beforehand so that eggs may be laid immediately, and thus hives may, at any rate, be stocked in April.*

The impact on beekeeping of Dzierżoń's and Huber's observations, and those of the Reverend Lorenzo Lorraine Langstroth's 1851 – similarly described as the father of modern apiculture – often-claimed discoverer of bee space[67] (see *Art thou a skeppist*), and of their moveable frame hives, were to banish the skeppist approach to keeping bees. Quite quickly apiarists reliance on issued and captured swarms was replaced by a strategy of active swarm control and prevention.

Langstroth [1810-1895] was the first to clearly articulate the importance of very practical moveable frame hives for modern beekeeping practice. The 1862 Massachusetts advertisement for his Langstroth Hive articulates this momentous leap (see *Bee space and bee hive architecture*):

Langstroth's position on the control of swarming is ambivalent as was the then contemporary practice of hiving swarms as a means of increase. Swarms were a source of new queens and the only means of establishing new colonies. So the notion that swarming should be encouraged prevailed and presaged the active raising of new stock by the likes of Henry Alley[68], Gilbert Doolittle[69] and Charles C. Miller. They used trimmed comb containing worker eggs to start queens in temporarily queenless hives to establish new colonies, a practice that would usurp the traditional reliance on swarm capture.

The contemporary beekeeper perspective of swarming is usually one of practical minimisation rather than outright prevention. The shift in thinking came in the late 19th Century and early 20th Century where the beekeeper set out to entirely thwart the natural process of swarming, exemplified by Charles Miller in his 1911 book *Fifty years among the bees*[70]:

> *If I were to meet a man perfect in the entire science and art of bee-keeping, and were allowed from him an answer to just one question, I would ask for the best and easiest way to prevent swarming, for one who is anxious to secure the largest crop of comb honey.*

Swarming takes many forms and it is not only for the sake of setting up a new nest do the bees leave home. While bees venture out to harvest stores, to take out the garbage, to take cleansing flights after long periods of confinement and to make orientation flights these are not swarming events though beekeepers sometimes mistake them as such.

Bees also abscond, that is abandon their nest. They may do so shortly after being hived because the nesting cavity is too small or is contaminated by oils and chemicals. Many species and races of honey bee abscond when they are raided by predators, ravaged by disease or face starvation. Some, such as many Asian and African honey bees, abscond deliberately and innately on a seasonal basis, a few migrating great distances.

We start by examining the pattern of swarming where bees depart to found a new colony.

Reproductive swarming

Ecologist Mark Winston's bee-oriented perspective[71] is one of equating swarming to reproductive colonisation. Swarming occurs under a narrow window of near ideal conditions, a combination of intrinsic demographic (colony) and extrinsic (environmental) factors that enable successful establishment of new colonies[72]. Winston and coworkers[73] have demonstrated the importance of queen substance[74] secreted by her mandibular glands[75] in regulating swarming comprising 9-keto-2-(E)-decenoic acid, the two stereoisomers of 9-hydroxy-2-(E)-decenoic acid, methyl p-hydroxybenzoate and 4-hydroxy-3-methoxyphenylethanol. The effect of these simple organic compounds is further enhanced by coniferyl alcohol, methyl oleate, hexadecane-1-ol and linoleic acid also produced by queen bees[76].

While many manipulation practices work well to arrest swarming, many of the reasons put forward as to why they actually work are buried in folklore. Much of what we actually know about swarming is based on the founding studies of James Simpson[77] in the 1940s and 1950s[78]. He established that a combination of queen age, a crowded brood nest and inadequate honey storage space along with seasonal conditions are the main factors initiating swarming. Control strategies involving open brood nest architecture, giving the queen ample room to lay, and the timely supply and arrangement of honey supers and stores certainly delay swarming. Conversely providing more space than bees can comfortably manage, especially in times of dearth or in winter, will arrest colony development and make bees prone to stress-related disease.

His specialised studies apart, Simpson was fully cognisant of the very practical problems swarming presents to beekeepers and the amount of effort required to control swarming. He noted presciently[79]:

> *Since many colonies do not swarm, particularly if they have unrestricted space for bees in their hives, some estimate of the proportion which will do so must be made before the value of attempts to eliminate swarming can be assessed.*

To get this handle on the propensity of honey bees to swarm Simpson assembled a large swathe of records of hives that would have been requeened regularly and generally well managed. He found that between 10 and 40% of colonies would swarm in an average year, that is if they were given excess hive space and otherwise left alone. He also examined records of 300 hives that had been requeened the previous summer that showed that honey bees are very active in managing their queens (Table 1.3). About half the hives had made some attempt to replace their queens or were at risk of failing. With a different dynamic honey bees regulate their drone population.

Queen focussed assessment of hives	Percentage of hives
Had unsatisfactory queens that were given new ones	5
Became queenless before beginning queen rearing and were given new queens	4
Began queen rearing, but stopped without swarming	24
Began queen rearing and subsequently may have swarmed or might have done so if not given new queens	18

Table 1.3   Requeening propensity in managed hives.

The most successful swarm control measures – the Demaree operation is but just one – are those that closely mimic natural swarming, notably hive splitting techniques where the new colony suffers an extended period of broodlessness. There the split off colony develops an emergency queen and, in most other ways, behaves like an issued swarm. By the time this replacement queen commences laying most brood will have emerged and will be more than ready to nurse brood. Meanwhile the old stay-at-home queen lays into a colony depleted of much of its stores and workforce and closely resembles the parent colony that has swarmed. In a swarmed colony, the old queen will depart, at least with the prime swarm, so that in the practice of hive splitting, the swarming condition is not perfectly replicated. In the normal course of events the original colony will abandon any attempt to swarm. For the two colonies, one newly founded, to remain productive they must both recover and reestablish themselves in time for the main honey flow and to survive another winter.

The experienced beekeeper will, if needed, simply reunite the split colonies after almost all risk of swarming has passed and do so at the beginning of the main honey flow. As we shall see, the practice of splitting works best where the colony is manipulated before the colony commences preparations to swarm since split colonies with retained swarm cells may still swarm.

Other successful swarm control measures are focused on queen and, less likely, drone management. A necessary condition of swarming is the presence of an additional gyne, namely a potential replacement queen: queen cell cup eggs, queen larvae, queen cells, virgin queens and, in not infrequent instances, laying workers termed gynaecoid females. Without another gyne, that is without an alternative laying queen – the normal colony condition – a hive cannot swarm successfully and will not do so. The importance of the queen will be further elaborated on once we have examined the role of the worker bee.

As already intimated, some control strategies are focussed on management practices aimed purely at preventing swarming, working against rather than with the bees.

The role of worker bees in swarming

A nuanced approach to controlling swarming requires an understanding of the underlying behaviour of worker bees. This necessitates an appreciation of not only the signs and conditions that signal imminent swarming but also of the ever-changing roles and task orientation of worker bees. You cannot just *Tell the Bees* (Figure 1.23) that their brood nest is crowded or that there is insufficient space to put honey and pollen into storage[80]:

> *This is the common belief that bees must be made acquainted with the death of any member of the family, otherwise these intelligent little creatures will either desert the hive in a pet, or leave off working and die inside of it.*

However the notion that these intrinsic hive conditions are chemically signalled to worker bees and presage swarming warrants very close attention[81].

John Hogg, in reviewing the extensive literature of what is known about swarming, provides a cogent interpretation of the swarming phenomenon[82]. Earlier[83] he reported on the practical findings of Bernard Möbus[84]: Möbus's genius and keen observation of bees, made idle by brood-bee imbalance (too many young emerging nurse bees to be accommodated in the nursery) and too many nectar receivers and wax producers (more than needed to receive incoming nectar) or lacking space to store and ripen nectar, explains the swarming phenomenon remarkably well. The Möbus narrative tells us seemingly most of what we need to know about how swarming is initiated and practiced. His and Hogg's ideas are elaborated in the following.

Hogg has designated idle bees as a temporal swarming caste. Young workers, freed of cell cleaning, brood tending and thermoregulation of the brood nest, and isolated from queen pheromone that suppresses queen and drone cell building and ovarian development[86], take control of swarm preparation. If conditions such as weather change suddenly, as they often do in spring, worker bees in close contact with the queen will tear down queen cells, continue to feed the queen and prevent her departure from the hive. Under favourable conditions, however, this swarming caste not only prepares a colony to swarm, but also orchestrates the actual process of swarming. They first cease feeding the queen, then they eject her from the colony, conduct actual swarming, locate a new nesting cavity and quickly construct comb in the colony budded off and established in a new cavity. These swarm caste workers also appear to micromanage after-swarms and accompanying virgin queens.

Figure 1.23   Telling the Bees – Charles Napier Hemy's painting, The Widow (1895)[85].

Convincing lines of evidence for the existence of idle workers, apart from the practiced eye of overly high bee numbers of newly emerged workers on brood comb or bearding outside hive entrances, is the observation that – as swarming preparations get underway – many bees are packed-to-the-gills with nectar and hang off combs outside the brood nest. These worker bees facing insufficient super storage space are well known to act as repletes, that is – much like honey ants[87] that have a temporal caste that engorges honey stores – act as short-term honey storage vessels.

Such bees, being no longer actively engaged in receiving and processing incoming nectar, building comb, and processing and storing honey – apart from acting as a living honey repository – are as useful to the mother colony as the idle couch potato. Similarly, very young workers, forced out of the brood nest to facilitate ventilation, and that are surplus to requirements for brood nest duties, will congregate outside the brood nest and are isolated from circulating queen pheromone normally achieved by trophallaxis, that is the regulated exchange of food. Large numbers of emerging workers leads to a bee-age imbalance and this is reflected in their being surfeit to immediate colony requirements.

This theory, explaining the culmination of brood rearing, honey storage, brood bee-balance condition and favourable environmental conditions external to the hive, also explains the success and occasional failure of other tried-and-true swarm control measures.

Preparation for swarming is also indicated by a sudden cessation of wax production and comb construction as Demuth observed[88]. His classic treatise anticipated better methods for swarm control if not prevention.

> *Gradually methods were devised for the prevention of afterswarms, and systems of management were worked out whereby the actual working force of the colony is not divided by the issuing of the prime swarm. During more recent years methods have been devised by which swarming is either prevented entirely or the act of swarming is anticipated by the beekeeper, which permits the control of swarming without constant attendance. This has made it possible for a beekeeper to operate a series of apiaries without an attendant in each to watch for and hive the issuing swarms.*

His 1921 *Farmers Bulletin* description of the swarming sequence is hard to better:

> *About a week after the issuing of the prime swarm the first of the young queens in the parent colony emerges from her cell. Instead of destroying the other young queens and establishing this first emerged young queen as the new mother of the colony, the bees usually swarm again about eight days after the prime swarm has issued, this after-swarm being accompanied by one or more young queens. Other afterswarms, each one smaller than the preceding, may issue with an interval of one or two days between until the colony is so reduced in numbers that further swarming is given up and all but one of the remaining young queens are killed. About 10 days after emergence the surviving young queen usually begins to lay, and normal brood-rearing is again established in the parent colony after an interval of at least 16 days during which no eggs have been laid. Each after-swarm establishes itself in a*

new abode, begins building its combs, and the young queen begins to lay about 10 days after emergence. If sufficient food is available such colonies may build up to normal strength for winter. This is the natural method of reproduction of colonies in the honeybee.

Several years ago I was fortunate enough to be called a number of times by a 'neighbour of a bee owner', hardly worth the title a keeper of bees, who allowed his bees to swarm freely. This presented a rare opportunity to observe how unmanaged bees perform. I collected a number of small swarms from this neighbour's property. Presumably bees had exhausted prospective cavity nesting sites as two of a number of swarms I collected had built nests in the open (Figure 1.24).

The role of the queen in swarming

A colony deprived of an incipient queen may still swarm but without the queen. That swarm, itself being queenless, and having no potential to survive, will almost invariably return to the parent hive[89]. Since the swarming propensity is not lost by removal of queen cells and stray virgin queens – and even just one of these is easily missed – colonies will often subsequently achieve their pre-programmed goal of swarming. The well established practices of clipping queen wings and removing all queen cells therefore would appear to do little to prevent swarming and usually only delays it.

Queen status is a critical factor in determining the readiness of a colony to swarm. In the normal course of events, the queen is the singular mother of any and all hive progeny, adult as well as laying workers, drones and that special caste of female bees we have already visited, gynes. Workers, under the influence of queen mandibular pheromones, remove potential usurpers and even remove drones and drone larvae, if present, in a failed honey flow. Under conditions of dearth, they may also cannibalise worker larvae surplus to colony needs.

Figure 1.24 Regular and a long resident swarm at the same location at Waniassa in the Australian Capital Territory in late October 2016.
Photos: Alan Wade

Occasionally a prosperous colony preparing to supersede its queen may swarm (Figure 1.25) but this is ever only likely to occur when good conditions coincide with those likely to favour swarming.

While regular removal of all queen cells and all virgin queens will always prevent a colony swarming, culling all cells at the very least every ten days is essential since poor queens can be developed from worker larvae even older than three days of age. Further missing one such hard-to-find queen cell will anyway result in swarming. My repeated take home message is not to rely on cutting out queen cells to prevent or even delay swarming though many beekeepers rely on this measure. This said, the successful swarm control technique of removing the old queen and leaving only one swarm cell, one advocated by the famous section comb honey producer Eugene Killon[90], is testament to the importance of the physiological status of both the queen and worker bees.

These findings can be employed to understand other hive phenomena that occur in the lead up to swarming, notably idle bees disengaged from hive activities being those that encourage and engage in building of drone comb and the phenomenon of swarm queen cells usually being found at the periphery of the brood nest.

Figure 1.25 Typical supersedure induced swarming at Jerrabomberra Wetlands in early November 2016.

Other genetic and environmental factors influencing swarming and the special case of non-reproductive swarming, absconding, will be discussed in terms of its importance for honey bee survival. Non reproductive swarming has special adaptive value across all honey bees. Its role in optimising survival of tropical bees and amongst species of honey bees other than European *Apis mellifera* in the phenomenon of colony migration provides further insight into underlying factors controlling swarming.

Environmental factors affecting swarming

There are other genetic and queen-age related factors that affect swarming and these also need to be acknowledged even if they are not yet fully comprehended. Further the influence of seasonal factors: day length, temperature, rainfall and nutrition are by no means well understood.

Hepburn and Radloff denote a pattern of rainfall and flowering influencing swarming amongst African honey bees. Sufficient to say swarming will be abandoned in periods of dearth while additional swarming often follows unusually heavy rainfall. For sub species such as *Apis mellifera scuttelata* and *Apis mellifera adansonii* there are multiple swarming seasons. Hepburn and Radloff summarise these findings simply[91]:

> *The phenology of reproductive swarming in African honeybees closely corresponds with local climate and weather and the availability of forage…*
>
> *Flowering, a pollen dispensing mechanism, in turn drives brood production and reproductive swarming. The well-defined sequence of colony events forms a linear chain of relationships: peak rainfall –> peak flowering –> maximum brood-rearing –> reproductive swarming.*

One other intriguing facet of African honey bee behaviour is swarming in small colonies of *Apis mellifera capensis*, *Apis mellifera scuttelata*, *Apis mellifera adansonii* and *Apis mellifera jemenitica* triggered by pollen flows. That pollen trapping might be a means of swarm control in commercial apiaries is being investigated by Victor Croker and David Leemhuis at Australian Honeybee.

More thoroughly studied are European races of the Western Honey Bee. Young queens are well known to have a higher fecundity – that is, they lay more eggs – and produce more queen pheromone than do older queens, no doubt replicated across all honey bees in the *Apis* genus and indeed also amongst stingless bees. With the potential of a larger workforce to replace diseased bees and to care for brood, and for old overwintered bees to become foragers and less attendant to early brood rearing, the whole spring buildup strategy is advanced and swarming can occur from, under local conditions, from any time from late winter.

The pattern of production of worker bees also varies between honey bee races and in response to pollen availability and honey stores. Oliver[92] stresses the importance of bee nutrition, particularly the role of pollen stores. Pollen is converted to the egg yolk protein vitellogenin stored in fatty tissue by diutinus (overwintering or dearth) bees. This protein maximises bee longevity and is important for late winter brood rearing. High levels of vitellogenin are also found in bees preparing to swarm but its role in regulating swarming is not fully understood[93].

Large differences in the propensity of different races to swarm are fairly well known, at least amongst races of European honey bees. For example, the Carniolan Bee (*Apis mellifera carnica*) has a long-standing reputation[94] of swarming easily and also building up quickly under favourable conditions. This race is widely cultivated in Europe but, as Brother Adam[95] notes from his travels, there are strains of this race that are much less swarm prone. He also notes that the cultivated Syrian Honey Bee (*Apis mellifera syriaca*) has an excessive propensity to swarm.

The widely cultivated and well studied race of African Honey Bee, *Apis mellifera scutellata* (from the central eastern region of southern Africa[96]) is reputedly a poor comb builder and this may be a factor in its being a prodigious swarmer.

Non reproductive swarming

Here we start with the scenario where the colony does not swarm but is swarmed into, the bees being usurped.

Usurping swarms

There is some evidence that late swarms are not primarily destined to be reproductive swarms and may, instead, be issued for the express purpose of taking over existing colonies[97], an idea first proposed by Mangum[98]. Hence bee colonies unlikely to survive supplant less fit colonies exploiting their stores and bee population. Such an adaptive strategy is not dissimilar to that of the already-described parasitising Cape Honey Bee.

Absconding – Non reproductive swarming in response to predation and limited resource availability

When a whole colony abandons its nest, it does so largely to survive. While it is rare for European honey bee races to abscond – they usually starve if conditions are poor – they will leave a cavity if conditions are unsuitable at the outset as sometimes occurs when a large swarm is hived into too small a nucleus box. Even amongst cavity dwelling species, there is considerable variation in their propensity to abscond: tropical ecotypes of *Apis cerana* and *Apis mellifera* more readily abscond than do their temperate counterparts[99].

Propensity to abscond is commonly a response to variable resource availability, whereas for temperate races of *Apis mellifera* and *Apis cerana* reliable annual honey flows and long winters favour a strategy of storage. More narrowly Hepburn and Radloff[100] describe the behavioural disparity between European and African races of the Western Honey Bee:

> *Perhaps the most striking differences among the lineages of Apis mellifera involve the extent of resource investment in the reproduction and mobility of the colonies.*

African races of the Western Honey Bee and tropical Asian honey bee species, including a number of races of *Apis cerana* and its close allies, *Apis koschevnikovi*, *Apis nuluensis* and *Apis nigrocincta*, regularly abscond. Indeed all swarm both reproductively and non-reproductively with greater regularity than does the European honey bee.

With absconding there is, of course, no increase in the number of colonies. Open nesting Apis species (*Apis dorsata, Apis laboriosa, Apis nigrocincta, Apis florea* and A*pis andreniformis*) not only also swarm reproductively in a fast and furious manner, an extreme strategy based on taking full advantage of the good conditions, they all abscond when and as conditions dictate. They also leave nesting sites in response to heavy or repeated predation, important factors for species that nest on a single comb, the giant and the dwarf honey bees, in the open. There is less investment in storage and comb building and more investment in nest defense and in moving the nest to a more sustainable food source.

Little known is the fact that colonies that lose their queen may swarm, that is abscond, rather than raise a new queen. While a frame of young brood is often added to a freshly hived swarm to help 'hold the bees', the alternative scenario where a queenless split has brood but no queen the situation is quite different. Mark Winston[101] argues that while absconding will disadvantage bee survival – they lose built comb, stores and the windfall of developing bees that might add to their numbers – the workers themselves have no queen pheromone to exert the cohesion required to maintain normal colony functioning. Sometimes colonies raise a new queen, the course normally followed when strong colonies are split. At other times they may swarm in response to queen loss. A further and common outcome is that, after being queenless for two to three weeks, laying workers will commence laying unfertilised eggs on worker cell walls and the colony will become almost impossible to requeen. The signature then becomes one of scattered domed drone brood in worker cells, old worker bees and as previously noted numerous small drones.

Migration – Non-reproductive swarming in response to climatic and regionally predictable resource availability

Further examination of the swarming behaviour shows that the phenomenon of 'failed local condition-induced' absconding may extend to seasonally programmed colony migration across the *Apis* genus[102].

Oldroyd and Wongsiri[103] provide a wide ranging review of the biology and evolution of Asian honey bees including an overview their reproduction, swarming and migration. For example *Apis dorsata* will never use wax from old comb even if returning swarms nest within centimeters of deserted comb whereas *Apis florea*, while not reestablishing nests on old comb, may nevertheless scavenge old wax. Such strategies involving a tradeoff between employing wax, a valuable resource, and not using old wax to minimise the risk of disease and mite transmission. This behaviour is in strong contrast to that of cavity dwelling species that readily occupy comb in old nest cavities. There the reward of much larger amounts of wax may enhance their chance of survival and reproduction. There are also nuanced differences between cavity-dwelling honey bees (*Apis mellifera* and *Apis cerana*) and the open-nesting giant and dwarf honey bees in normal reproductive swarming behaviour. There appears to be evidence that at least some open-nesting species fly directly to a new nesting site and that this may be orchestrated by bees performing orientation dances inside the parent colony prior to swarming: in other species dancing associated with location of a new nesting site has not been observed. These behaviours may be an adaptation to open-nesting species not needing to locate a scarce nesting hollow.

A parallel perspective of the pan tropical Western Honey Bee subspecies by Hepburn and Radloff[104] suggests parallel evolution of swarming and migration in *Apis mellifera* bees of African origin. The patterns of swarming are very complex, vary between species and vary regionally even within the one race of bees. For example they state pointedly that:

> *The phenology [natural cycling or timing] of reproductive swarming in African honeybees closely corresponds with local climate and weather and the availability of forage.*

The giant honey bees (Apis dorsata [Apis dorsata dorsata], Apis breviligula, Apis laboriosa and Apis dorsata binghami)

Giant honey bee colonies occupy a single comb, where colonies may cluster communally. Robinson[105] as well as Underwood[106] and Woyke and Wilde[107] have described the cyclical annual swarming and migratory behaviours of these species. Like migratory birds, colonies often return to the same nesting locations but not to their original nest. Their movement patterns vary widely so that, while lower altitude populations of *Apis dorsata* may migrate over large distances, up to hundreds of kilometers, Himalayan Honey Bee (*Apis laboriosa*) migrations are elevational and colony relocation distances are relatively short. The overall annual pattern is one of colonies expanding rapidly and swarming to establish new colonies, then abandoning their nests and migrating. Robinson also describes a bivouacking (staged migration) behaviour of *Apis dorsata* and the importance of preservation of these sites to protect the integrity of populations of this bee. These bees are famously raided for their honey stores[108] and are important wild pollinators. The behaviour of these bees is considered in greater detail under *The giant honey bees*.

The cavity dwelling honey bees (Apis cerana, Apis koschevnikovi, Apis mellifera, Apis nigrocinta, Apis nuluensis and Apis indica)

Koeniger, Koeniger, and Smith[109] provide a broad interpretation of the swarming behaviour of this group of honey bees in evolutionary terms:

> *...cavity-dwelling honeybees like A. cerana and A. mellifera could extend their*

distribution to subtropical and temperate conditions, where colony reproduction would require large swarms that are forced to locate a nest cavity within a short time. Under tropical conditions, however, small swarms resulting from high colony reproduction may survive for a longer period outside and such honeybees could exploit seasonal differences in nectar availability. At the end of this development, the transition to open-nesting forms might have taken place to avoid the limitations of which result from the scarcity of nest cavities.

Hepburn[110] is more specific about the nature of honey bee migration:

...Seasonal migration is characteristic of Apis cerana, Apis florea, Apis andreniformis, Apis dorsata, Apis laboriosa and African (not Eurasian) Apis mellifera... [a] behaviour [that] is dampened at colder latitudes.

Little is known of the migration patterns of the other species closely allied to *Apis cerana* namely *Apis koschevnikovi, Apis nigrocincta, Apis nuluensis* and *Apis indica* other than the fact some some are sympatric [occupying the same area] and either maintain breeding barriers or show variable habitat preference, partly in response to competition from each other.

Swarming and absconding of African races of honey bees including *Apis mellifera andansonii, Apis mellifera scutellata, Apis mellifera monticola, Apis mellifera litorea* and *Apis mellifera nubica* and the influences of resource availability, pests and diseases on these behaviours is outlined by Mutsaers[111]. Schneider and McNally[112] provide detailed accounts of colony growth and impacts of a wide range of factors on migratory behaviour of African honey bees demonstrating the fine-tuned responses of honey bees to resource availability and climatic conditions.

Koetz[113] provides an in-depth overview of the swarming behaviour of *Apis cerana*, particularly of the Javanese strain that has spread across the Indonesian Archipelago and into Irian Jaya, Papua New Guinea and northern Australia.

The dwarf honey bees (Apis florea and Apis andreniformis)

Robinson[114] has made a detailed observation on a serially absconding *Apis andreniformis* swarm indicating some difference in this species behaviour to that of the much better studied *Apis florea* but the account is purely anecdotal. A substantial review of the biology of the two dwarf honey bees by Wongsiri and coworkers[115] signal some striking differences in their biology and absconding behaviour.

Apis andreniformis is more prone to absconding and over greater distances and is less adaptable than *Apis florea* in that it is restricted to undisturbed habitat. *Apis florea* decamps over relatively short distances sometimes returning to the parent nest location to harvest wax for comb building. Over much of its range reproductive swarming occurs at the end of the dry season and in many areas there is a secondary absconding swarming before the onset of the dry season[116].

Overall significance of swarming in honey bees

Swarming, the propensity of colonies to abandon nesting sites, either for colony increase under favourable conditions or to survive, in response to dearth and predation, are both responses to selective environmental pressures. Both enhance honey bee survival at a whole of population scale rather than just at the individual colony level.

Keeping the lid on the box

Replacing an old queen with a young queen, especially one with a lineage of low swarming propensity, is one of the most effective strategies for keeping bees together. There are now many well known swarm control practices the very best of which involve getting a new queen to work with the existing colony queen until the swarming season is over. The old queen can then be removed. These topics will be canvassed under *Requeening the Honey Bee Colony*.

Sound swarm control practices involve working with bees and their natural instincts rather than against them: measures such as reliance on removal of queen cells are counterproductive in that bees will immediately resume swarm preparations. Many Asian bees that swarm in response to dearth can be prevented from doing so if they are fed and stocking density of hives is controlled.

Most swarm control practices emerged at the end of the skep era. Their invention ensured that the productive capacity of whole apiaries would no longer fly out the front door. But what was discovered, what were the key swarm control strategies and why do they work? To answer these questions, we need to understand the underlying factors that trigger bees swarming. It is also instructive to record how and when the swarm control discoveries were made as the findings were those of very astute observers whose insights deserve recognition.

Honey bees have a natural proclivity to swarm and defend themselves: they have evolved to reproduce as efficiently as possible and to fend off marauders. However with most honey bee species, other than temperate strains of the Western Honey Bee (*Apis mellifera*) and some races of the Asian Honey Bee (*Apis cerana*), their propensity to swarm is so great that they are not amenable to managed honey production and provision of pollination services.

This does not mean that bees that swarm profligately are of no practical value. On the contrary, many are hunted for their honey and wax and all honey bee species, including widespread *Apis florea* and *Apis dorsata*, are important crop pollinators.

Demaree appears to be one of the first beekeepers to devise a scheme to keep bees under the one lid. He invented a technique that closely mimicked swarming but kept bees together, the details of which he presented in an 1892 oration (detailed below) to the Ohio State Convention in Kentucky. In so doing he attempted to put

Miller's well articulated concern about swarming to rest[117]:

> *When your apiary is as large as you want it, what would you give to be able, by a simple, practical manipulation at the beginning of the swarming season, to hold all your colonies in full strength of working and breeding force steadily through the entire honey harvest? You can do it beyond a doubt, by practicing my new system of preventative swarming...*

With the wisdom of hindsight we now know that this scheme may need to be repeated, notionally every six weeks or for as long as the swarming season lasts. He omitted to mention that queen cells, those that often develop in the queenless brood chamber, should be removed especially if emerging queens could escape the hive and mate facilitating queen swarming or supersedure. In very strong colonies, the scheme may benefit from a super containing some drawn comb placed above the excluder to further distance the queen from her brood.

The success of Demareeing, still celebrated today, does not reflect the difficulty that George Demaree had in refining a system to one that actually worked. The details of his earlier account of swarm control, published in 1884[118], are almost indecipherable. That all stated, we are often surprised that so few beekeepers today either understand or ever practice the scheme that Demaree eventually articulated.

Neither should Demareeing be practiced too early before the bees are in a strong condition and be at risk of swarming – this will put them back – nor should it be practiced too late. Elton Dyce, famous for his method of producing creamed honey, also made a very considered study of different swarm control methods[119]. Of Demareeing, he makes the very pertinent observation:

> *The raising of brood in the Demaree plan places a severe strain on the colony. Unless the honey flow is certain to last a sufficient length of time to insure the building up of another brood nest below, the Demaree plan will be a failure. In other words, unless the honey flow will continue at least three to four weeks after it is intended to raise brood, the plan should not be used. Strong colonies only should be Demareed as the shook would lessen the gathering activities of weaker colonies.*

Some schemes such as hive splitting are indeed effective in that they force bees to regroup and rebuild when they normally will abandon any attempt to swarm.

The tradeoff is always one of building bees as fast as possible while avoiding brood-bee imbalance conditions where bees naturally propagate by dividing the workforce. Practical measures directed at arresting swarming are vital if potentially productive colonies are to realise their potential. Presented here are conventional swarm control measures that focus on harnessing the full productive capacity of bees.

Super reversal and brood lifting

Super reversal

In this simplest of practices, an early season deep super reversal (Figure 1.26) is employed to simulate a light honey flow. This ploy keeps bees more fully occupied than they would otherwise have been. After exchanging supers, or manually placing most brood frames in the lower chamber, any honey not required by emerging brood will be moved up by the bees from the now bottom super B. This makes extra room for the queen to lay.

Brood emerging from the now upper super A will similarly make space for nectar storage above the lower brood nest. The honey stores, having a large thermal mass and narrow single bee space passages restricting air circulation[120], help insulate the brood below. Super reversal stimulates brood rearing but must be conducted judiciously as there must be enough bees to avoid chilling rearranged brood. Later in the season, as swarming conditions ensue, repeating this process or, more simply, moving frames of brood down will result in idle bees being put to work moving honey and expanding a now decongested brood nest.

Putting honey up and brood down is a reorganisation of colony resources that enhances colony expansion and delays swarming.

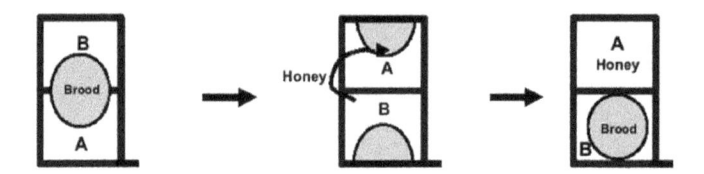

Figure 1.26 Super reversal.

The majority of Australian beekeepers appear to operate their hives with a single brood chamber. Provided excessive amounts of honey or pollen are regularly removed during a flow, and placed above the excluder, there will be ample room for the queen to lay even in a single eight-frame box. However queen excluders are usually removed from overwintered double box hives to enable bees to move brood up to stores in the upper chamber in late winter and early spring. Like many American beekeepers, I maintain double super brood nests throughout spring and through the honey flow to minimise the risk of congestion-induced swarming. Some recent poor seasons have suggested to me that this strategy may not always be wise at least when conditions dictate that the bees are being given too much space to rear brood.

Brood lifting

To control swarming while confining the brood nest to a single brood chamber, beekeepers, especially those employing eight frame gear, regularly move sealed brood and excess frames of honey and stores up above an excluder: the brood quickly emerges encouraging storage away from the brood nest and provides ample room for the queen to lay (Figure 1.27). Larger ten-to-twelve frame brood chambers may not benefit from a double brood chamber as a good queen will not fully lay out more than nine frames in a full brood cycle.

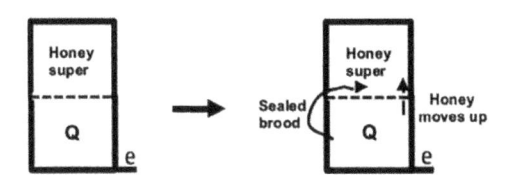

Figure 1.27 Brood lifting as an alternative strategy to brood box reversal.

The alternative hive management practice of under supering rapidly expanding colonies – nadiring – replicates the condition of bees in a spacious tree hollow. Ample room is supplied for the colony to expand and the bees move honey into comb from which brood has emerged at the same time expanding the brood nest downwards. A similar stimulatory effect can be achieved by brood spreading. This interspaces brood with empty combs, but should only be practised in very populous colonies and then very conservatively. However movement of the brood down, achieved by colony reversal, will facilitate rapid colony expansion and would seem to do more than just limit early swarming.

The colony split

The traditional colony split emphatically delays swarming simply because it mimics natural colony reproduction. In the definitive first step, the colony is divided into two separate parts (Figure 1.28a–>b). The old queen unit (OQ) now resembles the swarmed colony while the daughter unit on the original stand is now queenless (No Q). It raises its own new queen (NQ) from emergency cells and, once laying is commenced, is very much akin to the reestablished parent colony from which bees have swarmed. After supering, both units can then expand freely (Figure 1.28b–>c), the parent colony having avoided most risk of swarming.

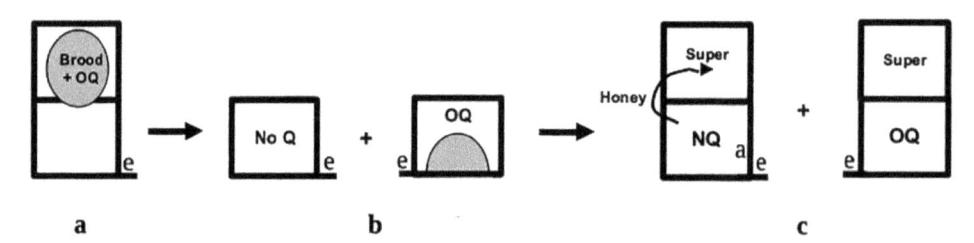

Figure 1.28 Colony split: e = entrance; OQ = old queen; No Q = No queen; NQ = new queen:
(a) strong colony at risk of swarming;
(b) colony is split; and
(c) queenless unit is either allowed to raise its own queen or a caged queen is introduced: the separate colonies are supered as needed.

Alternatively a new queen or a queen cell can be introduced to the initially queenless unit so that brood raising is minimally interrupted. Not too remarkably any idle bees in each unit will immediately revert to working bees intent on reestablishing their colony. When practiced well before the main honey flow, both colonies will have a good chance of building to produce a good honey crop.

The Demaree plan

Between 1884 and 1892 George Demaree developed a truly ingenious system of swarm prevention. The simple measure of separating the queen from all but a single frame of her brood, by using an excluder, simulates a near brood-free condition (Figure 1.29a –> 1.29b). Bees join the queen, but only as needed, while any excess honey located in the brood nest is moved up into the super to replicate the super reversal scenario described above. Bees can be shaken from all frames into the lower super or in front of the colony so that the queen does not need to be found.

Demaree's method, involving separation of the queen from her brood and forcing establishment of a new brood nest, diverted bees from swarming and kept all workers well occupied. In his 1892 public oration, Demaree describes his plan very lucidly and in splendid detail:

> *In my practice, I begin with the strongest colonies and transfer the combs containing brood from the brood-chamber to an upper story above the queen-excluder. One comb containing some unsealed brood and eggs is left in the brood-chamber as a start for the queen. I fill out the brood-chamber with empty combs, as I have a full outfit for my apiary. But full frames of foundation, or even starters, may be used in the absence of drawn combs.*
>
> *When the manipulation is completed, the colony has all of its brood with the queen, only its condition is altered. The queen has a new brood-nest below the excluder, while the combs of brood are in the centre of the super, with the sides filled out with empty combs above the queen-excluder.*
>
> *In twenty one days all the brood will be hatched out [sic emerged] above the excluder, and the bees will begin to hatch [emerge] in the queen's chamber below the excluder, so a continuous succession of young bees is well sustained.*
>
> *If my object is to take the honey with the extractor, I tier up with a surplus of extracting combs combs as fast as the large colony needed the room to store surplus. Usually, the combs above the excluder will be filled with honey by the time all the bees have hatched out [emerged], and no system is as sure to give one set of combs full of honey for the extractor in the very poorest seasons; and if the season is propitious the yield will be enormous under proper management...*
>
> *...The system above works perfectly if applied immediately after a swarm issues. The only difference in the manipulation, in this case is, that no brood or eggs is left in the brood-nest, where the swarm is hived back. – Read at the Ohio State University Convention. Christiansburgh, Ky.*

It is worth noting that both super reversal and the Demaree plan delay, rather than entirely avert, swarming. Indeed, and perversely, Demareeing practiced well before the brood nest is expanding rapidly can invite queen cell development even if swarming were to be some long way off. Further attempting to use the Demaree Plan too early in the season before the weather warms up will result in brood chilling setting the colony back.

Demaree does not mention the fact that queen cells may develop in the brood nest separated from the queen. While such cells – deprived of a flight entrance – are normally destroyed by the colony bees, they are relatively easy to locate in the upper hive chamber.

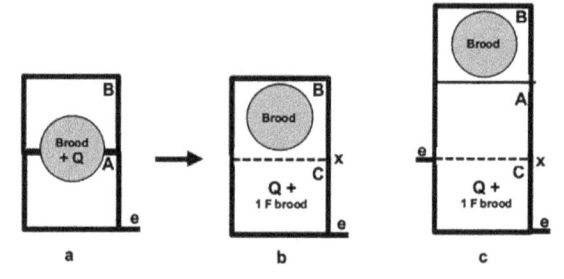

Figure 1.29 Demaree plan:
Q = queen; e = entrance; x = excluder; ds = double screen:
(a) strong colony in need of preventive swarm intervention;
(b) simple Demaree with queen separated from most of her brood; and
(c) enhanced Demaree with queen further separated from the brood to facilitate colony expansion.

An alternative setup (Figure 1.29c), accommodating parent hive frames with honey, will further separate the old queen from her brood. Any honey present will be moved up into the top super as worker bees emerge and will facilitate colony expansion while an additional entrance will facilitate free flight access for emerging drones. An excellent account of the Demaree plan using the additional super is provided by Snelgrove[121].

While the Demaree plan has been adapted to requeening, and indeed the colony may swarm or form an additional brood nest if virgin queens can find flight access, the primary purpose of the Demaree has always been to keep the lid on bees, that is to keep them building for the flow and to prevent their swarming.

At this point we should stand back and ask ourselves: 'Why does the Demaree system work well'? The likely reason is that, while the colony is maintained as a single functional unit, most of the post swarming conditions are replicated, as a vastly decongested brood nest and room for honey storage is presented by emerging brood. In the Demaree plan the queen, largely or completely removed from brood, will now be focussed on rebuilding her brood nest, while many nurse bees will be tending and thermoregulating brood elsewhere in the hive. At all stages of readjustment idle bees will be kept very busy and attendant to the queen, at least for some weeks, with the clear advantage of there being no break in the brood cycle.

Many beekeepers may not realise that a number of their swarm control measures achieve the same goal as the essential Demaree plan. For example, removing brood, young bees and some stores and placing them into one or more nucleus colonies will result in swarm-caste bees being reassigned normal hive and foraging duties in the respective parent and nucleus colonies. Since nuclei are unable to build good queens, any nucleus colony should be supplied with a young queen or a queen cell. Whatever swarm delay plan is adopted, once the main honey flow is imminent and the main swarming season is over, divided colonies can be reunited by simply removing the excluder (where the units are stacked) or by papering on separately located colonies. Apiarists have found that the new queen present in the top super will usually supersede the old queen below. Anecdotal records suggest that his occurs in about 80% of instances, a topic we will return to in examining the many ways of approaching requeening bees.

However, while large scale beekeepers may not always have time to find queens, most sideline beekeepers place high value on new queens and always search combs to remove the old queen. It is also sound practice to thoroughly examine remaining brood nest frames for the presence of additional queens especially during the swarming season or if there is evidence of supersedure in progress (Figure 1.30). Such supernumerary queens are remarkably common and may explain why queen acceptance is sometimes poor, especially in colonies preparing to either swarm or supersede the colony queen.

The net result of the full Demaree plan is the production of a hive much stronger than the original colony even had it never swarmed. A key advantage of the Demaree plan over colony splitting or colony weakening plans is that there is no interruption of brood rearing and that the colony is maintained as single functioning unit.

Figure 1.30 Repeated supersedure preparations in queenright colony (from which supersedure cells were previously twice removed) at Jerrabomberra Wetlands January 2017.
Photo Alan Wade

Enhanced Demaree plans

With the essential Demaree Plan, we discovered that the simple measure of separating the queen from all but a single frame of her brood simulated a near brood-free condition: the queen and her entourage faced conditions not dissimilar to those of a recently swarmed colony. Wedmore[122] provides a more detailed approach to application of the Demaree Plan especially in respect of its repeated use.

In a further variant of the Demaree Plan, the top unit in the strong Demareed double (Figure 1.31a) is offset to a separate bottom board as soon as queen cells are built and sealed, in practice ten days later. A queen in this offset colony will mate (Figure 1.31c) and tear down rival sister queens and become the new queen (NQ). This obviates any need to locate and destroy queen cells, removes any chance of swarming – timing is critical – and keeps the old queen laying uninterrupted. The original colony with the old queen (OQ), depleted of brood and supered will freely expand once brood begins to emerge as will the split off unit (Figure 1.31d). Again conditions are created where swarming is avoided and for the very simplest of reasons, the bees are kept in their early season brood expansion mode but now with the immediate prospect of a honey flow.

Alternatively the colony with the new queen can be papered onto the old colony and the resultant powerful single-queen hive will gather a good crop of honey.

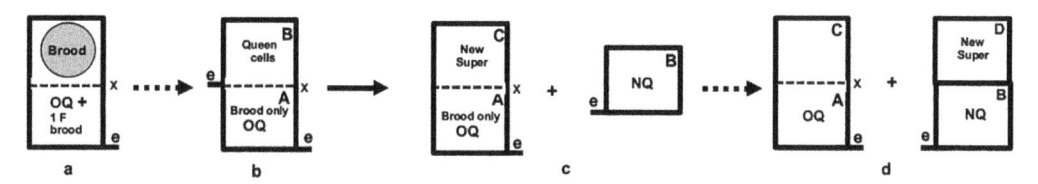

Figure 1.31   Advanced Demaree plan:
OQ = old queen; NQ = new queen; e = entrance; x = excluder:
(a) strong colony is Demareed;
(b) queen cells are developed above an excluder;
(c) the old colony is supered and the nuc(s) formed from parent hive are offset and a new queen is allowed to mate; and
(d) split colonies are supered for the honey flow.

In yet another Demaree scheme, the Lavender Hill plan (Figure 1.32)[123], the new queen offset can be papered onto the old bottom queen brood box forming a consolidated brood nest headed by two queens: the new queen in the stand alone new queen colony (Figure1.32b, super B) is papered on above the old queen (Figure 1.32b, super A) and the original honey super (C) is further papered over a second excluder. The resultant system is a very powerful colony that can be supered freely (Figure 1.32c). Hogg went on to develop a revolutionary section comb production system (Figure 1.32d) powered by two queens[124].

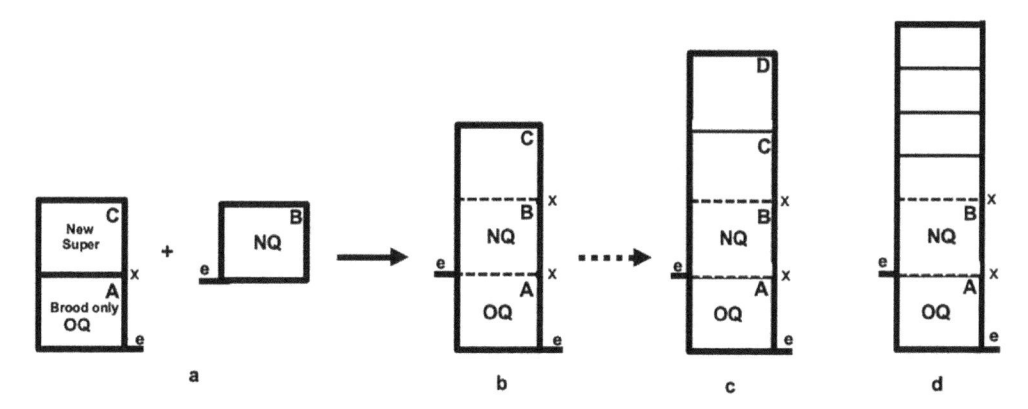

Figure 1.32   Lavender Hill plan:
OQ = old queen; NQ = new queen; e = entrance; x = excluder:

(a) new queen was formed as a result of Demareeing or nuc provisioned with a caged queen;
(b) new colony was papered onto the parent colony with old queen to form a conslidated brood nest colony with two queens;
(c) new queen was supered for extended flow conditions; and
(d) scheme as adapted to half-comb section honey production.

Near the beginning of short season flows or well before the major flow ceases, the lower queen excluder is removed to allow the colony to collapse to the natural single queen state or spilt into two single-queen colonies.

Shook swarming

Shook swarming involves shaking all bees from a strong colony preparing to swarm into new or irradiated gear. This system is a distinct alternative to hive splitting (where the colony is physically divided) and to the Demaree plan (where the bees are kept together but where queen is separated from her brood). The bees including the queen are simply pushed out the front door and housed elsewhere in much the same manner one would observe bees swarming naturally. Overall however shook swarming replicates colony abandonment where all bees depart.

The bees, having being artificially swarmed, are deprived of brood, stores and combs. As in package bee installation, the bees start again and do so with remarkable comb building and brood development zeal. There are many variants[125] of shook swarming both in terms of timing and the way in which it is performed. In practice the new colony may be fed or supplied with drawn and undrawn combs and stores to accelerate its reestablishment. In any event the bees are too intent on rebuilding the colony to swarm.

Planning the fate of the brood left behind as a result of shook swarming depends entirely on the condition of the parent colony. The remaining brood chamber may be left in place to receive returning field bees: the colony will then produce its own emergency queen in much the same way as a split colony operates. More often, healthy brood from a strong shook hive can be distributed to weak colonies and a weak colony can be located on the old stand and will be strengthened by the strong force of returning field bees. Suspect or diseased brood must be handled very differently as we shall now explore.

The process of shaking bees out of a colony is extremely valuable for purposes other than swarm control. Firstly there are very good reasons to shake bees where the colony has become hopelessly queenless. Such a hive is virtually impossible to requeen[126] and for one simple reason, laying workers – gynaecoid queens – ensure that they or their workers will ball and destroy a foreign introduced queen. The bees, mainly drones and aged workers, are of little or no value and can be simply shaken off frames well away from hive entrances and left to fend for themselves. Additionally, a drone laying queen, if present, or laying workers will be unable to fly and so will not re-enter the colony. The frames and gear can then either be stored, if needed treated, or more simply added to other colonies once a check is made to ensure the parent colony is disease free.

Secondly, shook swarming can be used to help regenerate hives badly affected by European Foulbrood, small hive beetle, *Nosema*, chalkbrood and even sac brood but *not* American Foulbrood[127]. Here shaking bees from a colony provides a means of separating heavily infected brood, stores and gear from largely healthy bees. Infected hive gear can be disposed of or separately irradiated. Meanwhile the bees can be reestablished in clean gear and fed with a much better prospect of overcoming disease burden, especially if requeened.

It seems almost inevitable that shook swarming will come into vogue with the arrival of the parasitic *Acarapis, Tropilaelaps* or *Varroa* mites. Separating bees from the parent colony forces a break in the brood cycle of parasites and is already a key strategy for controlling, though never eliminating, *Varroa* and certainly reducing its associated viral particle load. Breaking the brood cycle can be an effective means of eliminating *Tropilaelaps* at the apiary level but does not prevent re-infestation from feral colonies or from other apiaries. Swarming and drifting bees will always spread mites.

One variant developed by Leslie Bailey at Rothamsted involved rotating all combs out of a hive, obviating the need to shake all bees into new gear[128].

Shaking bees is widely employed in the packaged bees trade. Key differences between this practice and shook swarming include bees only being taken from healthy hives and their being shaken from many hives. The bulk bees are then funnelled into individual packages containing of the order of 10,000 bees, and shipped out with a caged queen and a heavy sugar syrup feeder. Establishing hives from package bees is not widely practiced in Australia as bees are relatively easy to maintain year round.

Swarm control in perspective

Many swarm control plans, involving simulation of the pre and post swarming condition, such as caging a queen to force a period of broodlessness, have a major downside. There is a loss in population attendant to the lack of a laying queen or brood and to the removal of bees. This reduces the rate of colony buildup and hence the capacity of colonies to take full advantage of every and any honey flow.

In the absence of disease, the brood cycle break problem has been averted in a number of ways. Not least amongst these is the practice of introducing a new queen to the initially queenless split colony unit or even introducing new queens to both splits. The alternative of using overly strong colonies to make up nuclei to introduce queens and then use these nucs to requeen hives in an apiary is a very practical swarm control measure and averts brood cycle interruption. Requeening an apiary using nucs involves extra work but improves queen acceptance, definitively replaces old queen stock prone to swarming and ensures continuous brood production.

Even with the simple colony split, performed 5-6 weeks before the main flow, it is important to again emphasise that the brood nests can be reunited at the beginning of the honey flow after almost all chance of swarming has disappeared. The result is a very powerful honey producing or pollinating unit.

Despite a great deal of evidence that sound management practice should avert swarming, it is important to realise that the natural tendency of bees to swarm and to replace queens means that swarming will still sometimes occur and often unpredictably. Swarm control measures simply improve the odds for colonies to successfully build up to full strength to coincide with major honey flows.

All this stated, it would be an oversight not to emphasise the value of regular requeening to minimise the risk of swarming, in my experience best conducted every 12-18 months. Queen age is a critical factor in determining the propensity of bees to swarm. Colonies headed by queens two years of age are three times more likely to swarm than with queens less than one year old whereas swarming was reported to occur in 10-40% of unmanaged colonies with young queens[129]. While queen breeders have been able to progress selection for strains of bees with a lower propensity to swarm, the take home message is that if you want to control swarming, make every effort to do so. We might sum up what is known about swarm management with Clarence Collison's clear statement:

Swarming is an instinctive desire of honey bees to increase their numbers by reproducing at the colony level, doubling [give or take the risks this process presents] their chances of survival.

Young queens have a higher fecundity – that is, they lay more eggs – and produce more queen pheromone. As we previously observed young queens produce more queen substance than do older queens. This arrests queen cell construction reducing the likelihood of swarms issuing.

Hence swarming signals a complex array of honey bee behaviours. As we have seen, swarming, critical to establishment of new colonies, may also result in queenright swarms taking over established colonies. Over time both strategies impact on honey bee gene flow and both confer long term survival of different bee lines.

In contrast, swarming involving colony absconding is a separate, but well recognised, response to adverse in-hive or environmental conditions. New colonies encountering poor nesting conditions may immediately seek more suitable nesting sites, that is abscond. For cavity dwelling species, this may be a response to inadequate space or hive contamination; amongst dwarf and giant honey bees temporary nesting or bivouacking may simply facilitate colony movement to exploit resources elsewhere. Most generally, absconding behaviour would appear to be a response to disease, to predation, to resource competition or to local resource depletion, where relocation of the bee colony and some of its resources confers an improved chance of survival. To this we may add the condition where bees are hived in a box far too small for its needs: I once introduced caged queens to a dozen well stocked three and four frame nucleus colonies under very favourable conditions in spring only to find the majority of them hanging from nearby branches several days later.

On the other hand, colony migration, where bees return to their original nesting locations on a cyclical basis, sometimes over large distances, would appear to be an adaptive response to regional and seasonal availability of floral resources.

III
Requeening the Honey Bee Colony

Having examined both *The honey bee queen* and *The honey bee swarm,* disparate expressions of the honey bee colony, we can now look in detail at practical ways to requeen hives, measures that influence swarming. We all know that bees, in replacing their queen of their own accord, undergo a process of wholesale renewal. While worker bees are replaced daily on a grand scale during the active bee season and can be viewed as a colony running repair exercise, the processes of colony reproduction and queen and associated drone replacement involve intergenerational renewal. In this light requeening can be seen as mediating the need for the colony to renew its own reproductive capacity and to facilitate swarming. With a clear understanding of these phenomena, beekeeper initiated requeening can be seen to be a means of maintaining colony vigour and an opportunity to control swarming.

Requeening schemes range from simple queen substitution through to sophisticated plans to build bees and arrest swarming. It follows that requeening honey bee colonies can be seen at two levels.

The first and simplest is straightforward colony queen replacement by the beekeeper, something bees do of their own accord when a queen dies or becomes infirm, here designated *Direct requeening*. At a higher and more nuanced level several requeening schemes, employing a second laying queen, have been invented that mimic swarming and that allow a new queen to automatically supersede the parent colony queen. This more sophisticated queen renewal practice I call *Requeening with swarm control*.

Swarm control employing two queens

We have visited various facets of swarm control and the pivotal role of the honey bee queen, not least those that a new queen imparts in having a high mandibular queen pheromone titre. Much as we look back nostalgically to the transition from skep to contemporary moveable frame beekeeping we can now look forward to the potential that judicious use of a second queen has on the potential to suppress the natural instinct of bees to swarm.

Direct requeening[130]

Queen substitution

Most beekeepers settle for the easy option, removing the old queen and installing a caged queen (Figure 1.33). This method of queen introduction was never going to be one hundred percent successful, but it works a whole lot better than opening the cage and trying to run the queen into the hive entrance and hoping the queen would replace an old and failing queen. It almost always fails. We have heard of commercial beekeepers simply swapping queens between hives in the middle of strong honey flows but give no credence to its applicability under more normal dearth conditions.

That not to say that the direct requeening method – after dispatching the old colony queen – does not work or that it is not the simplest means of requeening a paddock-full of hives. This said, any colony made queenless will almost immediately start building emergency queen cells so will become almost instantly queen receptive. There is no need to leave a hive queenless for long to have it accept a new queen although, occasionally, a new queen may fail in the delicate process of establishment.

When the old queen is found continue to conduct a quick search of other brood frames for a potential second supersedure queen that is sometimes present. Checking other combs for additional queens after that old queen has been removed requires minimal effort and should be standard practice.

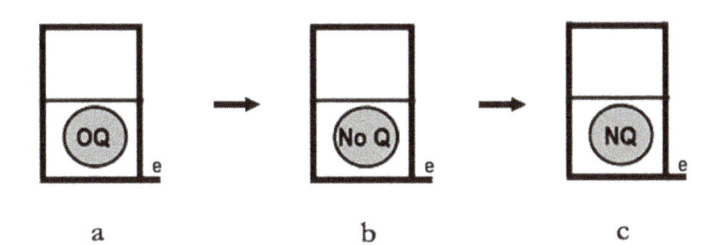

Figure 1.33 Direct requeening: e = entrance; OQ = old queen; No Q = no queen; NQ = new queen:
(a) find and remove the old queen;
(b) the colony is made queenless for less than 1-2 days; and
(c) a new queen (cell or caged queen) is introduced.

Locate the queen cage between top bars of frames of sealed brood (Figure 1.34) from which young nurse bees are emerging candy entrance facing upwards to prevent any escort bees blocking queen exit: some beekeepers simply lay the queen cage face downwards over central frames of the brood chamber. Unless the cage candy is rock hard don't adopt the ill-advised practice of punching a hole through it. Delaying the queen moving out into the hive improves her chance of acceptance so accelerating her release can be counterproductive. It only takes a few days for bees to chew out the candy and she is anyway tended by her escorts.

Alternatively a queen cell can be introduced. Because queen cells produce no queen pheromone, they do not need protection.

Queen acceptance is reputably more problematic when requeening a colony with a queen of a different honey bee race. Allowing such colonies to remain queenless for twenty four hours, rather than introducing the queen cage immediately after de-queening, may then be advisable. However, as a caged queen will take a few days to be released, for sake of convenience and at not much risk, many beekeepers requeen immediately upon dequeening saving the need to make a return journey to an out apiary.

Figure 1.34   Installing caged queen candy end upwards, Canberra Region Beekeepers' Jerrabomberra Wetlands Apiary 2 November 2016.
     Photo: Alan Wade

Every colony should be inspected after ten days to ensure that the new queen has taken and to ensure that she is laying. It is not essential to find the new queen as her presence will be indicated by the presence of eggs and young brood. Finding her however means she can be marked, a practice I follow but that most commercial beekeepers I know do not bother with.

An excellent alternative queen introduction approach is to remove the mailed queen (inside a butterfly net or a small bee-proof room) from the cage and introduce her onto emerging brood using a push-in wire cage (Figure 1.35). Mark the top bar and return a few days later and release her very carefully checking to make sure that she is laying. I like to mark queens using a water-based Posca pen as this is an ideal time to do so and as she will have increased in weight and normally won't be able to fly. Marking queens is a good way to adjudge whether a queen is subsequently lost to supersedure.

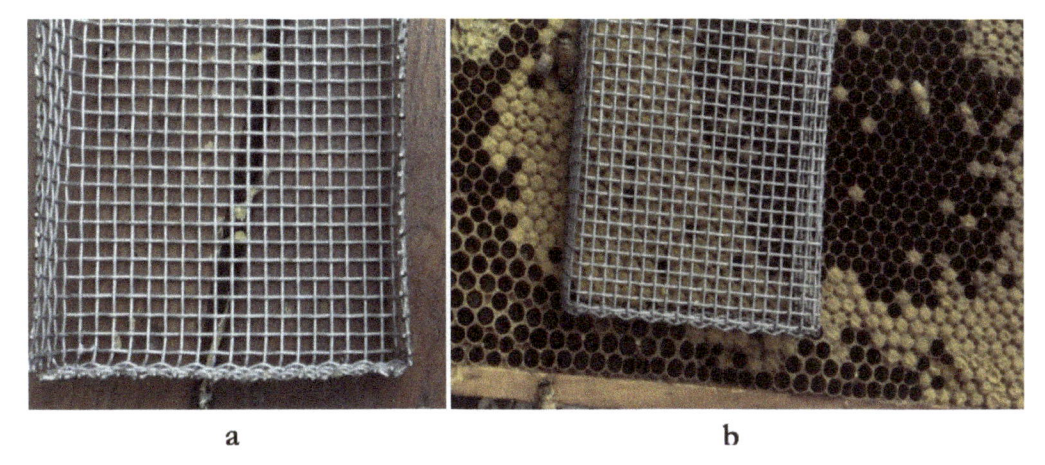

a b

Figure 1.35   Requeening using a wire-mesh push-in cage:
(a) cage profile (100x85x25 mm); and
(b) a mailing cage queen under a cage pushed into emerging brood.
Photo: Alan Wade

Queening into an offset nucleus colony

A far superior approach to requeening is to introduce a new queen to a nucleus colony made up from the parent hive frames and bees. To do this remove a few frames of sealed and emerging brood and a couple of frames of stores and put these into a 3-to-6 frame box along with adhering young bees.

This offset nuc method of queen introduction buys you time. Where a single nuc is made up (Figure 1.36) it can be inspected carefully to ensure that the old queen is not present and remains in the parent hive so does not have to be found.

Figure 1.36   Offset nucleus colony setup – Hillside Beeblebox Apiary 9 September 2020.
Photo: Alan Wade

Once the new queen is well established and has brood in all stages of development, about twenty one days, the old queen can be found and despatched and the nucleus colony papered on (Figure 1.37).

Alternatively a few frames can be removed from the parent hive brood nest and the nucleus can be slipped into the hive as a unit aided by a little smoke. In the absence of a light flow sprinkling or spraying a little sugar water into the parent hive brood chamber will help settle the bees.

If you want to establish a particularly valuable queen, say a breeder queen you have been gifted with, make up the nuc with a frame of emerging brood and young bees obtained by first gently shaking off older bees using frames taken from the brood nest. Such bees, less than several days old, will accept almost any queen and are ideal nurse bees.

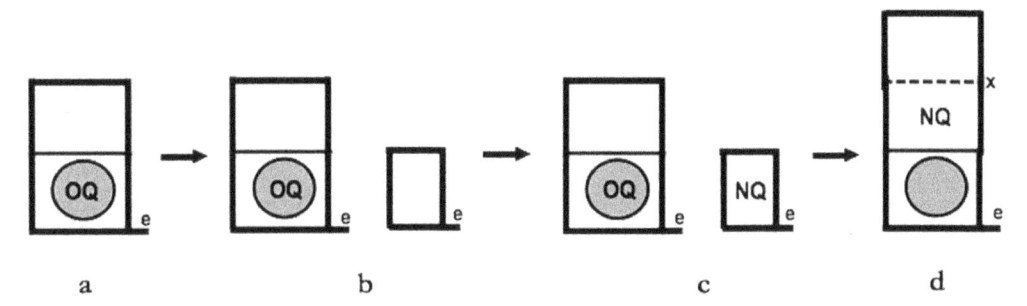

Figure 1.37 Offset nucleus requeening: e = entrance; x = excluder; OQ = old queen; NQ = new queen:
(a) colony needing requeening is identified;
(b) an offset nucleus with brood, stores and bees is established;
(c) the queen is introduced and is established over a period of 3-4 weeks; and
(d) the old queen is removed and nucleus with new queen is united with the parent hive.

The nucleus method confers a number of benefits over and above those of direct queen replacement:
- the new queen is more likely to be accepted;
- the old queen will continues to lay until the new queen is established with her first workers emerging;
- the inherent risk of losing both the old and the new queens is averted; and
- the colonies can be safely united after removing the old queen.

Here is a practical hint for anxious beginner beekeepers. Prepare queenless nucleus colonies a day in advance of the queen breeder advising that queens have been despatched express mail. This way my queens are put into hives the day they arrive. We find we can complete installation of twenty or more queens close to dusk when finding and removing old queens en masse is no longer feasible.

Colony split requeening

The principles for conducting a colony split are much the same as those employed for the offset nuc method, that is except where swarm cells are present. It is then advisable to use other hives to establish new queens and anyway it is better just to split or Demaree colonies rather than resort to removing cells every 7-10 days to avert swarming or attempting to requeen them.

With the split method, well suited to the difficult task of requeening very populous colonies, a new queen is introduced to the queenless unit.

The split also differs from the direct replacement and offset nuc methods in that the split can be performed several days to a week before the planned new queen introduction. It is the method of choice when requeening overly strong colonies. There are many variants to the procedure and, if desired, a new queen can be introduced to each half of the split to produce two newly queened colonies. Splitting colonies is particularly useful not only in planning requeening in a well-established apiary but also in controlling swarming.

To start the process, the colony is simply divided into two units (Figure 1.38) or is alternatively separated by a solid division (split) board (Figure 1.39). The only difference is that in the latter scheme all the gear can be kept together and that an extra cover and base are not required. If a double screen is employed instead of a solid division board the bees are still separated but the upper unit benefits from the warmth generated by the lower brood nest. In any case the queenless unit will raise its own emergency queen if not requeened shortly after the colony is split. A super – or supers – and an excluder need to be added as the colony builds.

This technique initially puts the parent colony back a week or so by removing some bees and resources, but this deficit is quickly made up for by the presence of a new laying queen. The two queens lay far more eggs than had the colony with the old queen not swarmed and been allowed to continue to lay uninterrupted.

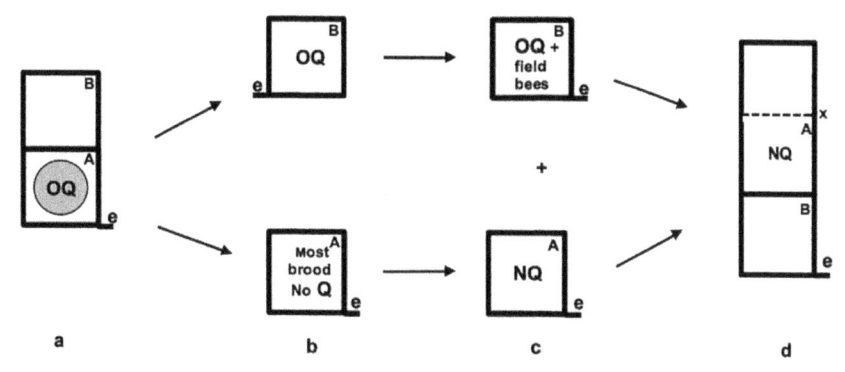

Figure 1.38 Queen introduction to split on separate stands: e = entrance; OQ = old queen; No Q = No queen; NQ = new queen:
(a) strong colony at risk of swarming;
(b) colony is split to stand alone units;
(c) new queen is introduced to queenless split; and
(d) old queen is removed and colony is united.

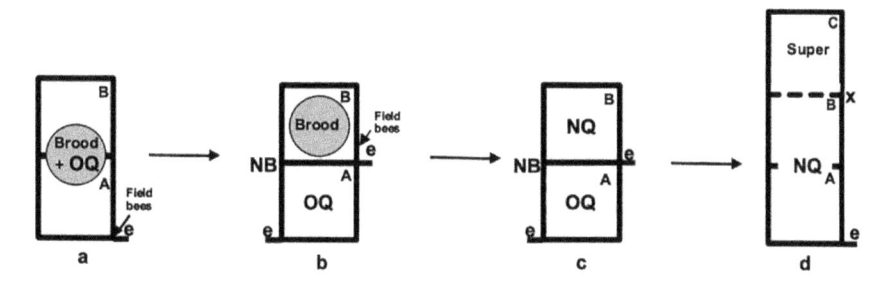

Figure 1.39 Queen introduction to split colony on parent stand: e = entrance; OQ = old queen; NQ = new queen; NB = nuc board:
(a) strong colony at risk of swarming;
(b) colony is split using split board or double screen, the top colony is given a new entrance;
(c) a new queen is introduced to the queenless split; and
(d) the old queen is removed and colony is united and supered.

Real world variants of these schemes (Figure 1.40) signals that the colony split system can be adapted to make maximum use of standard gear.

Figure 1.40 Hive divisions effected by removal of a super with some brood and bees, John Robinson's Apiary in Red Hill, 12 March 2021:
(a) offset split; and
(b) tiered split – note makeshift use of old carpet underlay and stick to provide the upper hive entrance.
Photos: Alan Wade

Faced with an urgent need to control swarming a couple of years back, we came across a club hive with a large number of queen cells. From all appearances that hive would have swarmed a day or two later. As we were in a hurry, we simply split the hive, supered each unit, stacked the hives and walked away (Figure 1.41).

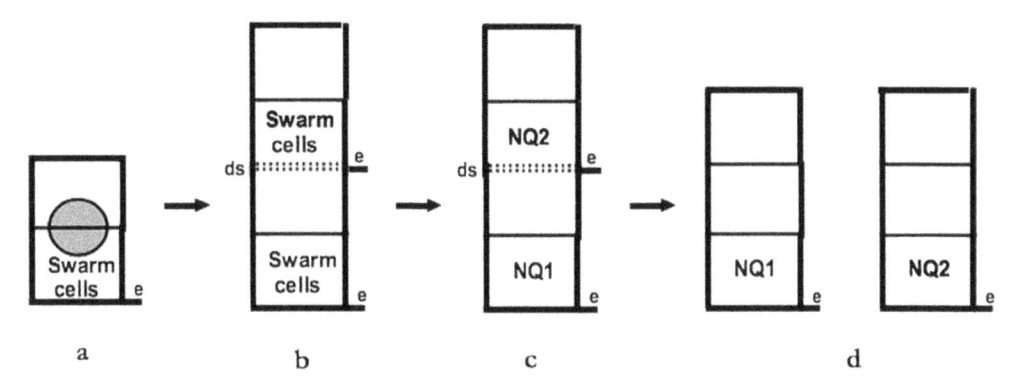

Figure 1.41 Opportunistic swarm control with self requeening at Canberra Region Beekeepers' Apiary:
e = entrance; ds = double screen; NQ = new queen:
(a) a colony was found with numerous sealed queen cells;
(b) the colony was split on the vertical pattern and supered to arrest swarming;
(c) two queens emerged and were successfully mated, one queen superseding the queen the colony would otherwise have swarmed with; and
(d) the top colony was offset and both hives were supered for the flow.

Not only did the hive give up swarming, but two new queens were found laying well when the piggybacked hives were inspected a month later. The old queen had been superseded and the queenless split also had a new queen. The swarm cells that had threatened to result in swarming were immediately repurposed to supply two exceptionally well raised queens.

Demaree swarm control with purposed requeening

Requeening occasionally occurs as a result of a simple procedure, one designed to thwart swarming, discovered one hundred and thirty years ago. As already detailed, George Demaree (1832-1915) came up with the novel idea of separating the queen from her brood – something that happens when the colony queen departs with the bulk of the bees when a colony swarms – but all conducted under the one hive roof[131]. Now and again the simple act of isolating most of the brood results in bees attending young brood present raising a few supersedure queens, a technique I have and many others have used shown in action here (Figure 1.42).

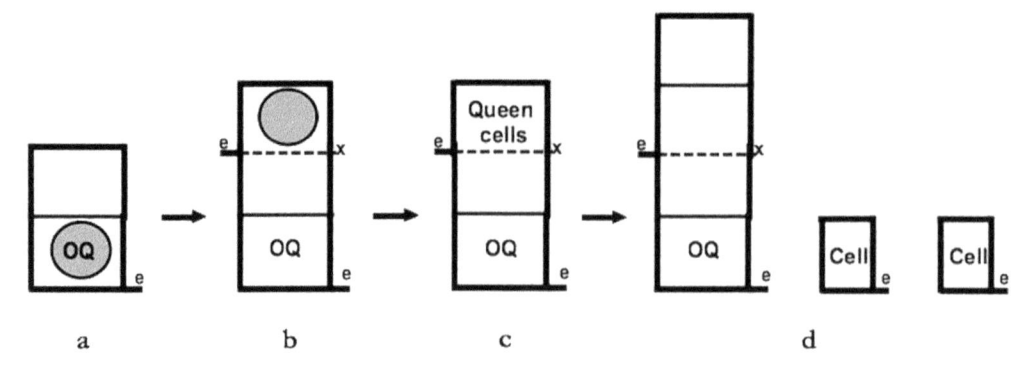

Figure 1.42 Demaree plan for swarm control with supersedure-raised queens: e = entrance; x = excluder; OQ = old queen:
(a) a strong colony at risk of swarming;
(b) the colony was reorganised where young brood was transferred to the top super and an additional super was sandwiched between, all conducted over a queen excluder. The old queen was left with one frame of brood and some stores in the bottom brood chamber;
(c) supersedure queen cells were raised in the top brood nest; and
(d) the cells were transferred to mating nuclei while the hive was retained as a normal supered colony for the honey flow.

Waving the queen farewell

Can any scheme of direct requeening be recommended above all others? This it is difficult to say as simple queen substitution will probably always be the method of choice if only the sake of convenience. However requeening into small nucleus colonies or into splits, that is staging requeening, makes common sense as does maintaining spare nuclei in any apiary.

Having spare queens and making the best of their productive capacity – controlling swarming apart – is the best recipe we know of to get bees to harvest and process nectar from trees and meadows dripping with nectar.

The nuances of queen replacement are examined more closely in *Requeening with swarm control*. There we will examine less direct methods of requeening where a new queen works with the colony queen to facilitate a more seamless transition mirroring, in most cases, natural supersedure.

Requeening with swarm control

During the 1950s an eminent group of American apiarists collaborated to present a coherent approach to spring hive management. They realised that the processes of honey bee population buildup, swarm preparation and queen replacement were essentially inseparable. They argued that these elements of bee survival should be viewed collectively as the natural processes of colonies making provision to requeen themselves, to reproduce and to build stores for periods of dearth.

To replicate the natural inclinations of honey bee colonies they asked themselves how they could devise simple plans to build bees continuously while at the same time requeen their hives. They wanted to copy what bees do well and that the bees had practised for millions of years. And to avoid swarming, they played a confidence trick on the bees: they divided the population at a critical juncture to imitate swarming and either supplied a new queen or got the bees to raise one themselves without curtailing the full laying capacity of the old queen.

The Killions of comb honey production fame conjectured that swarming was not the problem it is always made out to be, rather an opportunity to work with and harness the natural reproductive instincts of bees.

So what did these American beekeepers actually practise? Some of their schemes were adaptations of the Demaree swarm control plan where they also got colonies to raise or accept new queens. The new queen would invigorate the colony, head off queen failure, suppress the swarming instinct, and get bees to focus on brood nest rebuilding and on harvesting and storing honey and pollen.

A recurrent feature of their plans was the establishment of a new queen without having to resort to finding and removing the old colony queen. While old, diseased and patently failing queens heading struggling colonies were found and quickly despatched, every effort was made to continue to employ the laying capacity of any well-functioning colony queen that was at least until the new queen were well established and had her own retinue of bees.

A proactive, not a preventive strategy

Ready access to a bank of new season queens is good insurance against imminent queen failure, in practice either a supply of caged queens, ordered at the close of the last season to ensure availability, or cells raised in a strong colonies early in the season under ideal conditions.

However spare queens, lying idly in offset nucleus colonies, do little for apiary productivity unless their brood and stores are routinely raided to strengthen colonies. Extra queens put directly into service not only boost colony numbers but also supply an additional titre of queen pheromone. These queens, working the on either side of a double screen or nucleus board, were used subsequently to replace old and failing queens or double to supersede colony queens as soon as the services of the old hive queens were no longer needed.

Put more bluntly there was a ready awareness that honey bee colony queens wear out, just as do drones and workers: a second queen could be used to work in concert with a still well-functioning queen to fortify the colony and, in due course, replace her.

You might ask what these innovative beekeepers had in common in terms of their approach to using two queens laying in parallel? The answer is not a great deal. However they all recognised the inestimable value of having a second laying queen present instead of the standard new queen replaces old queen routine. Many, like Clayton Farrar and Winston Dunham, went much further. They were experienced two-queen hive aficionados and could reckon on their adjudging the relative merits of a newly grafted in queen versus that of the old parent colony queen working together in the same hive.

The goal here, however, was not to run hives with two queens but to run conventional single-queen hives at full tilt. Their shared trick was to employ a second laying queen to smooth queen succession and to avoid any interruption to the brood cycle during the period of buildup. As importantly they realised the serious downside of maintaining a second queen for any longer than were absolutely necessary. Two laying queens at the end of the main honey flow would equate to too much brood and an over supply of hungry bees, a common enough mistake made in two-queen hive operations.

In most of their schemes they introduced a new caged queen into a nucleus formed on top of the hive, that is above a double screen or division board, the nucleus being made up of bees, brood and stores taken from the powerful parent hive below. Once well established, and the main risk of swarming was over, this nucleus was united with the parent hive. No effort was made to find the old colony queen: in most instances the new queen successfully superseded the old colony queen obviating the need to search for and remove this queen.

Two queen schemes for automatic requeening, swarm control and crop increase

Here is a summary of the requeening schemes employing two queens designed to control swarming presented in the May 1952 and April 1954 issues of the *American Bee Journal*[132] broadened to include elements of advanced two-queen hive management set out in the journal in March 1953 where a second queen is commandeered to work in the same hive[133].

Professor Mykola H. Haydak from Minnesota: *The causes of swarming*

Mykola Haydak [1898-1971] sets out his approach to control of swarming with a self-evident observation[134]:

> *Swarming is a natural phenomenon in the life of bees. One reads a statement to the effect that if it were not for swarming, there would be no honeybees in the world today.*

Haydak follows with a nuanced explanation of the events that precipitate swarm preparation. He notes that, during the development of the hive in spring, and where the colony is crowded (27-35 brood nest bees per 100 cells), there is an increasing tendency of bees greater than 1-3 days old to be displaced from the brood nest:

> *In a strong colony, sooner or later, there will be a disproportion between the number of the displaced nurse bees and the space available for egg deposition…*

> *As soon as the queen is around the swarm cells the bees do not bother her. She approaches the swarm cups and deposits eggs in them. As soon as the queen larvae appear the displaced nurse bees can lavishly supply them with food. At the same time the bees stop feeding the queen. The latter feeds herself on honey, deposits less eggs and the size of her abdomen diminishes.*

He then went on to describe the loss of control by the queen over her progeny, the consequent swarm preparations and the subsequent absconding of unemployed bees. The solution to swarming, Haydak observes, is to resolve the supply and demand condition getting out of kilter, that is to keep all the bees well employed.

Other authors in these series provide the actual solutions, most employing a second queen to paper over the traditional disruptive approach to requeening that employs the emergency queen replacement impulse.

Mykola's notes on bee idleness being a key driver of swarming are enhanced by Scottish master beekeeper Bernhard Möbus's incisive observations on the impact of replete idle bees on their readiness to depart the parent hive[135].

Henry A. Schaefer: *Swarm prevention*

Henry Schaefer, a once president of the US National Beekeepers Federation[136], presents an unconventional approach to swarm control[137]. It is one that I had never heard of or read about except in the writings of Eugene Killion[138] and in Henry Schaefer's reference to Gravenhorst, the latter gentleman also famous for inventing a skep with moveable frames[139]:

> *A young queen mated before the nectar flow from the hive she later heads, plus ventilation and plenty of super room [won't swarm].*

In reading Charles Miller's famous book *Fifty years amongst the bees*[140] Schaefer records somewhat amusingly:

> *It was not long before I was very interested in what Dr Miller had to write about swarm prevention. Here it is: 'A colony disposed to swarm might be prevented from doing so by blowing it up with dynamite'. But, he says, 'that would be unprofitable'. He was seeking profitable swarm prevention.*

How did Schaefer link his swarm control efforts to requeening. First he reflected on his hive records of the excellent honey crop season of 1927. In that year, amongst his 200 colonies that produced an average of 91 kg (200 lb), there were two standout colonies each producing 194 kg (405 lb) of honey. He had used the standard Demaree swarm control protocol throughout the buildup phase but still pondered on what would constitute a more ideal routine, namely mating young queens from each hive before the honeyflow while also keeping the old queen laying.

He first toyed with using nucleus colonies, each furnished with a queen cell, established above a screen on top of each hive to establish new queens. Though successful in his goal he found this technique was too time consuming. Instead he switched to a scheme, very much the reverse of standard practice, where queens were raised below a screen underneath parent hives.

Here is what he did. Each strong colony (Figure 1.43a) was moved from its parent overwintered position and replaced with a new hive body to which was added a frame of eggs placed between a comb of honey and of pollen after first shaking bees back into the offset parent colony. He then added a couple of shallow supers or a single full depth hive body needed to accommodate returning field bees. To these he added a double sheet of newspaper and a screen with a rear entrance placing the parent hive on top (Figure 1.43b). As in other schemes, no queens needed to be found – Schaefer assiduously avoided finding them – and the parent colony, if it contained swarm cells, would normally be so depleted of bees that it would automatically destroy any present.

Ten-to-twelve days later all hives were inspected and two good queen cells found in the lower brood box were left in place (Figure 1.43c) where the best queen would mate and become established. As further contingency, if the upper unit with the old queen above the separator screen had been injured and queen cells had been built to replace her, it was reduced to a two frame nuc and left with one queen cell to avoid a high risk of swarming.

Then, by the commencement of the flow, the screen was replaced with a sheet of newspaper (the newspaper would only have been needed if a nuc board rather than a double screen were employed) to unite the two colonies (Figure 1.43e). The brood was consolidated at the next apiary visit to facilitate normal honey flow operation (Figure 1.43f) of hives.

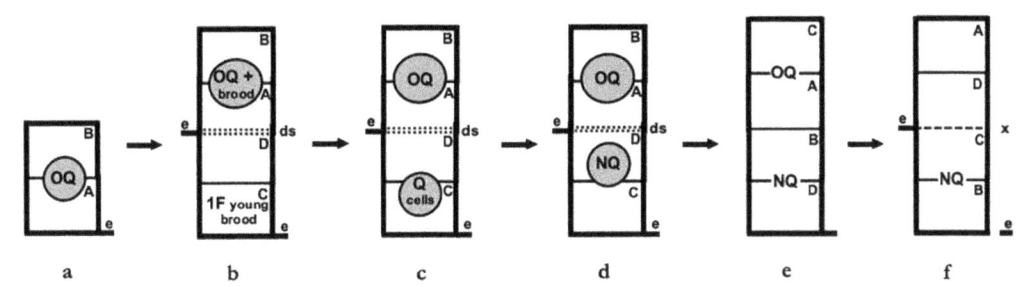

Figure 1.43 Schaefer requeening plan using two queens: e = entrance; x = excluder; ds = double screen; OQ = old queen; NQ = new queen:
(a) strong over wintered colony;
(b) colony is Demareed;
(c) new queen cells are raised;

(d) a new queen is mated and has her own emerging brood;
(e) the excluder is removed to allow the established new queen to supersede the old; and
(f) the brood is consolidated to prepare the colony for the flow.

Carl E. Killion and Eugene Killion: *Swarm control and queen rearing in comb honey*

The Killions, Carl [1875-1962] and Eugene [1923-2022], lived very long and, one might say, very fruitful lives. These famous father and son producers of comb honey describe a procedure[141] that almost fully controls swarming in populous colonies where crowding, essential to the successful drawing out of comb and filling of small honeycomb sections, was needed. As inspiration for their endeavour they refer to Geo Demuth's Farmers' Bulletin maxim[142]:

To bring about the best results in comb-honey, the entire working force of each colony must be left undivided and the means employed in doing so must be such that the storing instinct remains dominant throughout any given honey-flow.

They surmise that the provision of a young queen early in the season is not sufficient to prevent swarming under their crowded brood nest conditions instead observing:

We did find that any colony requeened with a ripe cell, in the swarming period, after being queenless for eight days, did not swarm nor swarm with a young queen reared and given instead of the cell, after the honey flow was underway.

In an apparent act of self deprecation they stated:

We hesitate when we say that we welcome the swarming season, for fear that any who hear us may think we are just bragging. But we are not: to us, swarming is just as necessary as the nectar and pollen that go into the bees.

While most beekeepers do not operate bees the way the Killions did – few produce section comb honey for sale these days – their technique of breaking the brood cycle and introducing a new queen allowed them to collapse already strong hives to single brood chambers. They could then super the bees with section combs where the bees would focus on comb building and honey production, not swarming or brood raising.

It has always been abundantly clear to me that the Killion technique, outlined in detail in *Honey in the comb*[143], would greatly facilitate the operation of the honey-tapping flow hive[144]. If you delve into Eugene Killion's famous book you will discover that they offset the old queen – in their case not so old girl – in a separate colony to build sufficient stores to provision the parent colony at the end of the season. After all you want the bees to fill sections – or flow combs – not to be diverted into the task of also provisioning the parent hive bees for a far distant overwintering.

However the Killions add a rider to the many who claim to have the solution to swarm prevention:

> *There are many methods of swarm control. We have tried almost everything recommended in the last thirty years. Most of these sound so very convincing on paper, but if some were left on paper and never tried out in the bee yard, the bees and beekeeper both would benefit.*

Newman I. Lyle: *Swarm control with the nucleus system*

Newman Lyle first reiterates the conventional swarm control measures[145] every experienced beekeeper is familiar with, large brood nests with ample worker and minimal drone comb and a young prolific queen. He then goes on to describe – much in the John Holzberlein style (see below) – the technique of introducing a young queen to a nucleus hive established on top of the hive by raising some sealed brood, stores and bees above a double screen. If nothing else the removal of brood from the strong colony puts back the potential of the bees to swarm but there is more to it than that. He had established a new nucleus colony and with a new queen.

Without getting into the detail of his scheme using this new queen to establish a two-queen hive – a topic for the rather more experienced beekeeper – it is worth noting that this newly established nucleus colony was subsequently used to simply requeen the parent colony. By simply shuffling the brood boxes, after taking the old queen out, he replaced the old queen with a new one in the process greatly reducing the incidence of swarming. Newman Lyle provides a number of other subtle hints, not least the potential to use the nuc to requeen other hives in greater need of requeening and – under good conditions – to draw more comb foundation to keep young bees active and less inclined to contribute to the swarming impulse.

John W. Holzberlein from Colorado: *Swarm prevention – not swarm control*

While the offset nuc method (or the equivalent hive split technique) is perhaps the most assured method for introducing new queens, Winston Dunham [1903-1980] (see below) and John Holzberlein [1902-1966][146] had many other ideas about the usefulness of the technique. They pioneered truly sophisticated plans, ones that integrated swarm control, requeening and colony build up for the honey flow. While both were great exponents of two-queen hive operation, they were equally cognisant of the many simple practices that used two queens to effect automatic requeening.

Instead of introducing queens to stand alone nuclei, they adopted the now familiar practice of moving brood and stores from strong hives to a new super located on top of a hive above a double screen (or split board). To this nucleus colony they introduced a new queen rather than a queen cell to accelerate its development. While in most respects the queening is identical to that of the queen introduction technique used by Clayton Farrar, Charles Gilbert and many others, John Holzberlein was focussed on seamlessly requeening whole apiaries. To paraphrase his technique:

> *Set a strong colony (Figure 1.44a) back and in its place put a new bottom board and an empty brood chamber. Quickly sort the frames filling out this empty chamber with unsealed brood and some frames of honey. On top of this place an excluder and above that a further empty super returning the remaining brood combs and stores to this chamber (Figure 1.44b) after first shaking bees off these frames at the hive entrance. The queen will be in the bottom brood chamber and nurse bees will move up through the excluder to tend maturing and emerging brood. A day or two later the excluder is replaced with a double screen, an empty super is inserted and a new queen is introduced to the older brood and stores in the top super (Figure 1.44c).*
>
> *Once the new queen is laying well and has her own brood some weeks later (Figure 1.44d), the two units are reorganised and united (Figure 1.44e). Newspaper will be needed only if a solid nucleus board is used instead of a double screen. The old queen is not found, the new queen superseding her.*

As Holzberlein notes the early set up process can be concertinaed if the queen is seen when sorting brood (Figure 1.44a –> Figure 1.44c) where the queen can be run in at the lower entrance as the excluder is not needed to isolate the queen.

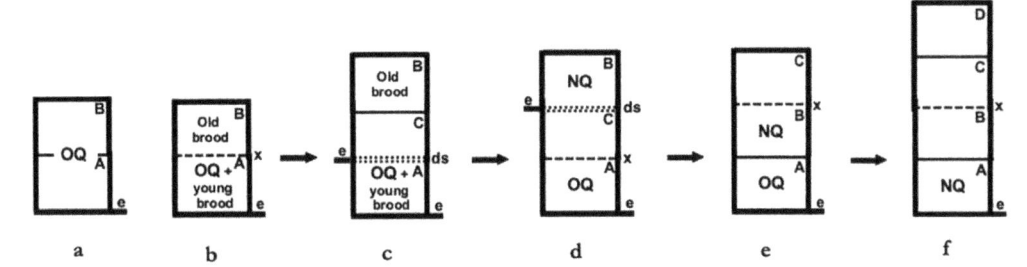

Figure 1.44 Holzberlein's requeening system employing two queens: e = entrance; x = excluder; ds = double screen; OQ = old queen; NQ = new queen:
(a) overwintered colony is reversed regularly AB->BA->>AB;
(b) frames are sorted and the queen is isolated below an excluder (a->b) noting that, if the queen is seen, this step can be omitted (a->c);
(c) the excluder is replaced by a double screen (or nuc board), a super is added and the queen receptive super with old brood is moved up;
(d) a caged queen is introduced and allowed to become well established;
(e) the colony is reorganised so that the new queen supersedes the overwintered queen; and
(f) the colony with a single queen is operated on the honey flow.

The merits of the requeening scheme include:

- never having to locate the old queen;
- forcing both colonies to rebuild their brood nests reducing the risk of swarming;
- minimising the risk of queenlessness;
- avoiding any break in the brood cycle;
- allowing the new queen to be evaluated prior to removal of the old queen; and
- facilitating seamless requeening where the old queen need not be found.

While contrived supersedure (allowing an introduced queen to replace an old queen) is never foolproof, Dunham and Holzberlein were forever watchful of queen performance and always replaced unproductive queens using spare queens they kept in nucs or raised in hive splits. It seems likely that their system of requeening mirrors Don Peer's remarkable success[147] in effecting supersedure of 80% of 4000 colony queens – introducing a queen cell to the top of hives to replace clapped out queens – late in the season.

If an old queen from the parent colony has had a good performance record and is still laying well, she can be later offset with a few frames of brood and stores as a spare nucleus colony. Such colonies are easily requeened, repurposed as spares or employed to requeen any ailing colonies.

If you only operate a handful of colonies and do not wish to risk of loss of newly purchased queens, you can of course first search for and remove the old queen rather than rely on supersedure to do the requeening for you. If finding that queen in a populous hive proves difficult there is a simple remedy. Offset the brood box on the bottom board with the old queen, move the unit with new queen to the bottom of the stack and search for the old queen once most of the field bees have returned to the parent stand (Figure 1.45).

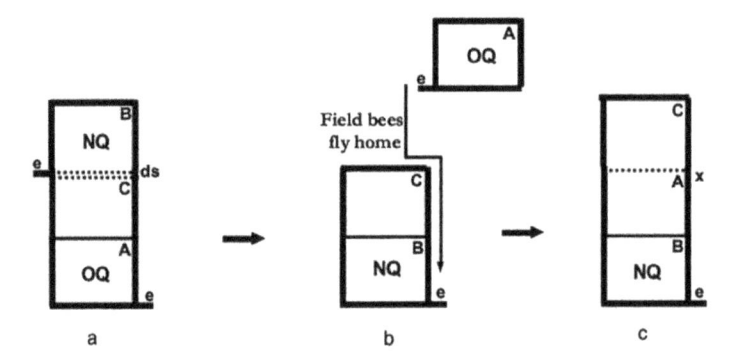

Figure 1.45   Offset hive technique to find old queen: e = entrance; ds = double screen; x = excluder; OQ = old queen; NQ = new queen:
(a) new queen is well established;
(b) the old queen is offset to facilitate her easy location and removal and where the chamber with the new queen is moved to the bottom of stack; and
(c) the brood is consolidated by reorganising boxes.

Elsewhere Holzberlein provides some very candid assessments of the swarming phenomenon. In framing an approach to swarm prevention[148] he noted that:

> *The phenomenon of the swarm is one which I doubt if anyone fully understands in all its phases. We know its basic cause. It is Nature's means for perpetuating the honey bee colony.*

To give substance to this observation he describes the many confounding conditions in relating his long experience of keeping bees, on the one hand feeling that he might line up with the 'experts' but on the other hand flummoxed by handling bees in a super swarming year:

> *But then a swarming year comes along and I know for sure that I belong right where I have rated myself all those years – alone with the learners.*

In describing the natural propensity of bees to swarm in beekeeping operations, Holzberlein notes dispassionately that:

> *Swarming is a natural instinct in bees but it has no place in modern honey production. Our great trouble [in hive] foundation is in trying to curb the urge once it takes possession of a colony. It almost parallels the sheep killing urge that develops in some dogs. The only way to stop it is to destroy the dog. Once a colony makes a firm determination to swarm, gets to the point of having sealed queen cells, with the old queen shrunken up ready for her first venture since her honeymoon days into the wide, blue yonder, there is nothing much we can do about it except to destroy the colony as such.*

John Holzberlein identified two effective cures for swarming:

> *…One is to swarm them, that is, some phase of dividing the old bees and queen from the young bees and brood. The other is to make the colony queenless and queen cell-less, causing it to raise a young queen from scratch.*

Neither he concedes is feasible in commercial operation as both are counterproductive to the building bees for the honey flow. His solution is found in a lesson that he learnt from the Nebraskan beekeeper Ralph Barnes, of Oaklands. Barnes mantra was 'Divide or requeen'.

Holzberlein's solution was to do both though, for the beginner, it may seem rather daunting:

> *The kind of dividing I am going to tell you about has no part in making increase. The divide is made all under one cover. The 'split' or 'divide' is set up over a solid or screened inner cover with an entrance of its own, and given a young queen. It is the beginning of the two-queen system, but right now it is a divide and the best little swarm preventer that you ever tried. Aside from being almost sure-fire swarm prevention it has the added advantage of getting and holding more bees in the field force of the colony than one queen could possibly*

produce. It keeps them all coming to the same hive, yet divides them at a time when the desire to swarm is almost sure to take over if nothing is done…

It is clear that Holzberlein understood that while two-queen colonies might produce rivers of honey in a protracted honey flow, the essence of productive bee management was to build powerful colonies and to avert the eventuality of bees swarming. He recognised that even the temporary presence of a second laying queen would achieve multiple objectives: less likelihood of colonies to swarm, much stronger bees and seamless requeening.

Winston D. Dunham from Ohio: *Dunham's modified two-queen plan*

Winston Dunham was an early architect of schemes to control swarming during the buildup phase[149]. Like others Dunham established new queens in nucleus colonies on top of strong hives early in the season, his Ohio modified two-queen plan removing the complicated supering arrangements that beset operation of hives with an additional queen during the honey flow. Without getting into the particulars of how he operated with two queens, his plan was simple and decisive. He concluded:

This plan is very versatile and allows also for preflow conditioning, making increase, and testing queens.

Figure 1.46   One of Dunham's Ohio modified two-queen apiaries.

Dunham's modified Ohio plan hit the sweet spot. His bees were ready – unswarmed, requeened and at full strength – to take advantage of the main Ohio district sweet clover flow where all bee effort was focused on honey collection not on a mix of harvesting and excessive brood raising (Figure 1.46). For southeastern Australia this would translate to uniting hives, those sporting an extra queen, in the middle of our Yellow Box and Blakely's Red Gum flow in mid summer, returning the bees to their easy to manage single-queen condition through to the end of summer and into autumn.

Charles S. Engle: *Automatic Demareeing*

Demareeing is an old chestnut[150] and is little more than asymmetric splitting of a colony and doing it under the one roof: most of the brood is relocated well away from the queen. She is left with minimal brood and some stores. In Charles Engle's approach[151] the queen and a single frame of brood are confined to a bottom brood chamber over which is placed an excluder while the remaining brood is lifted above a new empty super of combs and placed on top of the hive, the technique adopted by Henry Schaefer.

Engle's very early account of operating bees in the 1920s provides a nuanced application of this seemingly well-practised routine. Like all Demaree plans it upsets the division of labour in the hive and that stops the bees swarming: he could Demaree one hundred colonies in one day.

Amongst his many observations he found that supersedure cells raised in the top brood chamber rarely led to swarming or to the establishment of a second queen. This only occurred when virgins could find an escape hole, that is find free flight. Such new queens would turn a super of honey into a super of brood.

He experimented further by routinely providing a flight entrance to the top super, then employing any new queens he found to promptly requeen the colony after first removing the old queen. Given that he was operating apiaries of 100 or so hives he found locating and removing old queens too time consuming. So he resorted to shaking bees from the top nuc with the new queen in front of the hive entrance discovering that practically every new queen was accepted and had disposed of the old queen, in essence forced supersedure.

Engle had discovered a simple and efficient method of raising queens and requeening. Here it is worth highlighting a technique independently pioneered by David Leehumis and Victor Croker at Australian Honeybee in the southern highlands of New South Wales. They introduce a select queen cell to a nucleus made up of a few frames of young as well as sealed brood and stores established above a queen excluder but without bees: this obviates the need to find the colony queen: the young brood draw bees up through the excluder.

Since cells and virgin queens have no distinguishing pheromone, they found that a queen excluder sufficed to allow establishment of a new queen. By mating queens on strong colonies in isolated yards under flow conditions, most of their hives are successfully established as two-queen hives, the nucleus on top not only benefitting from rising warmth of the colony below but also from free access to colony resources below.

Australian Honeybee are migratory beekeepers so they do not then run their hives as two-queen units for honey flows. Instead they simply move the top super with the new queen to the bottom of the stack and move the bottom super with the old queen onto their truck. Most field bees drift to their parent stands and the truck with a load of single boxes with old queens is driven off into the sunset. Single brood boxes with such queens are depleted of field bees (see Figure 1.45) so are easily found. These colonies are re-celled in due course to provide a large number of strong nuclei with new queens.

G.H. (Bud) Cale: *Emergency swarm control*

Bud Cale, a then editor of the *American Bee Journal* and a prime instigator of the 1952 *Swarming round-up series,* provides an overview of the other author contributions[152] noting that all swarm control boils down to adopting emergency measures. Yet he concedes that the contributions of the likes provided by John Holzberlein suggest that swarming should be anticipated. Proactive measures such as splitting all strong colonies will change the mood of bees and keep them in that active expansionist condition essential to building colonies needed to gather large amounts of honey.

Like any beekeeper with a good grasp of the swarming impulse, Cale noted that:

> *The age of the queen has much to do with the amount of swarming because supersedure will be carried out at the same time swarming is usually imminent. The older queens therefore will be more involved in swarming than the young ones.*

The key swarm control strategy that Cale floats, one he highlights in *The Hive and the Honey Bee*[153], is to switch the location of strong and weak colonies, noting however that the strategy will only work if any started queen cells are first removed. It is a strategy that would be unwise to adopt unless you first checked your bees for brood diseases but it is a simple practice to adopt when you have a paddock full of beehives in need of emergency swarm control. Such hive shuffling is certainly unsuitable where parasitic mite infestation is not well controlled or where American Foulbrood is masked by use of antibiotics, a practice illegal in Australia.

In a later foray Cale outlines a variation of the traditional hive splitting method of swarm control (Figure 1.47)[154]. A hive with double brood nest with an additional shallow food chamber (Figure 1.47a, super C) is first established. By spring the upper supers (B and C) are occupied by brood when the supers are reversed (Figure 1.47b). In most instances the queen soon migrates to the now upper super A.

About two weeks later the colony is split and supered when a caged queen securely corked is immediately introduced to the most likely queenless split (Figure 1.47c, super B), the majority of bees drifting to the parent stand (Figure 1.47c, super A). Five days later, if hive B has no eggs, the cork is removed and the candy punctured to facilitate quick release of the queen. If, alternatively, the old queen is found laying in hive B, the caged queen is simply returned to the parent stand hive A where – after removal of the cork – she will be quickly released.

Cale's plan was simple and flexible: he did not need to locate the parent hive queen when making the split and introducing the caged and securely stopped queen. So he got the bees to tell him which of the splits had the laying queen but this required him to check for and remove any started queen cells if one of the splits were queenless.

Once the new queen was well established the colonies were united and supered, Cale noting that in most cases both queens both continued to lay for some time.

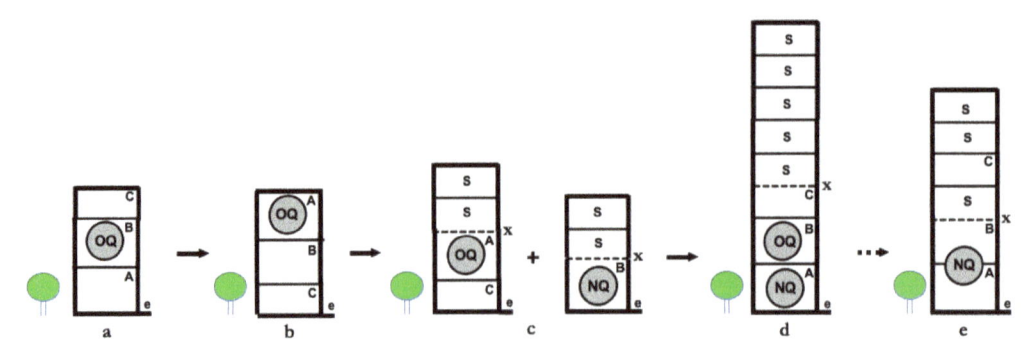

Figure 1.47 Cale split to arrest swarming: e = entrance; x = excluder; s = super; OQ = old queen; NQ = new queen:
(a) strong overwintered colony;
(b) the brood boxes are reversed and left for several weeks;
(c) the colony is split and the queen is introduced to the unit without eggs after five days (most likely brood box B);
(d) the colony is reunited and supered after the new queen is well established; and
(e) the new queen supersedes the overwintered queen and supers are extracted and returned to the bottom of the honey super stack or removed as the flow ebbs.

The scheme is notable for its extraordinary simplicity and any requirement to find queens.

Loren F. Miller from Minnesota: *Two queens to reclaim weak colonies*

In northerly Minnesota Loren Miller started his hives each spring from package bees in shallow brood bodies, feeding them heavily to establish them quickly[155]. He started colonies as single-queen hives in shallow hive bodies with a full frame of honey, a week later adding a shallow super where needed (Figure 1.48a). To recover weak colonies, and knowing that all colonies had young queens, Miller supered them above strong colonies – those that came away promptly – recognising that the two queens laying together (Figure 1.48b) would produce more honey than had the hard case colonies been left alone to flounder.

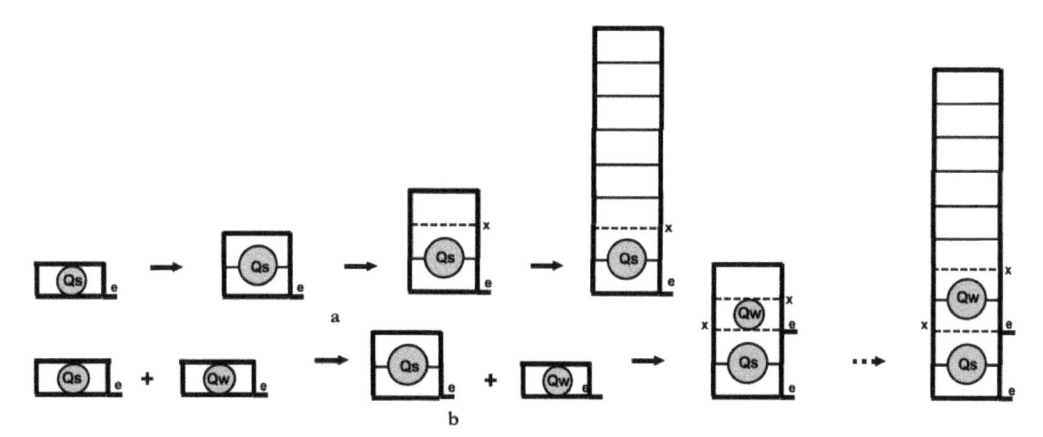

Figure 1.48 Miller package bee weak colony rescue plan: e = entrance; x = excluder; Qs = strong queen colony; Qw = weak queen colony: (a) course for normal single-queen colony expansion; and (b) employing weak colonies that come away slowly to further strengthen strong colonies.

In a more fulsome earlier article[156] Miller concedes that operating two-queen colonies can be problematic:

> *I found that all hives run as two-queen colonies are difficult to operate without costly extra visits and some disappointing results. Some swarm, some lose all their workers to one queen, some develop so much late brood they finally make a wonderful crop of late off-grade weed honey.*

Like others Miller succeeded in building his bees faster with a second queen though only under conditions where it proved profitable to do so.

Eugene S. Miller from Indiana: *Swarm control with extracted honey*

In reflecting on Loren Miller's and the very famous Charles Miller's [1831-1920] contributions to concomitant requeening and swarm control, it was hard to ignore the contributions of their namesake, once President of the American Honey Producers' League and early pioneer of commercial beekeeping in North America, Eugene Miller [1861-1958]. He had an uncanny grasp of effective hive management at an apiary scale (Figure 1.49).

Figure 1.49   One of Eugene Miller's bee yards 1918[157].

Eugene Miller provides us with what would now appear to be a fairly conventional approach to swarm control[158] for his era starting with a remarkable claim:

> *In the production of extracted honey, swarming can be prevented. It has been proved by more than six years without a swarm in a yard averaging about 50 colonies. In fact, the bees seldom started any queen cells. I believe this record can be duplicated by anyone who has a good strain of Italians and who will use the method which I will attempt to outline.*

He went on to describe how he did this:

> *Confine the queen to the lower story. If you have trouble finding her just shake off bees and queen and let them run in under the queen excluder. Next, use as a second story a deep super of drawn combs, which is to be left in place during the next twelve months as a food chamber. Then place on top of this what was previously the second story. Time required, about 15 minutes per colony.*

Miller's scheme avoids the classic labour intensive approach of regularly reversing brood boxes to delay the onset of swarming. Instead he used an excluder to confine the queen to a single largely empty lower brood box where the queen was forced to rebuild her brood nest, a de facto Demaree. He later conducts a more formal Demaree claiming that the overall scheme had entirely prevented swarming in his apiaries:

Then about June 1 [our 1ˢᵗ of December or quite a lot earlier given local ACT region conditions] or whenever the brood chamber becomes crowded replace the brood with drawn combs, moving to the top or fourth story all brood except one comb, which is left below so that the bees will not desert the queen.

He emphasises that this second operation must be conducted before the bees have any inclination to swarm and certainly well in advance of swarm queen cell construction. He points out that this measure removes the brood nest congestion inimical to colony expansion and Haydak's notion of supply and demand condition getting out of kilter.

Eugene Miller does not signal how he went about requeening his colonies but he did unite weak colonies and did set up his colonies with new queens – he lived in the US mid northwest – in August, that is in mid to late summer. This would have ensured that his colonies came away early in spring and be headed by young queens, a strategy that would have averted the need for spring requeening of his colonies.

Elsewhere Eugene Miller signalled a few other essentials, notably the use of young queens, but he claimed his system was essentially foolproof and that it was backed by years of experience. In a May 1943 article[159] he lists all the measures one should take to prevent or minimise swarming:
- queens should be bred only from the best non-swarming stock...;
- failing queens, and all queens over two years old, should be replaced, otherwise attempts at supersedure may tend to induce swarming;
- defective combs... should be discarded...;
- suitable ventilation should be provided with an entrance to the hive at least 7/8 inch (22 mm) deep by the full width of the hive;
- sufficient room [should be provided] for the expansion of brood and for storage of honey...; and
- ...the excess of young bees and the emerging brood from the combs to be occupied by the queen [should be removed].

Much earlier he had briefly noted a scheme for successfully starting a queen in the top super using a select stock queen cell in lieu of those raised by the colony[160]. This technique is now part and parcel of new queen establishment amongst time-starved commercial apiarists.

It seems likely that all these schemes are imbedded into commercial practice but are not well known amongst sideline beekeepers. To the extent that these schemes are a much simpler alternative to the operation of doubled and two-queen hives and that they achieve seamless requeening, they could make backyard beekeeping a more productive pursuit especially if this resulted in annual or biennial requeening.

IV
Honey Bee Keeping

Of the best ways to house and keep bees there are too many texts claiming the ultimate ways to do so. Abbé Émile Warré certainly had the needs of bees in mind when he developed his People's Hive. And while I still keep my bees in poorly insulated wood and plastic boxes, I have switched over to fully screened bottom boards – these work well in temperate Australia – and have packed all migratory covers – ventilated hive lids – with insulation. For those simple measures my bees now overwinter far better.

In good years, I get more honey than I can poke a stick at especially from my doubled and two-queen hives. But aside from brief excursions into two-queen beekeeping, I've instead included some topics that have helped me understand what influences beekeeping practice most. We start with an account that reflects the transition from traditional skep hives to frame hives and the enhanced benefit incurred if they are well insulated. We then take a fleeting glance at honey bee chemistry (most bee behaviour is mediated by bee pheromones and olfactory sensing) and dive into a few basics, making sure your bees are always well fed and having spare queens at hand to replace those that are patently failing. Floral resources and siting of hives aside, keeping your bees as safe as houses is about the most a keeper of honey bees can do to ensure bees thrive.

Art thou a skeppist?[161]

Part I The origins and practice of skep beekeeping

Here we introduce ourself to the lost art of skepping and its transition to the modern and robust frame hive. We first trace the origins and practice of skep beekeeping.

By the late 19th Century the continued use of the skep hive was in the balance (Figure 1.50).

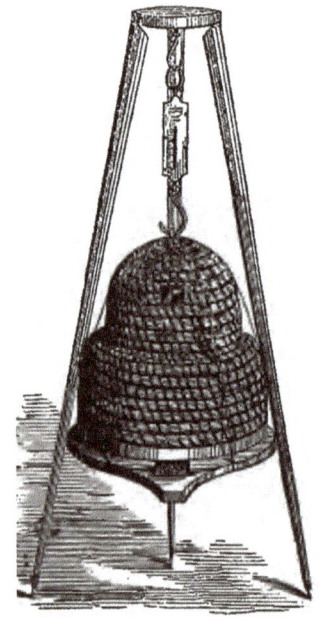

Figure 1.50   The skep in the balance[162].

Skeps[163], contrived from woven wicker or reeds, or coiled sheathes of straw daubed with mud and animal dung, are 'bee house baskets' (Figures 1.51 and 1.52) placed open-end-down. They date back to Egyptian civilisation some 2400 years ago[164].

In their simplest form, skeps had a single bottom entrance, their free-formed combs being fixed. Initially, at least, honey and wax could only be obtained by complete destruction of the hive. Like earlier clay pot and log hives they could be neither closely inspected nor their size regulated.

Skeppers were and remain skep builders. Many were skilled artisans producing a wide variety of skep designs ranging from simple domes to peaked structures with weather protecting hackles. Later flat-topped skeps, over-supered with honey-storing bell glasses or ancillary 'cap' chambers made of the same materials as the skeps themselves, came to the fore.

Skeppists were beekeepers, sometimes equally innovative and also highly skilled. Their hives were fashioned from natural fibres.

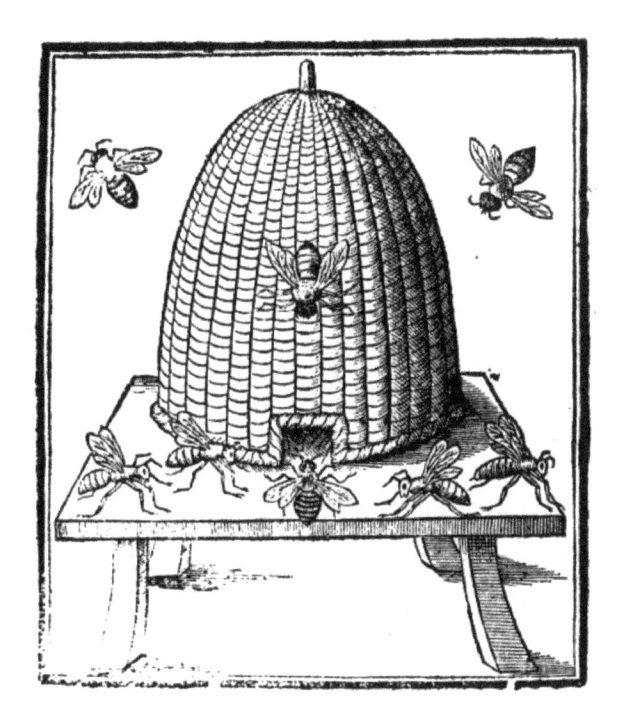

He who by Bees doth ever thinke to thrive,
Must order them, and neatly trim his Hive.

Figure 1.51 A 1634 skep setup[165].

Traditional hives can be viewed as replicas of the natural cavities, tree hollows and rock crevices, that honey bees employ in the wild. Those cavities, providing shelter, allowed honey bees to radiate from the tropics to temperate climes where they were insulated from extremes of cold. Indeed only two species of honey bees, *Apis mellifera* and *Apis cerana,* from the subgenus Apis have been widely cultivated and have extended their range into latitudinally colder climes. Thomas Seeley argues they have never been truly domesticated.

Their nesting boxes were also a larder, an extension of the colony engine. Honey bees garnering reserves of honey and pollen in brief plenty enabled them to survive over the long period of winter dearth. Combs attached to the roof were inaccessible to predators and, in some measure, from the predations of honey hunters whose garnering history goes back to the Mesolithic. This practice is depicted in cave drawings dating back at least 10,000 years[166].

Figure 1.52   Middle Ages skeps 1236-c.1250[167] and c. 1200[168].

To this day the skep hive symbolises not only the indefatigable bee, but also the industrious and busy beekeeper, the latter reminding us of the over-fawning ditty[169]:

How doth the little busy bee
Improve each shining hour,
And gather honey all the day
From every opening flower!

Isaac Watts

The skep also tells the tale of long gone beekeeping practice and, perhaps, the yearnings of beekeepers seeking a more authentic way of making apiculture more sustainable. For a charming accounts of old skep practice see A.H. Bowen's 1923 tale of skep beekeeping in England[170] and Granvenhorts 1907 moveable frame wicker hive (Figure 1.53) in Germany[171].

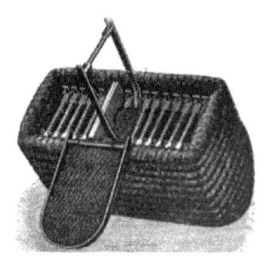

Figure 1.53   Gravenhorts' wicker skeps with moveable frames.

It is equally interesting to reflect on the cultural origins of the contemporary inclination of some natural beekeepers to allow bees to swarm freely. Traditional swarming beekeeping was integral to skepping. The practice encouraged bees to swarm engendering an annual swarm catching frenzy. Very active swarm collection, largely supplanted wild honey hunting, the raiding of natural tree hives often involving tree and hive destruction alike. The list of casualties might have also extended to include the occasional honey hunter misjudging the many hazards of raiding wild bees. In swarm beekeeping, some skeps were deliberately restricted in size to encourage swarming, cast swarms being eagerly chased and watched for much as a nomadic shepherd guards his flock.

In the skep era large numbers of swarms, equating to plentiful queenright colonies, allowed the progressive establishment of hives until the swarming season was over. The theory was that the more swarms one could collect the more honey one would get. Much of the putative success of swarm beekeeping can be attributed to late season blooms, notably of heather, a time lag that allowed honey bee colonies to build and become populous enough to be reasonably productive. End of season hive destruction was also a factor in encouraging the culture of promotion of swarming to produce new replacement colonies. This practice thwarted the natural inclination of bees to survive over winter to become very productive colonies each spring.

Since weak hives would not normally survive winter they were sacrificed as were the most productive hives, the latter so that all honey could be harvested. The practice of 'killing the poor bees' gradually drew wide-spread condemnation forcing a good deal of reform and innovation.

So by the early 19th Century skeppists were encouraged to limit hive destruction by preserving medium weight colonies likely to successfully overwinter. Such measures were complemented by skep modification practices that increased the effective size of the skep and which encouraged bees to store more honey.

The art of skep construction is a wondrous and varied tradition. For a glimpse of actual skep making I cannot but recommend highly enough the rustic Irish video *Hands: Of bees & bee skeps*[172].

With the benefit of hindsight, we can now see that skep beekeeping presented two intrinsic problems. Firstly beekeepers could not inspect comb for diseases and pests. Secondly as, apart from structural bracing, there is no internal structure provided for the bees to attach their combs. Honey combs were attached to the inside of the skep making harvest so difficult that it resulted in the entire colony being demolished.

Some very old beekeepers might still claim that skeps were all the rage when they were boys and that keeping bees this way was very fashionable. But only thirty of forty years ago, even owning of bees was contentious amongst the most avid naturalists in the New World. Then, though their reward to the agricultural community and our food chain was accepted, they were treated with suspicion as competitors to native birds and insects. We suspect that story has not changed amongst purists where honey bees are not native, though the public sentiment for 'rescuing bees' and fervour for actually having a hive certainly has.

Setting up and operating skeps

Charles Butler (1609) provides a comprehensive account of skep construction (Figure 1.54) and operation of skeps in his *Feminine Monarchie*[173]:

> *In some countries they use straw hives bound with briar: in some wicker Hives made of Privet, Withy or Hazel, dawbed visually with Cow cloome tempered with gravelly dust, sand or ashes.*
>
> *The strawne Hives when they are olde and loded, so visually sinke on one side, (specially if they take wet), and so break the combes and let out the hony; and then spleec them strong with a Cop, v. fitted to the top of the Hive.*
>
> *The Wicker Hives will still be at fault, and lie open, (if they be not often repaired), unto Waspes, Robbers & Mise. Any of these if they find a little chap, will dig her way in: and the Mouse (unless the twigs be close wrought) though she find none.*
>
> *Both these Hives, if they be not well covered, are subject to wet: which maketh them musty, and if it be much, rotteth the combes, and destroyeth the Bees. But*

the heat in the Summer, and the cold in the Winter, and the raine at all times doth soonest pierc the Wicker Hives: which is double cause to double dawb them.

All things considered, the strawne Hives are better, specially for small swarmes.

The Bees do best defend themselves from cold, when the hang round together in manner of a Sphere or Globe (which the Philosophers account for the most perfect figure) and the neerer the Hive commeth to the fashion thereof, the warmer and safer the Bees. But of necessitie the bottom must be broad, for the upright and sure standing of the Hive, and for the better taking of the combs: and the top must be rise some two or three inches from then just forme of the Globe, to stay the hackle, and to shunne the raine: which yet, where the hives are covered with panns, is not necessary. Otherwise let your Hives vary no more from this round figure, then needs must, as where it is in the top skirts to the skirts seventeene inches, in the middle or the widest place through the center fifteene inches, and at the skirts thirteene, after this forme.

Hives are to be made of any size between a bushell (35 litres) and half a bushell; that any swarme, of any quantity or time soever, may be fitly hived. v. less than half a bushell will not containe a competent stall; and more than a bushell is found too bigge for any company to continue, and thrive together.

The middling size of three pecks, or within a pottle, under or over, as fitly conteining the naturall quantity of a good stall, is most profitable.

Have alwaies Hives of all sorts (but most of the middling size) in store, lest they be to seeke when you should use them.

The best time for making them, whether they should be Strawne of Wicker, is in the three full moneths of Winter, Sagittar. Capre. and Aquar. v: for then the straw, briers, and twigs are best in season: and then is it best to provide them, because they are best cheape.

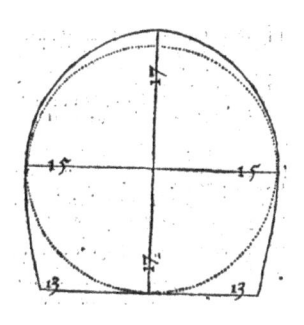

Figure 1.54 Skep dimension instructions: This forme with his dimensions will conteine three pecks: and the abating of one inch in each dimension abateth gawne in the content.

Other detailed instructions are recorded in widely circulated almanacs of good agricultural practice such as those of Edmund Southerne's 1593 *Bee Culture*[174] in a section entitled: 'The manner how to dresse Hives before you put in Bees':

> *… When you have trimmed your Hive, then take Sallow, Willow or Hasell stickes, which being cleft in the middle and cleane shaven, you may put five pearls into a Hive, that is to say, two a crosse foure fingers breadth within side, and the other foure parcels from the top to the mouth, being sticked fast betweene the squares of the cross stickes, is the best way you can splay your Hive, both for ease for the Bees, and also for splaying the Combes within side…*

To make a size comparison with the currently used Langstroth eight and ten-frame boxes in common use, we see that today's beehives are on the whole much larger. An imperial bushel has a volume of thirty six litres, so the recommended skep sizes were between eighteen and thirty six litres: the skep stood approximately 860 mm high and was 760 mm wide at the base.

The eight-frame box has an internal volume of thirty five litres, quite close to the maximum recommended skep size. It is also approximates the size of wild bee hives remarkably optimised to survive *Varroa destructor*[175]. In practice the set-up volume of a single eight-frame hive will be at least forty litres when allowance is made for migratory lid and bottom board space. For the recommended minimum two-box hive that volume increases to about seventy five litres.

The later invention of keeping bees in sturdy moveable frame hives meant that the colony did not need sulphuring (smoking to death) or drowning of bees to harvest the crop and that advances in queen breeding to produce more tractable and productive bees could be made. To otherwise harvest honey from skeps beekeepers had either to drive the bees out of the skep into a facing empty skep or, by the use of a bottom extension called an eke or a top extension called a cap, create compartments – the predecessor of the honey super – where honey alone was stored.

The driving of bees – drumming them from a mature skep of bees into an empty skep rather than always killing them to harvest honey and wax – was popularised by poet Thomas Tusser[176] in the 1550s:

Driving of hives and preserving of bees

Now burne up the bees that ye mind for to driue,
at Midsomer drive them and save them alive:
Place hive in good ayer, set southly and warme,
and take in due season wax, honie, and swarme.

Set hive on a plank, (not too low by the ground)
where herbe with the flowers may compas it round:
And boordes to defend it from north and north east,
from showers and rubbish, from vermin and beast.

In another Renaissance edition, Tusser[177] advocated the feeding bees to save their lives:

Saint Mihel byd bes, to be brent out of strife:
saint John bid take honey, with fauour of life.
For one sely cottage, set south good and warme:
take body and goodes, and twise yerely a swarme.

At Christmas take hede, if their hiues be to light:
take honey and water, together well dight.
That mixed with strawes, in a dish in their hiues:
They drowne not, they fight not, thou sauest their liues.

By the middle ages, skep design had advanced considerably, notably though the use of ekes, imps or nadirs, typically coils of wreathed straw made into rings (Figure 1.55) that could be 'undersupered' to expand their capacity to raise brood and store honey.

Figure 1.55
Eking, imping or nadiring of skeps[178].

But as we all know, these advances, with the ready transport of bees, came many maladies. While skep beekeeping did not allow for disease control other than by natural attrition, modern beekeeping has resulted in bee diseases being spread simply by interchanging gear both within and between beekeeping operations. Honey bee queen exports and long distance movement of bees for pollination services has resulted in global disease spread and mixing of the honey bee gene pool, the latter, for example, occasioning the killer bee phenomenon in the United States[179]. Despite claims to the contrary, the open mating nature of honey bee queens has not led to any significant decline in their gene pool, rather it is now more mixed and regional subspecies are no longer the pure strains they once were.

If you are not quite convinced about the unintended impact of free trade on bees, neither chalkbrood nor small hive beetle were even contemplated as problems in Australia thirty years ago. Yet most are now widespread and we are set to be affected by the likes of Asian hornets, the African large hive beetle and a large suite of arachnid mites, while American Foulbrood has become far more common. I am reminded of an apiary that I owned was quarantined following a suspected AFB outbreak back in the early 1980s. Fortunately that turned out to be EFB and all the hives recovered.

Welcome twenty twenties. Australia now has a likely unstoppable incursion of *Varroa destructor* though not of other mites such as *Acarapis woodi* or *Tropilaelaps mercedesae*.

Skeppists were often skilled and knowledgable beekeepers, more than capable of managing fixed comb hives far better than the many present day back yarders who own, but who all too often largely neglect, their bees. If you have any doubt about the skilled art of skepping, I also strongly recommend you watch the late 1970s video[180] of a 700-strong skep bee yard in Lower Saxony, NW Germany. It demonstrates the remarkable understanding of the swarm issuing season, the capturing and mating of virgin queens, skep surveillance to both unite swarms and to make queenless colonies queenright, and to successfully operate a large number of skeps.

Part II   From the skep to the modern frame hive

With the origins and the practice of skep beekeeping behind us, we now examine how this ancient way of keeping bees has morphed into the modern insulated and ventilated moveable frame hive (Figure 1.56).

Figure 1.56   Inside Alfred Neighbour's observation hive and skep apiary[181].

Beyond the skep

The invention of the moveable frame hive came very late in the history of beekeeping. Elsewhere we have discussed Sir George Wheler's 1682 account of swarm control where colonies were split to make increase outlined more fully in *Out of the box*. These skep-like hives were inverted, wastepaper basket style containers. Sticks supporting combs were suspended across the open face of the basket and covered with a lid, perhaps the earliest form of telescopic hive cover, so that some frames could be removed and transferred to an empty basket to split the colony.

By the following June 1683 an anonymous 'J.A.'[182] – whom Eva Crane attributes to the antiquarian, folklorist, archaeologist, natural philosopher and writer John Aubrey [1626-1697] – had provided detailed descriptions of a rectangular hive (excerpt Figure 1.57) that had top bar frames, namely sticks, evenly spaced, that could be interchanged with combs from, or split off to, other hives. These hives had vertical walls sloping inwards near the base, best visualised as modern Kenyan and Tanzanian top bar hives. The lack of appreciation of bee space meant that problems of cross bracing, more particularly the adhesion of side combs to walls, remained.

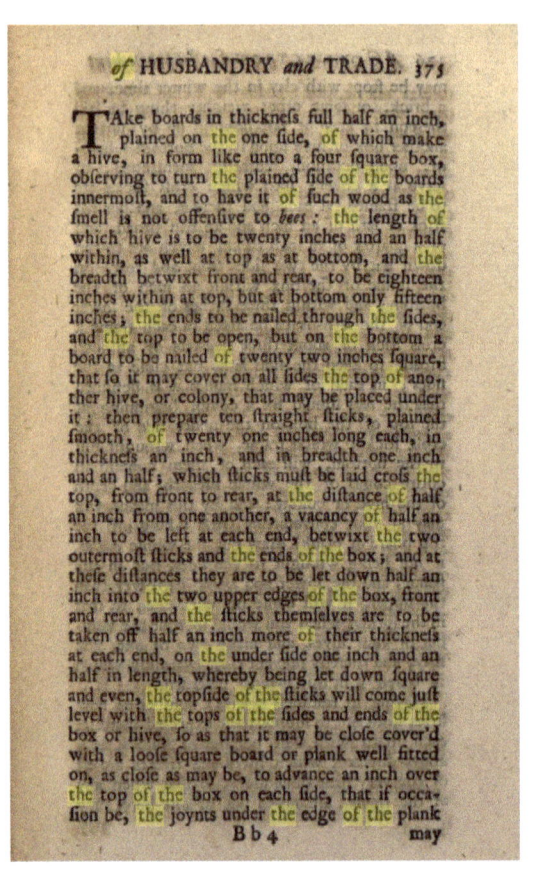

Figure 1.57 Excerpt of John Aubrey's instructions for construction of a moveable frame hive.

In an expansive 1927 review of beekeeping practice, Bevan discusses the tiering of such supers[183], as he called it storifying or:

the piling of hives or boxes upon each other... preserving a free communication between them; a method which enables the apiarian to take wax and honey without destroying the lives of the bees.

His earliest account of such supering he accords to a 20th of December 1653 account provided by Samuel Hartlib[184]:

My Appiary consists of a row of little houses, two stories high, two foot apart, which I find as cheap at seven yeares end as straw hacles, and far more handsome...

So it was not till much later that more robust full frame beekeeping became common place. That came around with François Huber's[185] [1750-1831] leaf hive (Figure 1.58), the first true frame hive, from which he made extended and detailed observations. His late 1780s study of the swarming process led to the discovery of the prime swarm containing the old queen and subsequent swarms headed by virgin queens. Rather later Johann Dzierzon[186] [1811-1906] and Lorenzo Lorraine Langstroth[187] [1810-1895] pioneered the spacing of frames so they could be easily removed. Dzierzon is further credited with discovering honey bee parthenogenesis.

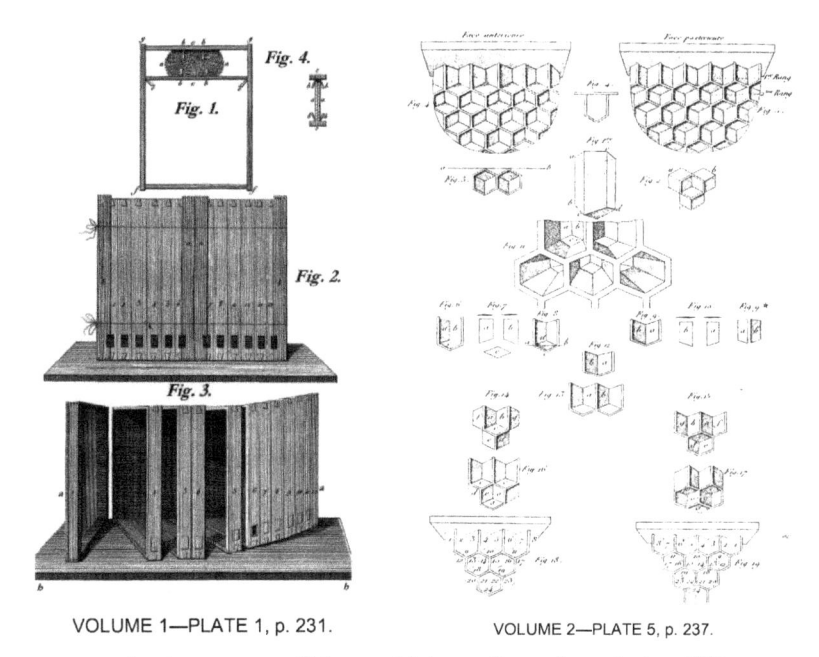

VOLUME 1—PLATE 1, p. 231.   VOLUME 2—PLATE 5, p. 237.

Figure 1.58   Huber's 1814 Leaf hive and his studies of comb detail[188].

The impact on beekeeping of Dzierzon's and Huber's many observations, and those of Langstroth's, meant that skeppist reliance on issued and captured swarms was replaced by a strategy of active swarm control and prevention. Langstroth very clearly articulated the importance of very practical moveable frame hives for modern beekeeping practice. He advertised his patented 1852 hive[189], one with a modicum of woodworking skills you can now build[190].

Bee space

In the introduction to early editions of Langstroth's *The Hive and the Honey-Bee,* Robert Baird, writing from New York, noted enthusiastically:

> *In my earlier life I had no inconsiderable experience in the management of bees, and I am bold to say that the hive which Mr. Langstroth has invented, is in all respects greatly superior to any which I have ever seen, either in this or foreign countries. Indeed, I do not believe that any one who takes an intelligent interest in the rearing of bees, can for a moment hesitate to use it; or, rather, can be induced to use any other, when he becomes acquainted with its nature and merits.*

Langstroth's penned prefaces to the 1853 and later editions makes no mention of bee space, rather alluding very much to his poor health – an often depressive state noted by Ron Miksha[191] – adding, as an afterthought, a hope that his hive design would prove serviceable to the community had begun.

In the 1863 edition, Chapter VII, *On the advantages which ought to be found in an improved hive* (pp.98-113), Langstroth elaborates sixty one requisites for the complete hive. In that listing he fails to make any mention of frame spacing, though many contemporary authors make claim that he published the discovery of bee space in the 1853 version even claiming that discovery was made in 1851.

In a revelatory publication, Johansson and Johansson make contrary the claim that other designs of the era – including British designs – had applied the same spacing principle and that Langsroth may not have indeed been the first to enunciate bee space[192]. This noted, it must be said that the Langstroth design has stood the test of time, and most beekeepers attribute the discovery of bee space to him. They note however that in 1861 Langstroth totally disavowed a claim to be the inventor of a moveable frame hive:

> *I will try to be the first to acknowledge that although an original inventor, I was not the first inventor of such a hive.*

The ABC and XYZ of Bee Culture nevertheless states clearly[193]:

> *Mr L.L. Langstroth, in the great invention which he gave to the world (the first practical moveable frame), made the discovery that the principle of bee space could be applied to design a hive, in which the frames could be removed.*

This authoritative manual provides very concise overview of the development of the Langstroth and subsequent evolution of such designs as the Dadant and Heddon hives. Most authorities have supported Langstroth's 'discovery of bee space', Miksha noting presciently:

> *He made wooden frames that held the wax combs, designing them so they dangled within the hive's box with their wooden edges always 3/8 of an inch (9 mm) from anything that might touch them: the lid, the interior box walls, the box bottom, other frames. Positioned like this, the bees neither waxed the frames together nor stuck them to the sides or bottom of the hive. The result was a beehive with movable frames. Combs could be lifted, examined, and manipulated. It was 1851 and modern beekeeping had begun.*

Eckert and Shaw make a similar observation[194]:

> *Talks of a bee space of 1/4 to 3/8 inch was made by Rev L. L Langstroth in 1851. This the discovery that bees would not fasten the frames or combs to each other or to the hive discovery revolutionized the beekeeping industry. The frames were hung in the hive so that they could be removed vertically and additional compartments could be added from above.*

We might sensibly resolve the issue of who discovered bee space by observing that the bees themselves had long invented honey comb architecture including passageway space many millions of years ago. Elsewhere we have noted Roger Morse's observation that giant honey bees (and no doubt also dwarf honey bees) maintained a bee space, or working space, between the curtain of bees and the comb (see *The giant honey bees*).

Charles Dadant's much revised 1913 version of *The hive and the honey bee* reduced Langstroth's *Requisites of a complete hive* (pp.137-139) to twenty six factors needed for good hive design, prefacing this listing with a more explicit discussion of bee space, the space between pre-Langstroth-design hive bodies, that allowed top bar hives to be split to make increase and to control swarming. He instances both cylindrical wicker top bar hive bodies (units we now call supers that could be tiered) and rectangular hives built in two parts (that could be split vertically) to illustrate how this was achieved.

Dadant notes pointedly that this did not resolve the issue of exact frame spacing that allowed individual comb inspection. This leaves one to wonder who made the first specific claim to fully comprehend bee space. Clearly Langstroth's design achieved this outcome as the (unstated) bee space rule was applied to all passages within his hive.

Transition to the frame hive

However skep beekeeping persisted, Alfred Neighbour[195] (1886) describing many maturing skep beekeeping practices, the driving of bees into empty skeps placed on top of inverted mature skeps, and extending the storage capacity of skep colonies coming into honey flows by superposing glass bells and other storage vessels onto the top of hives (Figure 1.59). He even describes a Ladies' Observatory or Crystal Bee-Hive: It was certainly a time of very fluid transition to modern beekeeping practice. Notably it was also a time where cottagers were the main class of beekeeper. Neighbour, however, was acutely aware of a range of moveable frame and observation hives as well as hive supering practice all signalling a shift away from fixed comb hives such as the skep.

But both Alfred and George Neighbour[196] were also familiar with the Langstroth's frame hive and bee spacing. They described and marketed many frame hive designs, recorded many decorative hive setups and were familiar with the intricacies of François Huber's leaf hive and even described the making of wax foundation:

> *Impressed wax sheets for artificial combs: These artificial partition walls for combs are sheets of genuine wax, about the substance of thin cardboard. They receive rhomboidal impressions by being pressed between two metal plates, carefully and mathematically prepared and cast so that the impressions are exactly the same size as the base of the cells of a honey-comb.*

BELL GLASSES.

To contain 10 lbs., 10 inches high, 7 inches wide.
To contain 6 lbs., 7 inches high, 5½ inches wide.
To contain 3 lbs., 5 inches high, 4 inches wide.

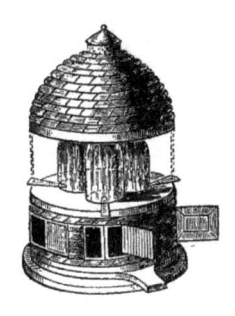

Figure 1.59 Neighbour's skep hives:
(a) Bell glasses: the largest for Nutt's Hive, the middle-sized is for our Improved Cottager Hive; the smallest glass is so very small that it is not often used;
(b) Improved Cottager Hive [£1.15sh] with observation ports and bell glasses; and
(c) Cottager's Hive No. 8 for taking honey in straw caps without the destruction of the bees.

There is also a history of protective hackles, weather-proofing witches hats and of bee houses, structures that are found in relic bee walls and elaborate bee shelters that persist across Europe today (Figure 1.60). Skeps are still employed occasionally to capture but not keep swarms in but are truly a thing of the past: Eva Crane and Penelope Walker certainly reflect on this almost dead art[197].

Figure 1.60   The last wicker skeps with hackles in Herefordshire.

The legacy of bee skepping

Skep beekeeping is now outlawed in western countries because frames cannot be inspected for brood diseases. So the question: 'Art thou a skeppist?' is purely rhetorical, despite the image of a skep as one of 'a hive of industry and exemplary work ethic' has carried over as a relic of its rich apicultural history.

Much was learnt from the keeping of bees in containers such as earthenware pots and natural log hives, not least from keeping bees in readily transportable skeps. Skepping sustained beekeeping through to both the Middle Ages and to the pre-Victorian era:

- providing detailed observations of the swarming behaviour of honey bees;
- recording the economic and cultural heritage of agrarian societies;
- leading to the first attempts at non-destructive honey harvesting strategies such as cutting out honey comb from hive back entrances to hives and driving of bees;
- developing the moveable frame hive;
- inventing supering, colony division and swarm control; and
- greatly facilitating the early culture of migratory beekeeping – skep hives were lightweight leading to easy transport and marketing of late swarms.

Recent advances in hive design would appear to be those that replicate conditions found in the most successful wild colonies, those deep inside tree hollows, that allow bees to best regulate the internal hive environment and that protect the nest from the vicissitudes of highly variable external conditions. These boil down to improved thermal and moisture regulation and better ventilation of hives.

Möbus[198] was one of the first researchers to show that bees circulate air vertically and that single bee space between honey frames and double bee space between brood frames appears to optimise heat retention and moisture and carbon dioxide exchange. Recent studies by Derek Mitchell[199] at the University of Leeds have shown that honey stores used to maintain bees through winter and on cold nights and the energy and hence stores required to ripen nectar are inversely correlated with hive insulation. Good insulation saves honey, stops bees having to cluster tightly, reduces their susceptibility to diseases such as *Nosema* and the *Varroa*-DWV complex, reduces bee wing wear and bee ageing and vastly improves nectar ripening and colony water cycling dynamics.

But these are topics beyond the scope of the window on skep beekeeping. Regrettably some skepping routines such as facilitating the old routines of swarming beekeeping (Figure 1.61), the lack of any effective disease control and the common neglect of beehives persist and are common enough amongst todays hive owners.

Figure 1.61   When 'swarming beekeeping' and skeps were all the go.
September in the Rohan Hours 14th Century[200].

Beekeepers with aspirations 'to save the bees' other than in skeps and to do so traditionally might start with Palladius's 350 AD adage:

The bee-yard should not be far away but aside, clean, secret, and protected from the wind, square, and so strong that no thief can enter it.

This was rendered in Middle English as *De opium castris. 145*[201]:

The Bee-yerd be not ferre, but faire asyde
Gladsum, secrete, and hoote, alle from the wynde,
Square, and so bigge into hit that no thef stride.

Bee space and bee hive architecture

In reexamining the homes of bees, mainly at the urging of Thomas Seeley, beekeepers are asking themselves whether hives, ostensibly designed to maximise honey production and produce powerful colonies optimising their pollination potential, are in fact those most suited to the well-being of bees or are indeed the best for beekeepers.

Bees in the wild make maximal use of the space afforded to them constrained only by the configuration of the cavity they nest in. They need corridors to undertake their nesting tasks and access to their stores and they pay much attention to home insulation, to propolising cavity walls and to hive ventilation and the best organisation of their brood engine and larder. Bee space governs the spacing of combs and efficient transport of materials and air circulation.

Good architecture is defined by factors such as comb loading and efficient use of space. Remarkably we need to learn much more about hive design from our cavity dwelling cousins.

Part I Bee space[202]

The discovery of bee space is elaborated upon in the previous section *Art thou a skeppist?* Many beekeepers make claim to understanding bee space (Figure 1.62) but in practice it almost always lands them in strife. Indeed it is one of the fundamentals tenets of beekeeping taught to beginners.

Anything bigger than a bee space will be filled with comb and anything smaller will be filled with propolis. Bee space is that magic 8-10 mm gap, in the old imperial measurement 3/8", needed to allow bees to move freely from one location to another in the catacomb that is the nest of the honey bee hive. Despite this known 'known' we are surprised that theory and practice are not very often on the same page.

Figure 1.62 No room in the inn – Bee space violation issue in migratory hive lid – Jerrabomberra Wetlands Apiary 9 January 2019.

An understanding of bee space tells us that with too much space bees will either increase the width of combs – increasing their effective storage capacity – or fill the void with burr and brace comb. When an examination is made of the sizes of available hive furniture (boxes, lids, bases and frames), we see a common mismatch of frame and hive body depth – not providing the exact bee space – that results, for example, in bottom bars being cemented – often with drone comb – to the top bars of supers below. This type of problem is a legacy of differences in how different bee gear manufacturers view bee space, a lack of common consensus on setting hive and frame dimensions and variable specifications adopted by governing agencies. This said an Australian full depth Langstroth super, typically 244 mm deep, might be trimmed to 239 mm to accommodate a fairly standard frame depth of 230 mm to provide a single bee space between supers. Notably the problem only arises between hive bodies and under hive lids. Building to hive walls has been overcome by correct spacing of frames and standardised 9 mm spacing between frame end bars and hive walls. The side attachment of combs to hive walls in natural tree hollows is replicated in attachment of combs to frame side bars, the 9 mm gap between end bars and hive walls allowing greater air circulation in hives than in wild nests is an anomaly that the beekeeping literature does not highlight as an aberrant feature of frame hives.

Provision of the critical bee space is similarly not a problem between the hive bottom board and the base of brood chamber frames. Indeed bees often fail to draw comb down to the base of bottom bar frames, a problem that Canberra Region Beekeeper's Dannielle Harden pointed out to me can be overcome by installing old-fashioned slatted racks that provide 'dark space' and sit on the brood chamber floor. Any inspection of a wild hive signals that bees need ample floor clearance to process hive detritus and to facilitate the vertical air circulation between frames that Bernhard Möbus identified by placing mini anemometers between hive combs.

The imposition of an industry standard, that is mandating a uniform 9 mm bee space and uniform measurements for all manufactured hive furniture, would be onerous and would likely meet strong opposition. However this is what will be required if many of the design mismatch problems are to be overcome.

As many of us have experienced, make the gap too narrow and you will give the bees a choice: they will either chew out the comb matrix to allow passage of bees or plug the gap with propolis. Leave out the hive mat under a migratory hive lid, and you will fill the roof space with comb honey with a maze of bee-ways.

We have already noted Scottish beekeeper Bernhard Möbus's discovery that not only is single space needed to prevent cross bracing of combs and to avoid attachment of combs to walls but also that this space is almost doubled between brood combs[203]. While this increased spacing allows bees to simultaneously tend both brood faces and facilitates closer packing of bees in the winter cluster, the very narrow bee-way between honey combs restricts air movement. This is important in maximising the insulating value of honey above brood. Möbus also suggests that the narrowness of this space may have been important in evolving the relatively small body size of cavity dwelling honey bees.

In *Part II Bee hive architecture* and elsewhere we will examine Morse and Seeley's finding of a distinct bee preference for a nesting cavity size of around 40 L (see also Figure 1.2 in *The honey bee queen*) but do not evaluate how bees utilise the space made available to them. It would seem that the topic of 'bee space' can be expanded to the notion of the total amount of space available to a colony of bees.

This gives us license to 'air our views'.

How bees apportion space in a tree hollow cavity

For an informed overview of bee nests in hollow trees it is well worth turning to the wisdom of Tom Seeley and his book *The lives of bees*[204] where he starts out his chapter on the honey bee nest with this insight:

> We shall see that when colonies live under a beekeeper's supervision in wooden hives, rather than in their own hollow trees, they are forced to cope with homes that are often jumbo-sized, poorly insulated and closely spaced.

To this we might add:

> Bees have to live with partitioned hive interiors that may be less than conducive to efficient use of the space available.

But in the wild, and short of gnawing loose any fibrous material and shoring up crevices with propolis and wax, honey bees are constrained by the walls of the cavity they have sought out. In attaching their combs to the cavity roof and adjusting their points of attachment to match the natural hanging surface, bees can fill out the whole space with comb, only making provision for their free movement and ventilation. Like the development of stalactites in a limestone cave bees build downwards but they control the space-filling architecture.

Suppose a swarm enters an empty full depth 8-frame Langstroth box of internal capacity of 34.8 L [35.4 L if the Australian Honey Bee Council recommendations are adhered to] (see Table 1.4). The volume of this box is just short of the optimal size sought by swarms, but this space will be increased by the bottom board riser depth.

To this box let's add eight equally-spaced foundation starter strips (that is to the underside of a close fitting telescopic lid) to simulate bees starting combs in a natural cavity. We might then expect the bees to draw out perfectly parallel combs leaving a single bee space around the four walls and at least this space between the hive floor and lower comb bottom bars.

In practice, of course, the bees create a cross-braced and impenetrable labyrinth. However, having opened many possum and bird boxes occupied by swarms,

I can attest to having found large sheets of vertical well-spaced combs, never well aligned, but following the general pattern of proper comb spacing.

So set up, we first notice that the bees leave a 9 mm gap between all combs storing honey and pollen (Figure 163a). We then observe a relatively large gap between sealed worker comb surfaces (Figure 163b), but rather less interstitial space where drone comb is present. For a full box this translates to a maximum of 25.4 L (69%) of the total space being utilised for honey storage but a rather less, 12.6 L (65%), if all the space available were given over to raising worker brood. In estimating this space for brood we are further indebted to Möbus who pointed out that capped worker cells spans about 25 mm across the comb septum.

| Single bee space | Single bee spacing between honey frames | Single bee space | Double bee spacing between brood frames |

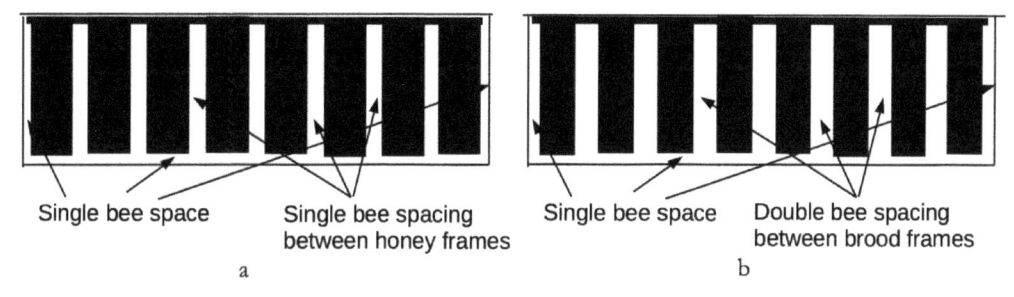

a   b

Figure 1.63   Bee space in free-comb built into a standard full depth Langstroth 8-frame hive body:
(a) occupied wholly by stored honey with 9 mm single bee space between sealed store comb faces; and
(b) fully laid out worker brood comb with ~13 mm space between sealed worker brood surface, the shaded black combs being ~25 mm wide.
Notes: Standard Hoffman frames are 35 mm across at the shoulder, and an 8-frame super is 309 mm wide internally. With 8 frames this allows a gap between each frame of ~4 mm.

As an aside, it is interesting to note that with an Iodice Kenyan Top Bar (KTB) hive[205] with the same 35 mm top bar width, and with combs built to within 9 mm of sloping walls and the floor with a single bee space between combs all round, comb occupies 40.3 L (70%) of the total 68.6 L working capacity. The greater theoretical efficiency of use of space – below the roof formed by top bars – arises from the fact that bees are able to form free-form comb. This efficiency is almost identical to that achieved where bees invade an empty rectangular box (as in Figure 163a).

We can now turn to the situation where the comb space is broken up. We start with the same scenario of a swarm settling into the same empty super and forming perfect evenly-spaced combs hanging from a flat hive cover. But, let us say, rather than the one full depth box, we now examine the space allocation where there are two half depth boxes.

Here we 'fake a false ceiling' by inserting thin wooden slats to start the lower combs. So rather than a single full depth super 244 mm deep (Figure 164a) we now have two half depth supers (Figure 164b) with the same equivalent total void space of 34.8 L (or 2x17.4 L). Notice, however, that there is now a 9 mm gap between the two half depth boxes. This reduces the overall usable space by an additional 3.7% (1.3 L). This small space penalty is always incurred where multiple brood boxes and honey supers are used.

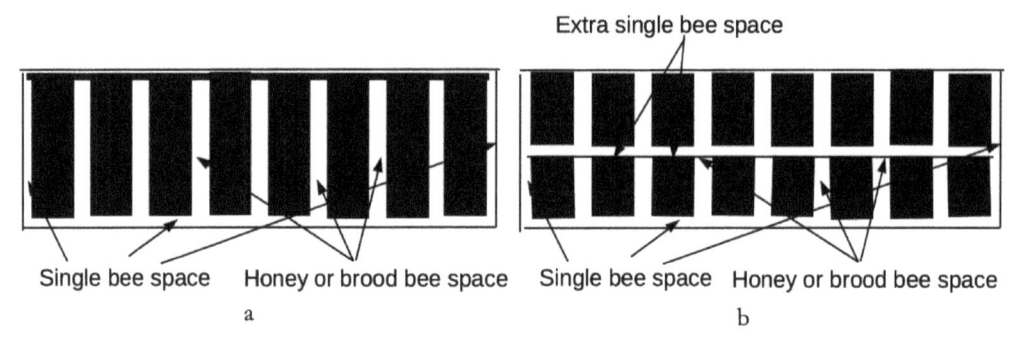

Figure 1.64   Bee space in free-comb built into a full depth Langstroth 8-frame hive body:
            (a) a standard full depth Langstroth 8-frame hive body; and
            (b) a double half depth Langstroth 8-frame hive body with additional bee space between supers.

Comb in wild hives also varies naturally in width. Dave Cushman records that spacing between comb centres of any reasonable size is usually 36-38 mm[206], although this may vary slightly where there are small pieces of infill comb. We can now turn to 'man-maid' hives to see how bees accommodate a built environment fitted out with standard rectangular frames.

Bee use of space in the frame hive

For the bees, what are the essential design tradeoffs between the cavity space and the comb area available for storage and raising brood? The only free space is the single bee space that that exists between combs of honey, or the roughly double bee space between adjacent sealed brood and the bee space that exists between bee frames and cavity walls. The natural constraints encountered in wild hives such as honeycomb stress loading – governed by the depth of comb from the point attachment to the roof – and the ventilation dynamics needed to supply oxygen and remove carbon dioxide from the hive cavity and maintain water balance, are largely overcome in any well-designed frame hive.

With regular frame hives we find we need only to discount the total volume of materials that make up the frame and the additional space lost between frame components (e.g. between frame edge rims) to calculate the usable space available to the bees. The overall space allocation requirements include the need to provide access highways (the amount of urban space taken up by roads and car parks), the need to provide perimeter access and, in the case of the moveable frame hive, the facility to make the frames movable (i.e. not attached to the roof) and avoiding any attachment of combs or frames to hive walls.

Bee space, first observed by François Huber[207], was put into practical effect by the Reverend Lorenzo Lorraine Langstroth[208] in his design of the modern frame hive. In examining the overarching features hive design – wide versus narrow hive body designs, ventilation features and the value of insulation – we examined these overall features but not the details of the internal use of space.

Hive dimensions

Hive body widths and depths vary considerably but are constrained by the natural limits of bees to manage the space available. Shallower hive bodies have significantly less effective brood rearing and storage space per unit volume while more closely packed frames offer more brood rearing area though reducing effective storage space.

For bees, the requirement of the colony to exercise climate control over the hive void must be traded against the available comb space area for raising brood and storing and curing nectar and pollen. Too much hive space – and poor insulation – and the bees will have unavoidable heating and cooling bills. Too little space to raise brood and to put away stores – a lack of real estate – will severely inhibit colony development and promote swarming or, under extreme circumstances, occasionally encourage absconding. Further a poorly configured hive giving little attention to bee space tolerances will result, as we have seen, in construction of burr and brace comb and the propolisation of gaps.

Frame design

There are two basic types of standard frame, Hoffman frames with tapered end bars, that rely on the frame shoulder width to space frames across the hive, and Manley frames with rectangular end bars (used mainly in honey supers) to minimise the amount of wax needed to seal honey). Manley frames space frames all the way down (Figure 1.65).

Manley frames are famous for being propolised, that is where the frame end bars meet, but appear not to be in wide use in Australia. However both frame types present the same usable comb surface area. Manly frames are 41 mm (1 5/8") wide, about 6 mm wider than standard Hoffman frames so eight frames cannot fit into our standard 8-frame hive body. However they are truly self spacing.

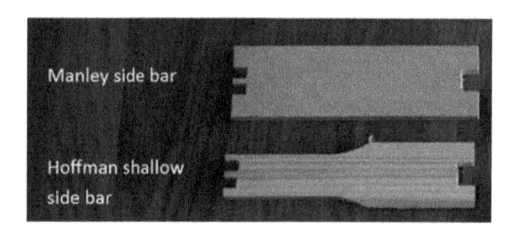

Figure 1.65
Manley and Hoffman frame side-bars[209].

Factory Hoffman frame widths vary considerably but range in shoulder widths (the top wide section) from 'shaved down' 32 mm (1 ¼") to standard 35 mm (1 3/8") and to Dadant 38 mm (1 ½") and are tapered top to bottom. Provided the frame spacing is fairly even (we estimate just over 4 mm additional spacing between each pair of Hoffman frames in an 8-frame hive body), there is little cross-bracing allowing combs to be interhanged freely. Bees adjust outside comb width to provide bee space adjacent to the hive wall.

And since flow hives are now popular, it is worth noting that these frames are 51 mm wide. As shown (Figure 1.66), anyone looking to modify their existing hive bodies to fit flow frames should take note that the 242 mm depth of a Flow® frame will be too large but a standard 8-frame super will comfortably accommodate six flow frames and sit on an excluder. In our apiary we have observed that these frames will be propolised between frame edge gaps and extend comb width to follow the bee space rule.

This brings us to the point that frames in well designed hives should provide a single space between supers and must clear the floor by at least a single bee space (8-10 mm). With frame top bars typically 15 mm deep and 25 mm wide, bottom bars 9-10 mm deep and 25 mm wide and with frame side bars typically 9-10 mm thick, the timber framework results – less shoulder support lugs and frame holder hive rebate – in about a 2 L (6%) reduction in total space available in an 8-frame full depth box.

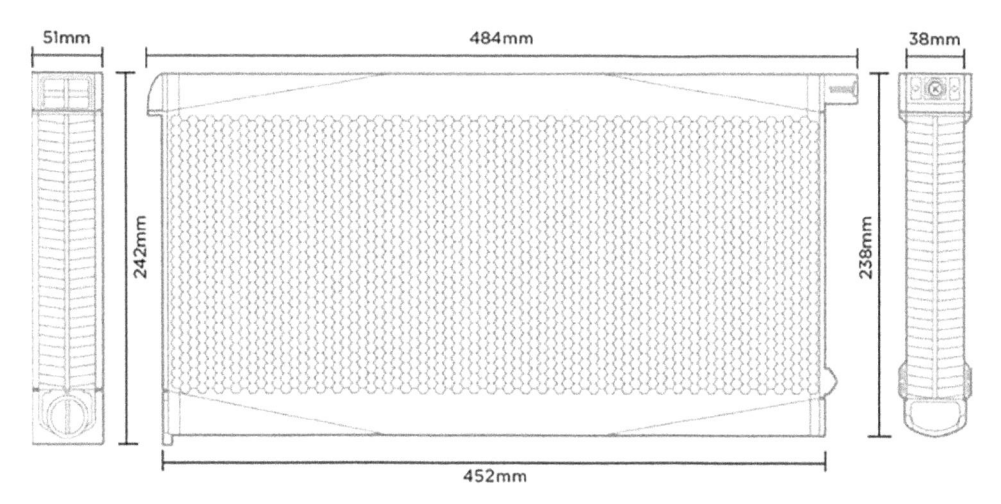

Figure 1.66 Flow® frame dimensions[210].

This leaves the vexed issue of leaving bee space adjacent[211] to a queen excluder inserted between supers. A flat wire excluder placed flush with frame top bars on a super below does not present problems provided that the bottom bars of supers-to-excluder below are spaced by about 9 mm. The queen and drone excluding wire spacing (4.2-4.3 mm) does however invite burr comb building into the actual excluder and has led many beekeepers to avoiding their use altogether. This problem is accentuated if screens are clad with a rimmed riser frame (Figure 1.67). These screens need routine cleaning to facilitate free bee movement and ventilation. In our experience they are best removed and placed in a solar wax smelter for effective cleanup.

The tradeoff between usable frame space and total hive cavity space is also influenced by frame configurations. Honey supers are often operated one frame (or even two frames) short but must then be very carefully spaced. Removal of frames does not significantly change the amount of honey that can be stored in a super as bees will draw out comb and still maintain single bee space between frames, though there will be at least one less bee space. These so-called 'fatties' are easier to decap but Australian Honeybee claims that this economy of frame use must be offset against the fact that the wider spacing will increase the amount of comb cross-bracing.

Figure 1.67 Framed queen excluder (left) and double screen (right) with build in flight entrances.

As we have already seen the use of shallower hive bodies and proportionally smaller frames, easier for the beekeeper to lift off than full supers of honey, imposes a small penalty on the amount of honey that can be stored.

Other factors affecting usable comb space include the relative proportion of worker and drone comb and comb defects (warping, holes, and space between the comb edge and frame bottom bar). These factors taken together result in reduced real estate for raising worker bees and storing pollen and nectar. This is a clarion call for regular comb renewal.

Table 1.4 and Table 1.5 relate hive dimensions to total available volume. Comb area on the other hand is correlated with the amount of hanging space. Where bees are allowed to construct free-form comb, say in an empty super, each comb edge reaches from its roof attachment to within 9 mm of the super below. In our free form 8-frame comb hive (Figure 163a) the total comb area is 1.74 m^2 (0.2171 m^2 per free formed comb).

In hives with frames the comb area is reduced by the depth of the top bar (15 mm) and the widths of the side bars and bottom bar (nominally 10 mm) and by the bee 9 mm space left between frame edges and hive walls. The total comb area for an eight frame hive is then just 1.40 m² (0.1746 m² per comb) or 80% of the equivalent free form comb area of 1.74 m² (Figure 163a) if comb is fully attached to the roof and walls and extends to within 9 mm of the super below it.

The amount of brood that any one super can hold is directly proportional to the number of frames. The normal practice is to fill out supers with as many frames as will fit. A notable exception to this rule is the use of eleven frames in a 12-frame brood box as in the British National Hive. The tolerances for fitting in twelve frames are so small that the almost universal practice is to operate eleven frames and insert a narrow follower dummy board to fill out the available space.

A further issue arises when double brood chambers are employed. The common North American practice of using two full depth brood chambers facilitates regular brood chamber reversal and is one I have long practiced. This stimulates colony expansion and inhibits swarming, but the downside is that the brood nest is split.

While running an additional brood box is not warranted where 9, 10 and 12-frame supers are employed, we have found that running a double brood box is helpful – at least during the spring build up phase – in 8-frame operations. Unless honey and pollen frames are regularly moved out of a single 8-frame brood box, i.e above the excluder, a prolific queen may not have adequate space to lay to her capacity.

There is, nevertheless, a clear space penalty for employing shallow frames instead of full depth gear for raising brood. Some beekeepers employ extra deep brood chamber boxes – with correspondingly larger deep frames – to ensure room for the queen to lay and to always confine the brood nest to a single bottom chamber. Spreading brood over more than one super often becomes a necessity when a single standard full depth 8-frame brood box is used.

Hive body	New South Wales Department of Primary Industries[212]			Australian Honey Bee Industry Council[213]		
	Outside dimensions (mm³)#	Inside dimensions (mm³)	Volume (L)	Outside dimensions (mm³)	Inside dimensions (mm³)	Volume (L)
Langstroth 8-frame FD	500x347x244	462x309x244	34.8	508x352x240	470x314x240	35.4
Langstroth 8-frame WS Pender	500x347x193	462x309x193	27.6	508x352x192	470x314x192	28.3
Langstroth 8-frame Ideal	500x347x147	462x309x147	21.6	508x352x144	470x314x144	21.3
Langstroth 8-frame Half (shallow)	500x347x122	462x309x122	17.4	508x352x125	470x314x125	18.4
Langstroth 10-frame FD	500x400x244	462x362x244	40.8	508x406x240	470x368x240	41.5
Langstroth 10-frame WS Pender	500x400x193	462x362x193	32.3	508x406x192	470x368x192	33.2
Langstroth 10-frame Ideal	500x400x244	462x362x147	24.6	508x406x144	470x368x144	24.9
Langstroth 10-frame Half (shallow)	500x400x244	462x362x122	20.4	508x406x125	470x368x125	21.6
Langstroth 12-frame FD				508x508x240	470x470x240	53.0
Non standard hives						
Paradise 8 (9)-frame	545x408x240	465x328x240	36.6	-	-	-
HiveiQ 9-frame*	545x408x240##	465x328x240	36.6	-	-	-

Paradise 10-frame	540x455x240	465x375x240	**41.9**	-	-	-
Warré 8-frame[214]	318x318x210	300x300x210	**18.9**	-	-	-
Kenyan Top Bar	Top bars 35 mm wide	1200x(240-360)x190.5	**76.6**	-	-	-
	A sample of other hives		-	-	-	-
British National 12-frame** Deep[215]	460x460x315	442x442x315	**61.5**	-	-	-
British National Standard	460x460x225	442x442x225	**44.0**	-	-	-
British National Shallow	460x460x150	442x442x150	**29.3**	-	-	-
Langstroth 12-frame Jumbo	448x448x286	410x410x286	**48.1**	-	-	-

Table 1.4   Nominal void space volumes for various hive bodies focussed on those in common use in Australia.
Notes
   #   Standard hive body timber width is 19 mm (¾"); plastic hives are of variable wall thickness but upper and lower rims are 19 mm wide and internal dimensions match those of the traditional wooden hive. Paradise hives and *HiveiQ* frames have a wall thickness of 40 mm.
   ## Other features of the *iQ* hive include a 5 mm telescopic rim (total hive body depth 245 mm) that locks supers together and prevents water ingress and improve air leakage seal, the now widely employed mesh bottom board to minimise floor debris buildup, a generous 35x280 mm landing board with a three-part fold-down gate system fencing a 270x12 mm entrance with a precise 22 mm gap between a 240 mm deep frame and screened floor – that bees do not build into – and facility to incorporate a simple under-hive pollen trap actuated by a lever.
   *   The new Hive*iQ* 9-frame hive® has all requisite bee space features including a rebate to accommodate a queen excluder that sits flush on frames below, provides an exact 9 mm spacing between the excluder and the bottom bars of the super above.
   ** Both British National and (now obsolete) Langstroth Jumbo[216]

hives are nominally 12-frame but normally take eleven frames and sometimes a space filling follower board and are often accompanied by shallower depth supers.

Langstroth Jumbo hives have fallen into disuse in North America but are included as a reference to similar hives found in Europe and Australia.

Table 1.5 shows the sizes of Hoffman frames and comb areas.

	Top bar (mm)	Bottom bar (mm) [ID†]	Side bar [ID†]			
			Full Depth [ID†]	WS Pender	Ideal	Half (shallow)
Length (mm)	482	448 [430]	232 [203]#	181 [152]	136 [107]	111 [82]
Timber width at shoulder (mm)	25	25	35	35	35	35
Timber depth (mm)	15	9	9	9	9	9
Comb area per frame (cm²)			1746	1307	9202	7052
Cells per frame (2 sides)*			6300	4715	3320	2545

Table 1.5   Specifications for Hoffman frames using NSW DPI recommendations. Values in parenthesis are internal or comb dimensions.
† Values in parenthesis are comb as opposed to outside frame dimensions.
# Most full depth frames currently in use are 230 mm deep. Assuming this depth and allowing for combined top and bottom or end bar thicknesses, comb areas are 430 mm x side bar depth less about 25 mm per side (15 mm top bar and 10 mm bottom bar) or double this taking into account two comb faces.
* Based on actual counts (cells per row (75) x number of rows of cells (42) for full depth frames I found there were 360 cells per 100 cm² on a single comb face (720 cells for an equivalent 10 cm x 10 cm section of comb).

In practice the number of cells for a given area of comb depends on the die presses used to produce foundation (5.73 mm centre-to-centre in the comb locally purchased) to produce larger cells than than found in wild or naturally drawn out comb. For different hive depths, the estimated number of cells per frame is scaled to the area of the comb employed. Worker comb cell size in natural comb is much smaller (5.0-5.3 mm dia.) and the number of cells varies widely (830-920 cells per 100 cm^2)[217], being at the lower end of the scale amongst African races of *Apis mellifera*. The effect of cell size on comb packing density can be simply estimated from cell area, in turn estimated from average cell width[218].

Consider the difference in comb space that occurs in running two half-frame (shallow) supers in lieu of a single full depth box. The two shallows exactly replicate the full depth super in overall size but their total comb area is reduced by about 20% where:
* total comb area for 8-frame full depth super is 2x8x430x203 mm^2 (1.40 m^2)
* total comb area for two 8-frame half supers is 4x8x430x82 mm^2 (1.13 m^2).

The volume lost includes an additional 9 mm spacing between the two sets of frames and an additional 25 mm for the thicknesses of an additional top and bottom bar.

Based on these measurements, a simple calculation shows that a vigorous queen laying 1800 eggs per day will lay out exactly seven frames of brood into perfectly formed full depth worker comb. Given that brood combs are always less than perfect and that bees naturally store pollen and honey close to brood, a good queen will consistently lay out nine combs before that brood emerges to free up new space.

Warré and top bar hive combs

Warré hives have internal dimensions of 300x300x210 mm^3 and are sometimes fitted out with cutdown (or especially made) rectangular Langstroth frames lacking a bottom bar. Assuming the same thicknesses and widths as materials employed in Langstroth frames, actual combs measure 280x195 mm^2. For an 8-frame super the total comb area is 874 cm^2 (2x8x280x195 mm^2). Since both Warré and top bar hives employ naturally drawn out comb, cell numbers will be higher per given comb section area. This free-formed or natural comb:
▸ does not completely fill out all hanging space;
▸ has comb with a smaller natural ~5.1 mm centre-to-centre cell size, i.e. more worker cells for any given area than a Langstroth frame; and
▸ constructs proportionately (10%) more drone cells.

Figure 1.18 (see *Heiress to the throne*) depicts a typical Warré comb. The cell count per frame is conjectural but I estimate about 55 cells spanning 280 mm or a 5.1 mm average worker cell width. Note the preponderance of drone cells on comb margins.

The amount of drone comb found in wild hives is generally reckoned at about 20% of total comb area, more typically 10% in managed hives where pre-formed worker foundation is employed. The amount of drone comb produced in the early stages of swarm occupation of a tree hollow is quite small – workers are needed to establish the colony – but can be quite high in large colonies and particularly at the margins of brood nests where the influence of queen pheromones is much reduced. We know this as inserting an empty frame into a strong colony will typically draw out nearly pure drone comb.

Bees expend a seemingly inordinate amount of energy and resources raising drones (and indeed also replacement queens) but we should see this as part of the natural calculus of honey bee survival. In managed hives the relative number of drones present is only poorly related to the amount of drone comb present: bees readily store honey and pollen in drone comb as well as in interstitial and burr comb, so the effective amount of storage space is likely best estimated from comb area rather than comb type.

But how much space does a highly fecund queen need lay out in a top bar or Warré hive? A queen laying at the same high rate in a of 1800 eggs per day in a Warré hive – those employing an open frame – would fill out one comb daily. Assuming a 20% of drone comb (drone cells are 6.6-6.8 mm wide) we can calculate the relative areas of drone (6.7 mm centre-to-centre) and worker comb (using 5.1 mm centre-to-centre) in naturally drawn comb.

So with drone cells occupying 20% of the 874 cm^2 of a fully filled out Warré frame (450 cells per frame) and worker comb occupying 80% of the same space (3103 cells per frame) a queen laying 1800 eggs per day of mainly worker comb would require twelve brood-only combs allowing for the fact that she will lay some drone eggs. Thus Emile Warré's allocation of two eight frame brood chambers makes common sense, especially since the Warré hive normally comprises 2-3 chambers. And since bees have free range of all supers in top bar and Warré hives comb with small worker cells and regular large drone cells will store honey and pollen roughly as efficiently as in foundation-based comb (less wax is needed to make large than small cells). This is corroborated be the fact that Warré hives rarely accommodate brood in more than two eight-frame chambers.

Many design features remain to be explored, for example rebating hives to more securely incorporate queen excluders, and the potential to design hive covers that provide a single or no bee space between lids and top bars. Many of these measures might resolve commonly encountered problems such as wax moth and hive beetles occupying space under hive lids.

We have probably only touched upon the many issues that honey bee space has presented: we have not canvassed space usage in other hives such as the square British National that we briefly touched on. When it comes to prizing frames and supers glued together by burr comb in the paddock, especially in the middle of a major honey flow, you will be equipped to rectify the problem or at least understand why such problems arise.

Part II Bee hive architecture[219]

In a quest to discover better housing for honey bees, those more suited to extremes of climate and weather, some beekeepers have turned to the bees for answers. More forward thinking beekeepers have realised that the common or garden plastic or thin-walled wooden hives and other design features are short changing bees natural ability to operate efficiently.

Wild about hives

If you were seeking an impartial arbiter to get the optimal design for a honey bee colony nest, you might go no further than examine the architecture of wild hives such as those found in tree hollows. Tom Seeley and Roger Morse did just that[220]. They dissected the nests of twenty one hives taking measurements to define the essential characteristics of honey bee nests, trying to define a fairly typical hive structure (Figure 1.68), one chosen if not dictated by the bees themselves.

They found that colonies have a general preference for a nest deep inside the tree cavity, those with a small (32-89 mm^2) bottom entrance and a nest cavity size between 30 and 60 L with a median capacity of 42 L.

Figure 1.68
Exposed nest structure for a cold climate hive some 1750 mm in height, Icitha New York.
Note knothole through left wall (see again figure 1.2); honey above, brood below; and drone comb at edges.

Only cavity-dwelling honey bees (of the sub-genus Apis) build multi-comb nests and then usually, but not always, inside tree hollows or rock crevices. Those built in the open for lack of available nesting sites normally fail to overwinter. This nesting environment gives bees a large measure of control of the internal hive environment, one that also protects them from predators such as honey hunters, bears, skunks, badgers, rainbow bee-eaters, cane toads, rodents and wasps. Temperate races of the Western Honey Bee (*Apis mellifera*) and of the Asian Honey Bee (*Apis cerana*) amass fairly large stores of honey to tide them over long winters.

Wild about natural beekeeping

One of the objectives of natural beekeeping is to mimic the conditions of a successful wild hive. The most common examples available of this are the Kenyan Top Bar (KTB) hive with sloping sides, the rectangular section Tanzanian Top Bar hive and the Warré hive.

The hive commonly referred to as the Kenyan Top Bar was developed in 1971 by Maurice Smith and Gordon Townsend from the University of Guelph[221] under a Kenyan project sponsored by the Canadian International Development Agency (CIDA). These and similar top bar hives have replaced many of the traditional African log hives where the nest was often destroyed to harvest honey.

Needless to say the more efficient and more productive Langstroth hive[222] has replaced top bar hives almost everywhere including Africa, though these hives remain a favourite amongst amateurs.

One variant of the Langstroth hive, the square British National, holds a notional twelve frames but, as noted, are always operated with eleven frames and a follower board. Others such as the Dadant hives[223] vary only in respect of hive dimensions, numbers of frames and hive depth but are likewise not fundamentally different to the Langstroth design. Michael Bush summarises some differences quoting imperial dimensions here converted to the nearest millimetre:

Dadant hive: 296 mm (11 5/8") deep and 505 mm (19 7/8") square;
Langstroth hive: 413 mm (16 1/4") by 505 mm (19 7/8")
   and of various depths:
Langstroth deep 244 mm (9 5/8") deep;
Langstroth medium 168 mm (6 5/8") deep; and
Langstroth shallow 146 mm (5 3/4") deep.

Again these measurements are not closely regulated, the standard full depth Australian Langstroth super being nominally 240 mm deep though even this measurement varies a little nationally, a topic canvassed separately under *Bee space*.

A significant problem with all natural comb hive management systems is the need for comb destruction to harvest honey. This may be good for generating wax if that, and keeping bees for the sake of keeping them, are the key objectives of fostering honey bees. However it is expensive in terms of honey production as it takes around five kilograms of honey to produce one kilogram of wax. While frequent rendering of combs is a certain measure for reducing hive pathogen loads, experienced beekeepers regularly replace comb in the process of removing old and damaged combs. This can further limit the number of combs stored over winter subject to wax moth and small hive beetle attack. Nevertheless comb reuse greatly increases the capacity of bees to store more honey.

Highly nuanced arguments for running bees as naturally as possible have gained considerable kudos from Tom Seeley's *Darwinian Beekeeping*[224]. Well practiced natural beekeeping reflects the real needs of bees, if not those of those of the honey seeking beekeeper, such as the balance of drone numbers and worker bees in a colony and allowing bees to swarm freely, both important survival strategies adopted by wild bees.

Top bar hives

Running top bar hives provides for a close-to-natural window observation of the wild hive, a feature not possible in the physically wood-armoured tree hollow. Of these the Kenyan and Tanzanian Top Bar Hives are the best known. KTB (Kenyan Top Bar) and TTB (Tanzanian Top Bar) hive combs – though fragile – can be removed for inspection, their free-form construction closely replicating that of comb attached to the roof of wild hives.

If you have spare timber and good woodworking skills, you may be able to construct a top bar hive from scrap timber for a song. They are, however inordinately expensive to buy and critical clearances – read bee space – must be incorporated into their every design feature to avoid excessive gap filling by bees. In practice free-hanging combs take the same form as they do in a tree hollow where, however, cross-bracing is the norm. Control of this maze is one of the many skills required to operate KTBs and TTBs successfully.

When built on stands, or with supporting legs, top bar hives can be comfortably inspected and are particularly popular amongst beekeepers with limited mobility. One saving grace of the top bar hives is that all frames are immediately accessible, that is without having to remove supers.

Employing standard full depth Langstroth frames, in in lieu of free-formed top bar combs, facilitates more ready frame handling while preserving the essential architecture of the top bar hive. Though less natural, Langstroth frames are far more robust and lend versatility to hive management as frames can be interchanged with those of any other hive.

Despite some attractive features found in top bar hives, I argue that any system of keeping bees in a fully horizontal arrangement has a significant downside. In the wild, combs are attached to the roof. This top-to-bottom, rather than side-to-side, mode of operation arises out of brood being raised on the lower comb perimeter and honey being stored in cells from which brood has emerged above. Then, as the honey flow ends and as brood raising declines with approaching winter, the bees move upwards consuming honey and pollen to maintain hive homeostasis. In any wide-brood chamber setup, say a conventional twelve-frame full depth brood chamber, some honey is stored laterally. That honey is not warmed by bees and cannot form part of insulating honey mass above the winter cluster, a condition exacerbated in top bar hives.

Conventional tiered hives are more thermally efficient than top bars for one simple reason. Heat rises rather than being advected horizontally and the hive cover area is relatively small. The vertical arrangement conserves energy and facilitates more ready access of bees to stores.

That bees in horizontal hives often migrate towards stores to one end of the hive, leaving them unable to reach stores at the other end during winter, is further evidence that the horizontal arrangement is less than optimal. Émile Warré, in singing the design merits of his hive with built-in top-insulation, also emphasises the importance of minimising the hive cover surface area and of having a thermal blanketing store of honey above the brood nest:

> *In winter, there is no risk of the bees getting cold, provided that the stores are above the cluster of bees.*

Bees can thrive in horizontal setups, but one might take cognisance of Horr's[225] experience with horizontal hives where he notes disparagingly:

> *The long hive is no longer a bee hive, but may become a recycled planter box with a glass top for early spring or winter vegetables. I cannot say it did not work, the bees did fine but the beekeeper came up short of a crop. There is no way I know of to economically keep a horizontal three-hive body colony of bees...*

In an even less complimentary tone an Australian National University researcher, Ian Wallis, noted wryly:

> *This KTB movement arose out of beekeepers reading grass roots magazines, keen to demonstrate that you can keep bees in any old wooden box or tea chest even if it does have moveable frames.*

Another natural limiting feature of KTB hives is that their full capacity rarely exceeds that of an equivalent of two full depth eight-frame Langstroth supers – most KTB combs being considerably smaller than the standard full depth Langstroth comb. Nevertheless the skilled KTB operator, employing follower boards to optimise the space occupied by a colony of any given strength, replicates the standard practice of adding and removing supers.

From these observations you might assume that I am averse to keeping bees in horizontal hives? This is not the case. The horizontal expansion of hives is the only really practical alternative to tiering up supers. Indeed we have experimented widely with extremely large brood chambers especially those containing two or more laying queens with bee populations of up to 100,000 bees. There, as we shall see, a mix of horizontal and vertical architecture makes incorporating a horizontal element an attractive alternative to the skyscraper hive[226].

Why Warré?

Abbé Émile Warré (1867-1951) operated an apiary of 350 hives. He experimented widely inventing the *People's Hive* [*ruche populaire*] or, as we now know it, the Warré hive. Warré wanted to create a practical hive that would not only assist a declining French beekeeping industry but also to replicate the features of wild beehives that would make built hives more productive. He published his findings[227] as *L'Apiculture Pour Tous* [*Beekeeping for All*] in 1948. Warré, despite his noble objectives, was not averse to preaching:

> *Furthermore, beekeeping is a moral activity, as far as it keeps one away from cafés and low places and puts before the beekeeper an example of work, order and devotion to the common cause.*

Unlike the Langstroth hive, even better designed to maximise colony strength and maximise honey production and housing very robust frames, the Warré hive was designed to be a natural hive that was nevertheless productive. Warré hive covers are insulating and moisture regulating while their combs, though constrained to minimise wall comb attachment, are free-formed allowing a natural balance of worker and drone comb. Furthermore the overall capacity of two Warré supers (2x18.9 L) closely approximates that of the cavity size of wild hives (~40 L). For comparison an eight frame Langstroth super has a 42 L capacity, a standard Australian eight-frame Langstroth super having internal dimensions of 240 mm (deep) x 470 mm by 368 mm (41.5 L) with clearance below the bottom brood box of 4 L. Warré's recommendation was to operate three storey hives (each with internal dimensions 210 mm deep and 300 x 300 mm wide and long), allowing for two brood chambers and a single honey super.

He advocated a minimal wall thickness of 20 mm – employed in the hives we have seen – though he did recommended 24 mm timber to improve hive body rigidity. In colder climates 38 and 50 mm thick timber has been employed to improve insulation.

To further replicate the natural hive condition, Warré hives are operated excluder-free, a minimalist scheme that many beekeepers claim avoids restricting queen laying potential. For the most part, brood is infrequently found in upper supers during a honey flow, though the use of excluders ensures that queens can be more readily found and are less likely to be lost when harvesting honey in Langstroth hive operation. However, and for these reasons, we find queen excluders helpful and use them except in winter where all bees need free and full access to stores.

So for the beekeeper intent on keeping bees for their own sake, the Warré has a wonderful attraction. Skilled Warré hive operators are keen bee observers and their practice of nadiring (under supering) colonies replicates the natural hive instinct. Because of their setup, Warré colonies can overwinter on more limited stores, and in a smaller cluster, than can bees in uninsulated Langstroth hives.

Do I have the same fatal attraction for the Warré hive? Probably not if only because I find that conventional hives are more readily managed and available. I prefer the Langstroth hive model over the Warré hive because I believe it better reflects the space requirements of the very productive bee queens now available, because frames of bees, stores and brood are much sturdier and less subject to cross bracing, and because frames of bees, stores and brood are more readily exchangeable with standard hive gear. Further Langstroth frames are readily extracted and returned to the hive and can be reused even when bees are not inclined to build comb.

Why Langstroth?

A key question remains: 'In mimicking the naturally occurring hive, do we get a good hive design?'

This all comes back to what ones beekeeping aims are. In their natural state bees swarm profligately to replace colonies lost to predation, disease and starvation, trading this strategy with building only just sufficient stores to overwinter. However nature is hard taskmaster and swarmed bees, likewise parent colonies, may not survive. This occurs more frequently when summer honey and pollen flows are poor and where winters are particularly severe.

Some wild hive design features, those shared by bees kept in skeps or that occupy a wall cavity, are entirely unsuited to contemporary hive management. More space is needed and combs that are moveable have become essential hive design components. Following the post-skep era, typified by Alfred Neighbour's *The apiary: bees, bee-hives, and bee culture* visited in the skep essay, a multitude of new-fangled moveable frame designs emerged.

Into this melee stepped Langstroth.

Firstly Langstroth very clearly articulated the importance of very practical moveable frame hives for modern beekeeping practice. In advertising his patented 1852 hive[228] – he noted:

> *Each comb in this hive is attached to a separate, moveable frame, and in less than five minutes they may all be taken out, without cutting or injuring them, or at all enraging the bees... New colonies may be formed in less time than is usually required to hive a natural swarm; or the hive may be used as a non swarmer, or managed on the common swarming plan...*

Secondly the original Langstroth hive incorporated all the critical comb and space requirements that bees employ, notably those featured in François Huber's[229] early leaf hive and improved by Johann Dzierzon[230]: the Langstroth[231] hive has, as we have already noted, stood the test of time.

However some changes were needed as the modern Langstroth hive bears only superficial resemblance to the original hive design. Back then the Langstroth comprised a set of crates that carried the frames, the crates being inserted into a box-like outer cover. But the standard frame width was, and will aways be, 27 mm (1 1/16") while most frames are 480 mm (19") long.

The relative merits of eight, ten and twelve frame hives and their shallow, ideal, WS Pender and full depth supers – there are many variants internationally – will always be contentious. To these we can add the Flow™ hive, a 2015 Australian invention of Cedar and Stuart Anderson. The super in this hive uses special frames with especially formed honeycomb cells – larger than worker cells – in the super. A tap, actually a lever mechanism, on the outside of the hive splits the cells when it is turned on. The honey flows through channels and drains into containers. It is not a new hive, well in the sense that it is essentially a Langstroth hive that has been adapted, but its design is novel. But like all beehives, it is a hive that needs bees that are well attended to. I was talking to a commercial beekeeper recently and he related a tall story of a beginner who thought you just turn on the tap to get honey. That beekeeper, if you can believe the yarn, was surprised to discover that this hive actually needed bees.

However, while the issues of frame design and bee spacing have been fully resolved, there is less consensus about actual hive width, the optimal number of equivalent full depth frames a hive body should hold.
From a bee perspective, and from an ease of handling perspective, a nine-frame full-depth frame hives would appear to be not only an optimal configuration, but also represent the equivalent normal maximum number of frames (in good condition) that a queen can sustainably lay out. So it would seem sensible not to employ nucleus sized hive bodies (e.g. six frame supers) for strong colonies with a queen in full lay or overly large brood chambers. Indeed Simpson has signalled that restricting the horizontal dimensions of the brood nest, i.e. artificially reducing space for the queen to lay, is an important factor in promoting swarming[232].

Overall the actual depth of hive bodies is less critical provided that frame size leaves 9 mm bee space. However full depth brood chambers, where the brood frames are a single entity and not broken by a vertical space would appear preferable. Many beekeepers use full depth gear if only to facilitate practices such as brood box reversal and, where hives are operated employing a single brood-chamber, to rotate of combs out of the brood nest.

For ease of lifting and handling, many beekeepers employ shallower supers for honey storage: manual decapping of smaller frames is certainly much easier. An extreme example is the employment of mini-nucs for queen mating. These may be stand alone units or incorporated into regular hive design (Figure 1.69) gear. Such hives are not self-sustaining and must be topped up with nurse bees shaken off brood frames.

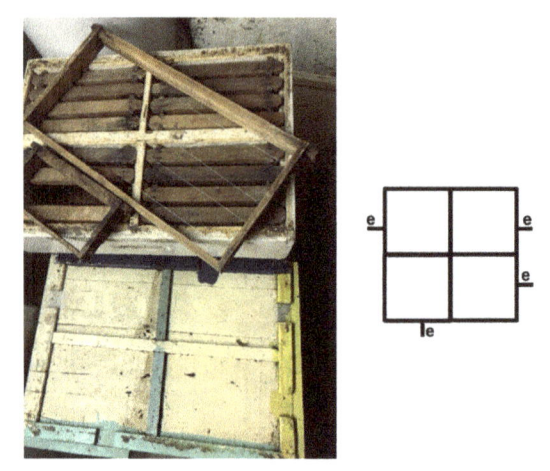

Figure 1.69   4-Way mini mating nucs on a shallow multi-entrance floor board pattern: e = entrance.
Photo: Alan Wade
Note: Image shows mini mating nuc frame alongside a standard full depth frame.

With wooden Langstroth hives, there have been two recent improvements in their overall design. One has been to fill out traditional vented migratory lids with insulation.

Infrared imaging (Figure 1.70) shows convective heat losses through migratory cover air vents and seams between supers and conductive and irradiative thermal losses near hive handles. Migratory covers were invented to facilitate air circulation for hives during transportation. Since they are also flush with hive body walls they can be more easily stacked – as shown in the images – more closely on a truck tray.

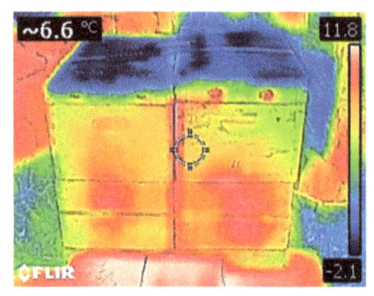

Figure 1.70   Mid winter imaging July 2020 through the lens of:
(a) an infrared camera; and
(b) a visible spectrum camera.

I have employed R 1.1 value extruded 30 mm polystyrene sheeting gluing cut sections into the roof with filler adhesive silicon and sealing off the foam with 3 mm ply to prevent bees termiting the polystyrene (Figure 1.71), a scheme savvy top bar hive owners are similarly turning to. The other major advance has been to modify or replace solid bottom boards with screened boards discussed below.

Figure 1.71 Simple retrofit insulation for standard Langstroth hive migratory lids.
Photo: Alan Wade

The issue of insulating hives better segues into an unanswered question about the operation of Langstroth hives: 'What are the relative merits of horizontal and vertical hive modes of operation?'

The horizontal vs vertical hive

Vertical beehives replicate the hollow tree advantage of rising brood nest warmth and the aforementioned annual cycle of colonies expanding downwards and contracting upwards. Tower systems are – for equivalent horizontal systems – far more productive and less problematic at least as far as the bees are concerned. Indeed the use of minimal sized boxes is inimical to the raison d'être of the Warré hive design. This said any hive operating with more than 4-5 standard full depth supers suffers from some measure of physical instability.

It is of course possible to marry horizontal and vertical hive element as this doubled hive (Figure 1.72) illustrates.

Figure 1.72 Horizontal or vertical architecture? – Actually both.

Optimal hive architecture is not just limited to the problems of handling multi-queened colonies where hive size must be up-scaled dramatically. In principle, hives might be set up in the traditional vertical arrangement (Figure 1.73a) or in a ground-hugging orientation (Figure 1.73b).

                              a   b

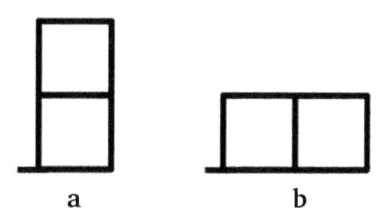

Figure 1.73   Vertical and horizontal hive modes of operation.

Note the partition(s) dividing the supers may be a regular hive wall, a perforated, double screen or a queen excluder and there will be a separate hive entrance where there is a separate colony with a second queen.

For the vast majority of Langstroth hive operations, the vertical hive, limited to no more than three of four stories employing standard full-depth supers, or their equivalent of shallower supers, is a no brainer. But for exceptionally strong colonies, building in some horizontal element would appear to both practical and wise especially where standard honey supers (sometimes two or more per queen) are employed for honey storage. Unless hives can be abutted or accommodated in divided extra-large brood chambers – as employed in double or two-queen hives outlined elsewhere[233] – the requirement for non-standard gear to operate horizontal Langstroth hive systems rules horizontal setups out for widespread adoption.

Yet hives containing around twenty full depth Langstroth frames were of common occurrence through from about the 1880s through until the early 20th Century and have even persisted amongst queen breeders in Australia: our famous Gretchen Wheen employed them for queen raising. Commercial beekeepers have found that, by inserting vertical queen excluders, or solid division boards, they could start and finish queen cells in a single strong large hive body. Figure 1.74 emphasises the disparity between tower and horizontal hive architectures.

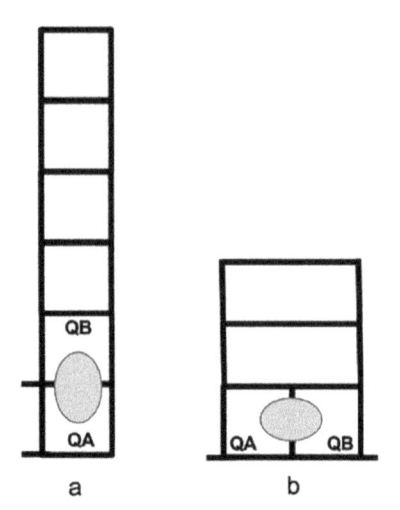

Figure 1.74 Multi super hives – for two-queen operation – showing alternative hive orientations (for simplicity queen excluders not shown):
(a) likely more thermally efficient tower operation; and
(b) more physically stable scheme employing either large supers or doubled single supers at each level.

Note that, in these setups, additional hive entrances can be employed during the main flow to facilitate direct return of nectar to honey supers.

Horizontal hives with two queens

In practice such monsters, either as towers or as compounded horizontal hives, have proved too cumbersome to move and are only suited to permanent apiary operations. We acquired a sixteen-frame brood chamber hive body (Figure 1.75), recently refurbished with a centrally located and removable queen excluder, and have used it, supered each way with standard eight-frame boxes to run a successful two-queen colony in the club apiary.

a b

Figure 1.75 Sixteen-frame double brood chamber setup – Jerrabomberra Wetlands apiary summer 2019-2020:
(a) divided brood box with screened bottom board and division board replaceable with a vertical queen excluder; and
(b) two-queen hive with central vertical queen excluder installed and tiered up with honey supers over horizontal queen excluders.
Photos: Alan Wade

Elsewhere fellow beekeeper Frank Derwent began experimenting with a 68-frame horizontal Langstroth two-queen hive (Figure 1.76). This hive accommodated separate single-queen brood nests, one on each wing. These brood nests flanked a common central honey chamber employing vertical queen excluders to separate the compartments. However, while Frank found that the bees worked well enough together, the colony functioned and performed like an equivalent pair of single-queen hives: only some of the central honey frames were utilised.

Figure 1.76   Derwent sixty eight frame horizontal hive with with two queens and two vertical queen excluders. The cement-filled bucket is a counterweight to facilitate ready hinged lid opening of hive.
Photos: Frank Derwent

A horizontal four-queen hive

A little wiser Frank has now modified the original structure. It is now a 65-frame full depth frame hive accommodating four queens (in three sixteen and one seventeen frame brood chambers), each nest divided from its neighbour by a vertical excluder (Figure 1.77). All parts of the hive are well-insulated overcoming in part the problems that bees have in working in a horizontal configuration.

Above these generous 16+ frame brood chambers Frank installed hinged queen excluders, supering the whole giant brood nest with a 69 F ideal frame coffin-box fitted with a new (lighter weight) hive covers. However, the mechanics of lifting the coffin super and gaining ready access to the four queen brood chamber has proved to be a singular challenge.

The past (2022-2023) season was very slow, but this colony was already easily outperforming equivalent four single-queen hives. By the early spring 2022 building phase the hive was already storing honey well (Figure 1.78b). And by early-to-mid summer of 2022-2023, ideal frames in the coffin super were filling out quickly (Figure 1.78a) while hives Canberra-wide were still building due to an exceptionally cool and wet spring.

a   b

Figure 1.77   Construction details of the horizontal four-queen hive:
(a) internal detail of hive showing screened bottom board, brood chamber compartments separated by vertical ventilation boards (these are interchangeable with the vertical queen excluders) and hinged horizontal queen excluders; and
(b) assembled hive profile showing four compartment entrances.
Photos: Alan Wade

Figure 1.78 Early stage establishment of horizontal four-queen hive showing exposed super ideal frames after supering with 69-frame ideal coffin super:
(a) Frank Derwent examining late autumn 2022 condition of recently established hive with four new queens superable with a single ideal coffin hive body; and
(b) early signal of bees building stores in ideal frames.
Photos: Alan Wade
Notes: The ideal 69-frame ideal super was added in early summer 2022.
Vertical queen excluders separating queens are seasonally alternated with vertical ventilation boards to allow colonies to overwinter separately as abutted single-queen units.

By the 19th of December 2022 – southern hemisphere mid summer – the 69 frame super was over half full. The first summer harvest was made on 2 February 2023 yielding 56 fully capped frames (140 kg) of honey. Of the remaining few (13 ideal) frames, 8 were full but uncapped, the remaining handful of combs being half full. Frank has just reported a final late season harvest (22 April 2023) of an additional 40 ideal combs of honey, another 100 kg, or a total of 240 kg of surplus honey for the season. The individual 16+ frame brood nest chambers were also well filled out to more than suffice their overwintering as separate single queen colonies. That one of those four queens is now failing was signalled by the fact that frames above one nest were only partially filled out.

This contrasts with the spectacularly poor performance of most single-queen beehives in Canberra this season, not least mine. I was forced to unite many hives and fed the residual eight hives a total of 150 kg of sugar made up as a heavy syrup simply to maintain a base stock.

Other multiqueen hive configurations

So how do bees fare more generally in these and similar behemoth hives and what designs are possible? There have been many variants to the design of multiple-queen hives (Figure 1.79) mostly formulated in the 20th century.

> *Behold now behemoth, which I made with thee; he eateth grass as an ox.*
> *Lo now, his strength is in his loins, and his force is in the navel of his belly.*
> *He moveth his tail like a cedar: the sinews of his stones are wrapped together.*
> *His bones are as strong pieces of brass; his bones are like bars of iron.*
>
> **Job 40:15-18**
> King James Version of the Bible

There are many inherent problems in operating multiple-queen hives – size, weight and portability aside. Most difficulties arise in their design and operation of multiple queen hives. While many linear (long) and vertical (tower) two-queen hives have been operated over much of the 20th century possible configurations exist for abutting colonies (Figure 1.80) though there are formidable design issues of frame support and bee space design – and even frame sizes and shapes – that would need to be overcome. Supering these brood chamber setups would involve considerable design and structural challenges and require some ingenuity.

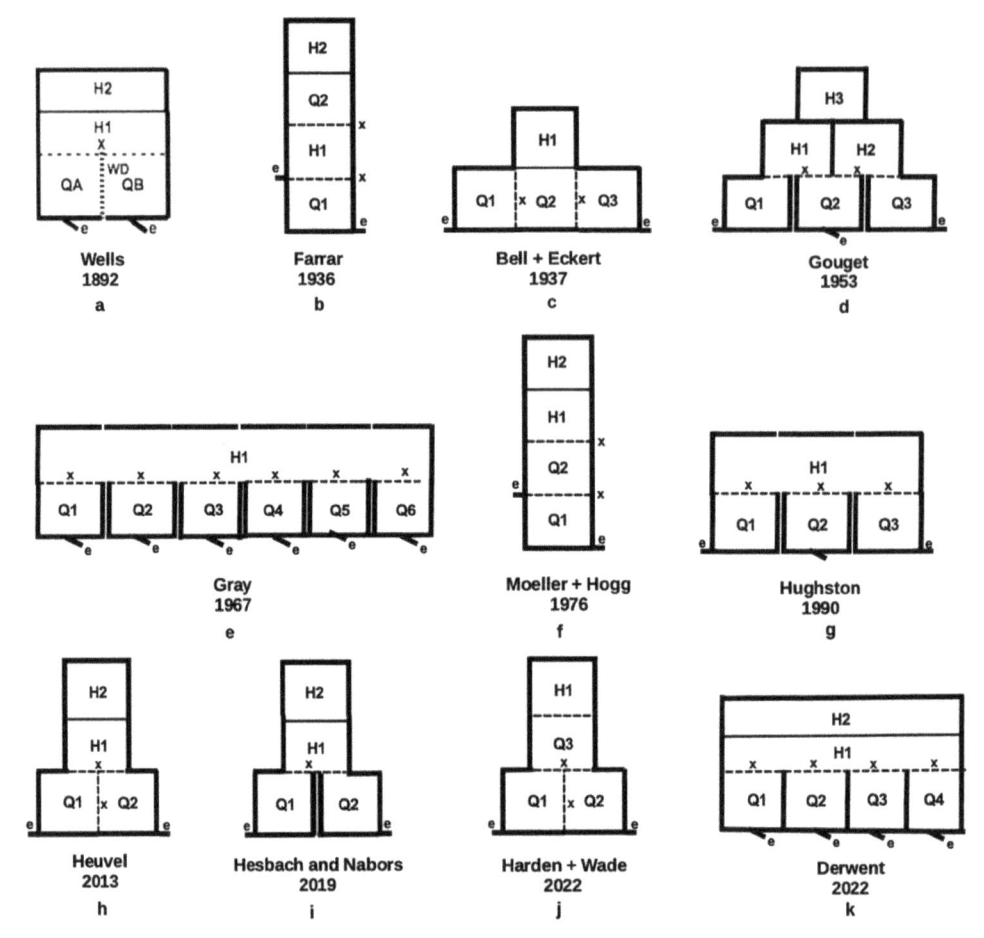

Figure 1.79 Representative design features for multiple queen hives: e = entrance, WD Wells Dummy; x = excluder; Q = queen; H = honey super:
(a) George Wells' 1892 doubled hive[234];
(b) Clayton Farrar's 1936 top and bottom two-queen hive[235];
(c) Leyroy Bell's and John Eckert's 1937 duo hive[236];
(d) Charles Gouget's 1953 tripled queen pyramid hive[237];
(e) Ken Gray's 1967 sextupled queen hive[238];
(f) Floyd Moeller's and John Hogg's 1976 Consolidated Brood Nest two-queen hive[239];
(g) Stringy Hughston's 1990 tripled hive;
(h) Bernhardt Heuvel's 2013 horizontal two-queen hive;
(i) Bill Hesbach's and Nabors 2015 doubled hive[240];
(j) Dannielle Harden's and Alan Wade's mixed horizontal and vertical 2022 tripled hive; and
(k) Frank Derwent's horizontal four-queen hive (currently operated with a single coffin super).

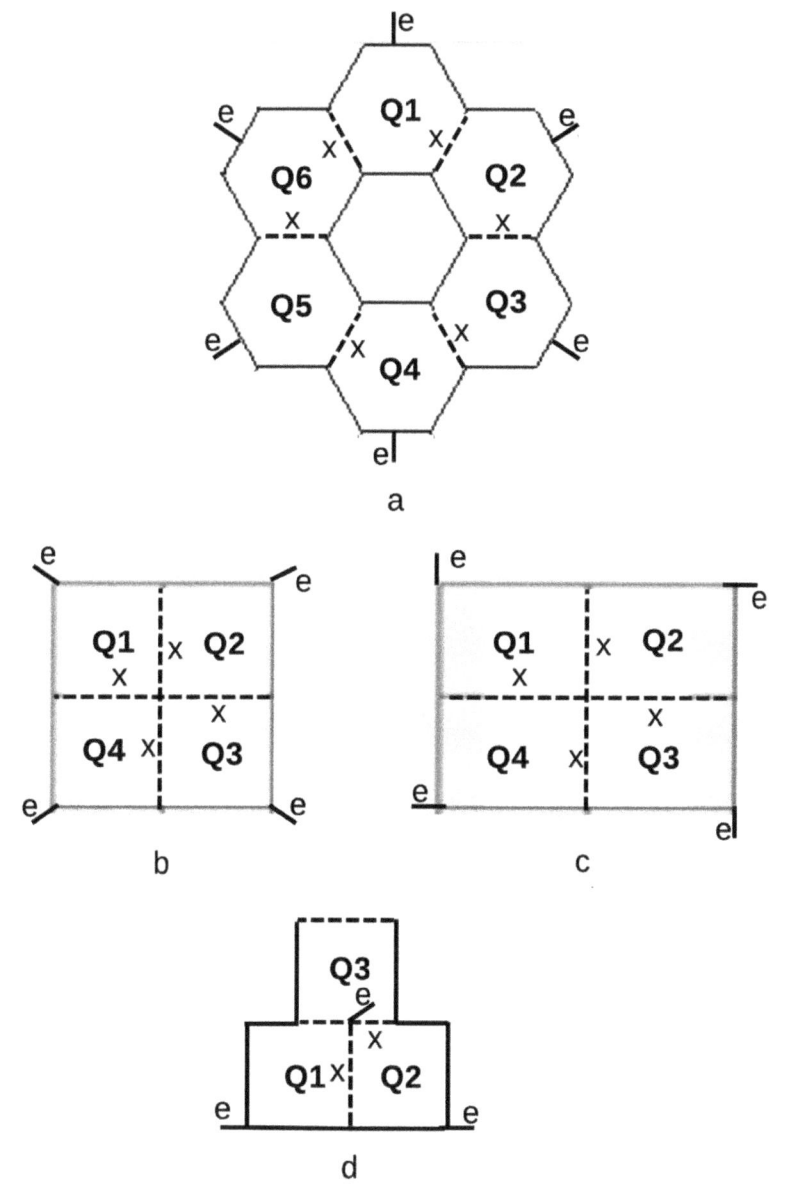

Figure 1.80 Notional configurations for multiqueen hives: e = entrance; x = framed excluder; Q = queen:
(a) hexagon-shaped brood chambers with variable frame length and problematic end bar support;
(b) square-section brood chambers with queen excluder window internal walls;
(c) equivalent rectangular brood chambers; and
(d) vertical hive setup for three queens.

Operational issues include migration of hives, employing only queens with a similar pheromone titre (queens should be of similar laying capability), failing queens not being replaced by the normal supersedure response, supers and brood nests filling quickly with honey preventing queens laying to capacity, exceptionally high honey and pollen demands in the early stages of hive population build up and excessive brood development if multiple queens continue to lay towards the end of honey flows. From this perspective, any successful establishment of multi-queen hives should be treated with some caution: production of honey bees is very resource intensive though hives with very large populations can produce rivers, if not tidal waves, of honey.

Design features exigencies of providing entrances to each brood nest to facilitate drone flight, complex supering arrangements and cumbersome manipulation and brood chamber alignment requirements suggest they are unlikely ever to be built. Similarly vertical hive systems with more than two queens will be likely beset with brood nest honey storage and tower stability problems.

Clearly any linear arrangement of brood nests is also possible though in practice these distil to separate interchangeable brood chambers connected by shared honey supers, either coffin chambers or offset supers.

An interesting facet of horizontal hives operating in the 1890s was their wide use as doubled queen hives. This comprised a jumbo twenty frame full-depth Langstroth brood chamber, partitioned by a thin reinforced board precision drilled 3 mm holes at 25 mm intervals – the so-called Wells Dummy. George Wells, the originator of the doubled hive, initially employed a fourteen frame brood-chamber (i.e. a partitioned two side-by-side seven-frame chambers), but added shallow brood supers to allow each queen to lay to her full capacity.

This allowed side-by-side colonies to share brood nest warmth and exchange of colony odour. In early spring the hives were united and supered over a large excluder using twenty frame shallow coffin supers that could either be used to produce extracted comb or section comb honey. At seasons' end the powerful double queen hive was simply returned to well-provisioned side-by-side single-queen hives where the central division board allowed shared warmth and better overwintering capacity than stand alone single hives.

Hive homeostasis and hive architecture

Homeostasis The brood nest of the honey bee is an extra 'ordinary' entity. Its role in raising large numbers of bees quickly – laid egg to emerging bee in twenty one days – is dependent on optimal hive nutrition, controlled temperature inside the working hive and the elevated temperature of the actual brood nest. While sealed brood can be insolated by so-called heater bees[241], the nest as a whole is regulated at 35 ± 0.5 °C by clustering or fanning bees, a thermal condition common to the whole *Apis* genus. For example the Giant Honey Bee, *Apis dorsata*[242] orients its large single comb to minimise the impact of wind and rain while the brood itself is warmed by a curtain of bees (at the same 35 °C). Tan and coworkers found that cavity dwelling Asian Honey Bee, *Apis cerana*, operates its brood nest at a slightly lower temperature[243]. In colocated hives they recorded brood nest temperatures for *Apis mellifera* (34.1 ± 0.05 °C) and for *Apis cerana* (33.1 ± 0.05 °C): elsewhere Jones et al recording a slightly wider brood nest temperatures for *Apis mellifera* (34.5± 1.5 °C).

Insulation: Mitchell[244] has investigated the value of year round hive insulation, pointedly noting that wild hives tend to be far better protected than even the most heavily insulated man made hives. But insulation – installed primarily to minimise stores consumption – is only a small part of the equation. Mitchell calculated that bees consume between 25 and 50% of the nectar delivered to the hive in transforming it to storable honey. Well-insulated hives improves not only colony productivity but also results in much less wing wear extending bee longevity.

The full benefits of improved insulation extend to effective regulation of the colony water budget, fewer water gatherers, less condensation on hive walls, earlier bee flight on chilly mornings and less bearding of bees at hive entrances on summer evenings. Further, better regulation of the internal hive environment results in lower incidence of stress diseases such as *Nosema*, European Foulbrood and chalkbrood.

Ventilation: Somerville and Collins[245] examined the impact of screened bottom boards on hive ventilation. The role of screened flooring in removing insect frass, comb debris and chalkbrood mummies means that spring cleaning of hive bottom boards has become a thing of the past. Screened bottom boards have also reduced mould problems and removed the breeding ground for small hive beetle and lesser wax moth. Nevertheless sheltering bees from wind and away from wet areas are sensible if not hive placement maxims.

Hive materials: It is difficult to express any preference for construction materials beyond stating that insulated gear, built in or added, greatly improves colony performance. All hives deteriorate over time and normally either rot or deteriorate in sunlight (some plastics are photo stable). Some are readily fractured if dropped and several are easily punctured or subject to wax moth and rodent attack. Most, though not plastic, hives require regular painting or treatment. (I have copper naphthanated wooden hives that have survived for forty years). My advice is to repair hives promptly or retire gear that lets in the weather or that weakens hive defence.

Hive frames: I have some preference for wax over plastic foundation though clip-in plastic foundation is quick to install but must drawn on a honey flow in, or just above, the brood nest. Undrawn plastic comb, like warped or regular wax foundation and combs with large gaps, equates to unused real estate. These hive dead spots can't store honey and pollen and can't be used to raise brood.

I also have a strong preference for wood over plastic frames, though the latter appear immune to wax moth attack. Most plastic frames have notches and crannies that fill with wax and propolis that are nearly impossible to clean. Further they curl up in hot water and in solar wax melters. Plastic gear, touch wood, is made of plastic.

So what is the best hive?

This it is hard to say, well not with any certainty or conviction. As to the hive choices my beekeeping pals have made, we are agreed we would all have acquired different gear to that we now own, and would have kept bees in a different way, that is had we now known what we know.

What we do know, however, is that wild bees normally select a snug home and would appear to have a strong preference for a hive architecture that reflects a fairly roomy vertical defendable hollow rather than a spread out McMansion.

Going overboard[246]

Part I Hive dividers

Ever thought that most beekeepers were a kangaroo short in the top paddock? Then try breaking into a beekeepers' shed to confirm your long-held suspicions.

We might start with the discovery of Zorzi's honey uncapping machine (Figure 1.81) in one old shed. In extolling its merits, the editors of the *British Bee Journal* noted dryly:

> *The uncapping machine, the first of the kind we believe ever made, is the invention of Count R Zorzi, of Bologna, who deservedly received a gold medal for it at the Exhibition*[247].

Intrigued, I obtained detailed operating intructions for this must-have piece of gear:

> *It will be seen by the illustration that it consists of wooden stand, to which is attached a large wheel worked by a handle. This, by means of a strap, actuates a pulley fixed to a shaft, having an eccentric slot cut, which in turn gives a horizontal motion to a sharp steel blade. This is set at an angle and is fixed by means of thumb-screws...*

Figure 1.81   Count Zorzi's 1886 uncapping machine.

Now, just over 130 years on, there is little call on such hand operated machines. The hot knife is just as effective and a good deal cheaper. But the world has changed in other ways and mechanised de-capping machines will be found in any large-capacity honey processing plant that you might care to visit.

In a routine sort of 'our club apiary' shed, I dragged out an equally weird assortment of gear. I have simply call these items flats (Figure 1.82). For what purpose would anyone want to keep most of these space wasters? Or could it be that, like the finders of the wondrous uncapping machine, we are sadly mistaken and that most of the gear is more or less indispensable?

Figure 1.82   Bee shed assorted flats.

We note that most of these items are modified bottom, top and division boards that exert control on the bees to varying degrees. As we shall see a few even help to dispose of hive floor rubbish, wonderful self-cleaning devices.

Now there are many and varied 'inventive gizmos' that one has read about but will probably never lay eyes on, things like the Snelgrove Board[248] (Figure 1.83). These items are now largely consigned to the dustbins of history, simply because there are easier and better ways of carrying out their functions: others extol the virtues of the board and Snelgrove's inventive genius. But, hang on, beekeepers, as much as bees often surprise: I have only just discovered three very competent beekeepers who swear that the Snelgrove board is the best thing since sliced bread.

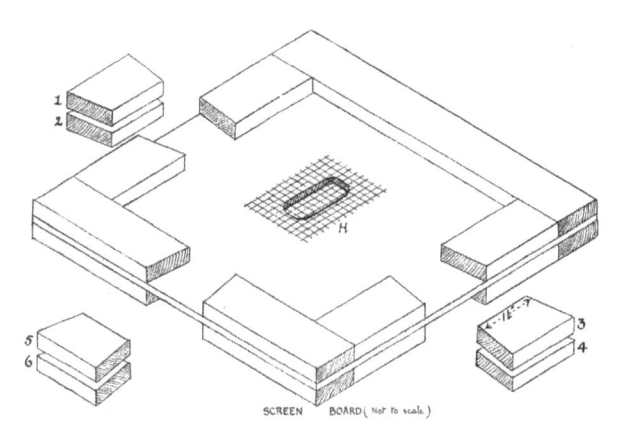

Figure 1.83 The Snelgrove Board, a multi-entrance division board.

Another example of an 'odd find' is the carbolic acid bee repellant mat. Carbolic acid, better known as phenol, is truly nasty. Bees understandably desert their precious stores to escape their suffocating fumes, and you would too risking life, limb and lung in their use.

But let's turn to many flats to explain their purpose and let you be the judge as to whether they will clutter your shed, empty your wallet or let the gear become bosom friends. We will focus here on those flats that fit between hive bodies and come back to lids and bottom boards.

The queen excluder

An old standard and beekeeper companion, one that surfaced in the 1880s[249] is the queen excluder. Way back then they were made, as today, of 'perforated zinc' aka galvanised iron sheeting (Figure 1.84), here slotted sheeting. But they only became really popular after passing the pub test for reliability in 1891[250]. Even today, they tend to remove bee legs and wings. Gap width is between 4.1 and 4.4 mm.

Figure 1.84 'Excluder zinc' invented in the late 19th Century.

Wire queen excluders (Figure 1.85) have largely overcome this problem by eliminating the sharp edges but always trap drones and sometimes bees. Wire excluders are either welded, brazed, or die cast. They are reliable, sturdy, and are one of the most commonly used hive accessories.

a　　　　　　　　　　　　　　b

Figure 1.85 Standard galvanised wire queen excluders with reinforced edges: (a) eight frame excluder scraped to remove burr comb; and (b) new ten frame excluder.

Queen excluders serve the simple purpose of preventing queens from entering overlying honey supers. In any large scale operation the queen excluder becomes an almost indispensable tool in ensuring that frames of honey are brood free. This said, many hobby beekeepers operate their hives excluder-free claiming that excluders are 'a honey barrier' and that they deny bees optimal use of hive space. In practice, and since honey is stored above brood, the small-scale operator can selectively harvest frames containing honey that are anyway mainly located above brood.

A simple variant of the flat excluder is the rimmed excluder (Figure 1.86) with a flight entrance that allows drones their normal access to any part of the colony.

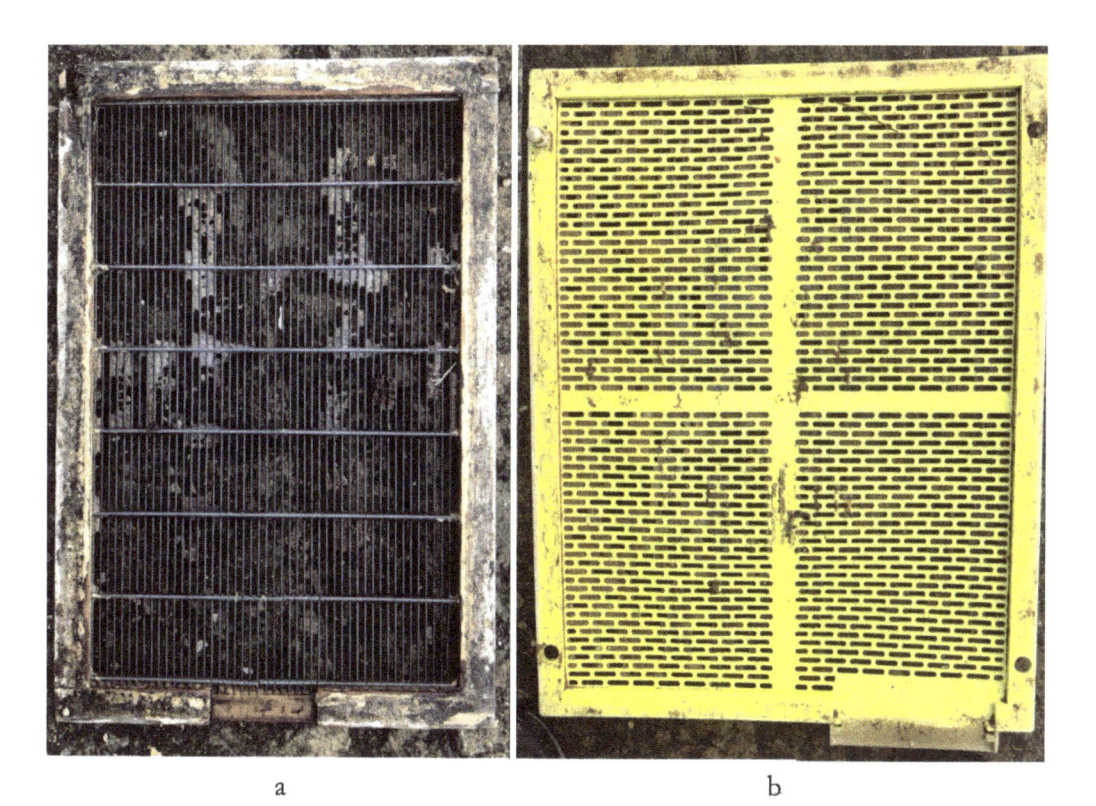

a   b

Figure 1.86   Rimmed queen excluders:
  (a) rimmed 8F excluder with 8 mm risers above and below excluder with central cut out flight entrance; and
  (b) moulded Technoset slotted 10F excluder with offset flight entrance and landing board with 8 mm bee-space riser.

Of course it is possible to combine a simple riser rim (Figure 1.87) with a standard flat queen excluder (Figure 1.85) to achieve the same function.

Figure 1.87 Riser rims with flight access often used in conjunction with queen excluders:
(a) 8F timber rim with slotted flight entrance; and
(b) 10F cross-braced Technoset rim with offset entrance.

Riser rims are often employed alone to provide flight access (in lieu of the old practice of boring a circular hole in super walls) during strong honey flows: the additional entrance also allows emerging drones flight access. I also use them in two-queen hive setups where the upper brood nest requires a second, or even a third, entrance.

Vertical queen excluders were once widely employed in queen raising operations, that is to partition oversized brood boxes. We have used them in two-queen hive operations (Figure 1.88) to keep queens apart.

Figure 1.88 The vertical queen excluder:
(a) oversized full depth brood box with central slot accommodating either a solid division board or a vertical excluder; and
(b) vertical excluder lubricated with food-grade vaseline to allow for easy removal.

A parting note on queen excluder design and condition. Most cheaper plastic and un-clad excluders are easily damaged and hardly worth purchasing and are items you toss when they become un-serviceable (Figure 1.89).

Figure 1.89 Unsheathed excluders showing:
(a) damaged wire excluder; and
(b) warped and often photo-degraded plastic excluder.

Importantly queen excluders should not be left on a hive over winter. Bees located in a bottom brood box in autumn will move up into stores in late winter (Figure 1.90) so leaving that excluder in place risks stranding the queen below the excluder and prevent critical raising of young bees coming into spring.

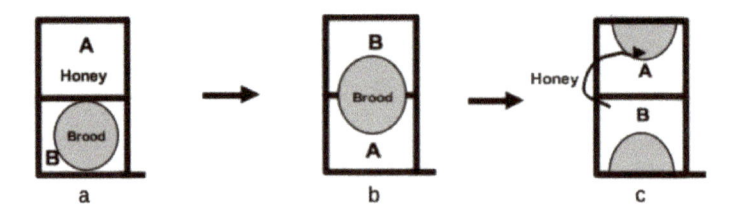

Figure 1.90 Brood box reversal:
(a) colony readied for overwintering in late autumn;
(b) brood moves up into stores in late winter and early spring; and
(c) brood boxes reversed every few weeks to stimulate brood rearing.

The split board

After the queen excluder, perhaps the most useful piece of flat we've come across is the nucleus or split board (Figure 1.91). Its primary function is to divide a colony on the parent stand. This 'hive split' obviates the need to establish an offset hive requiring both an extra bottom board and an extra lid.

So the nucleus board is useful in splitting colonies (an emergency queen will be established in the queenless portion) and in swarm prevention where dividing the hive simulates and helps prevent swarming.

A ready supply of split boards are also a much better alternative to sacks, old sheets and pieces of cardboard to cover supers of exposed frames when honey is being harvested and where robbing is likely to ensue.

The split board finds still other uses. It can serve as a temporary bottom board or lid or indeed, in a deep and well sealed version, as a drip tray when extracting or transporting supers of honey.

Figure 1.91 The split board:
(a) upper face showing 9 mm riser rim, closable pivot flight entrance and landing board; and
(b) lower closed face with 9 mm bee space riser rim.

The double screen

Double screens (Figure 1.92) operate in much the same way as a split board. They fully divide the upper colony from the hive below on the parent stand. The screened board confers two advantages over the simple split board. The rising warmth from the brood chamber below accelerates the development of the upper colony while the mingling of hive odours allows colonies to be united without the traditional use of smoke and newspaper. Additionally, double screens can be employed as temporary screened bottom boards.

Figure 1.92 Double screen with approximately 8 mm screen spacer and flight entrance on right edge.

The escape board

A further variant of the hive separator is the escape board (Figure 1.93).

Escape boards are used to clear a honey super of bees. Simply smoke your bees, lift off the one or two supers you want to extract, drop on the escape board – hole side up – and pop the supers full of bees back on. Come back a day later and the honey supers will be bee-free. A simple warning: bees won't leave brood so they really only work reliably where you use queen excluders.

Figure 1.93 Bee escape board showing escape holes and non-return ports employing:
(a) Porter escapes; and
(b) improvised home made escapes.

Using an escape board obviates the need to shake and brush each comb and is a life saver when extracting in autumn when robbing can make honey harvesting a perilous exercise. Of course beekeepers now use high volume low velocity bee blowers. Even cheap battery powered blowers quickly remove bees from supers.

We can now explore the myriad top and bottom hive cappers, the gear that also keeps bees on frames housed in open boxes out of the elements and keeps them snug and away from marauders.

Part II Hive lids, bottom boards and feeders

Natural beekeepers eschew the use of hive-dividing devices we have described, claiming that they interfere with the preordained organisation of the colony and settle with open frames and a hive with a floor and cover. To that extent, at least, not using fancy gear sets natural beekeepers apart from the commercial and most amateur operators. Of course one can go further, settle for an old drum or a more natural container with a hole to let the bees in and out but it is hardly legal to keep bees that way.

> *A box without hinges, key, or lid,*
> *Yet golden treasure inside is hid.*
>
> **The Hobbit**
> J.R.R. Tolkien

Western and Asian honey bees nest in a cavity, a hollow tree or a rock crevice. However managed hives by law must be operated with moveable frames. For owned hives one needs boxes (we call them supers) with free-hanging frames though, in top bar frame hives, this may just be a single closed cavity with simple drop-combs reminiscent of images of the single comb hives of the largest of all bees, The Giant Cliff Honey Bee, *Apis laboriosa*. Of course we also need lids and bases to keep the bees in and pests and the elements out. The exceptions to nested honey bees are the dwarf and giant honey bees that nest in the open, but even these need a roof or a twig or branch to hang from and some protection from the elements.

Little wonder then that in all the years that humans have kept bees[251], finding them homes has been a classic case of necessity is the mother of invention.

Now, getting back to the humble flat dividers we started with, we now describe the caps, or lids, and hive bases. Lids divide the honey bee colony from the heavens and opportunistic raiders, while bottom boards keep bees away from mother earth, the dirt and often damp underfloor. To conclude we will look at a few other contraptions called feeders used to artificially supplement the honey bee diet.

Hive lids

Standard covers

The essential functions of the hive lid are to cap the hive nest and to facilitate ready access to moveable frames. Lids come in a menagerie of designs (Figure 1.94) but in two distinctive styles:

- the migratory cover (Figure 1.94a) – so named because these are the lids that many Australian commercial beekeepers use transporting their hives from one floral source to the next. These lids sit flush with the sides of the hive boxes so that hives stack together efficiently. The lids provide ventilation and free clustering space under the roof to help prevent overheating when hives are moved; and
- the telescopic lid (Figures 1.95) – these fit over the top hive body in much the same way as a cosy fits over a tea caddy. Most have a flat top, but some have a sloped or gabled roof above to ward off sun and shed rain. Telescopic lids are not ventilated and are typified by the standard Flow© cover.

The popularity of the migratory lid among Australian hobby beekeepers, those that do not transport their hives, has had an unintended consequence. Bees quickly propolise these vents to prevent winter heat loss.

Figure 1.94
Hive covers:
(a) standard 8F migratory lid with end ventilation ports;
(b) commercial insulated 8F migratory lid; and
(c) insulated 10F Paradise lid.

Figure 1.95 Telescopic lided 3F nucleus box.

Émile Warré illustrates telescopic lid detail (Figure 1.96) for a much simplified sloping roof version of his hive cover. We've also used telescopic lids (in conjunction with a more standard migratory lid) where super sizes do not match lid sizes to cover exposed brood nest frames (Figure 1.97).

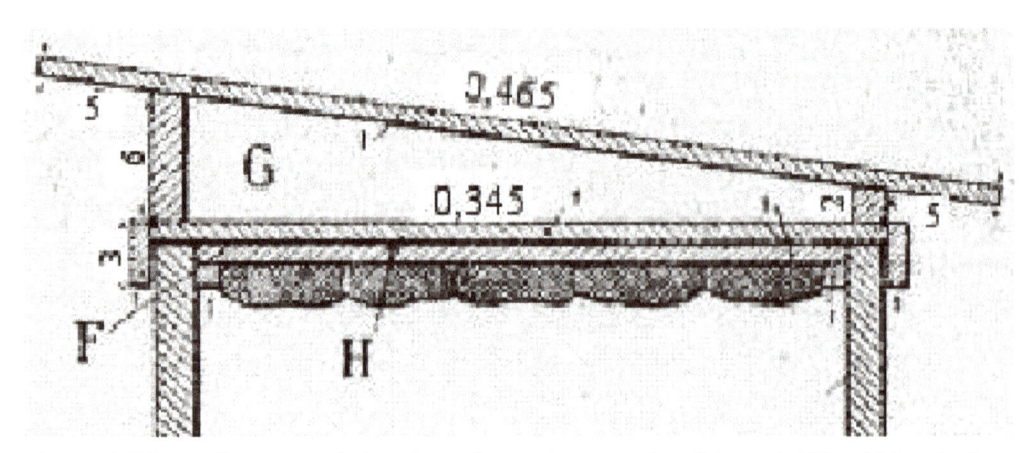

Figure 1.96 Sloping roof showing telescopic cover detail in early Warré hive design.

Figure 1.97 Telescopic side frame covers on doubled hive brood nests:
(a) side covers showing side cleats upper and handles lower; and
(b) telescopic side covers in use in combination with central honey supers with a migratory lid on honey supers.

Insulated covers

In a natural tree hollow bees aerate their space vertically venting excess moisture and waste gases through a low side or bottom entrance. They rely on a solid roof and side walls and upper blanket of honey to keep their nest snug. Images of hives from outside Australia show there is a strong preference for the telescoping lid style of hive covering. In conjunction with a top blanketing inner cover, this lid seals and insulates the hive better than the migratory lid.

Insulating beehives is by no means a recent innovation. In their bid to banish the icy fingers of winter, higher latitude northern hemisphere beekeepers have either brought their hives into sheltered bee barns or bee houses, wrapped them in heavy duty insulation, built double-walled hives or more recently built their hives from insulating materials. Such insulated hives function well over hot summers obviating any need for shade.

Beekeepers running traditional wooden hive furniture can improve colony performance by installing insulating material inside migratory hive lids or by purchasing insulated lids (Figure 1.94b and 1.94c above) from equipment suppliers.

Émile Warré was an early proponent of incorporating roof insulation[252]. His colony supers are topped with a box of moisture absorbing and insulating wood shavings, a telescopic cover and a gabled roof (Figure 1.98a): the natural tree hive nest is encased with a thick wood buffer. The flow hive lacks the insulating top-box but does incorporate an insulating inner cover, a telescopic cover and has a similar closed gabled roof (Figure 1.98b).

a   b

Figure 1.98   Gabled hive roofs incorporating telescopic cover:
(a) Warré roof structure with wood-shavings insulator box below; and
(b) Flow roof structure incorporating insulating inner cover under lid.

Hive mats placed under lids (Figure 1.99) are employed with traditional migratory lids and can be made of almost any material. We have tried thin ply, corflute, linoleum, tarred insulation foil and carpet squares. We've found that insulation foil is chewed out while carpet tends to become heavily propolised.

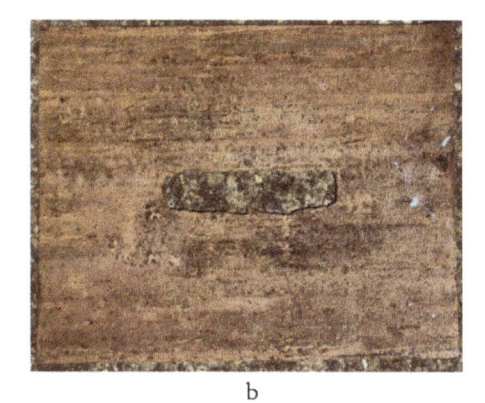

Figure 1.99 Hive mats:
(a) plywood mat; and
(b) linoleum Mercer mat with central top feeder slot.

Hive mats help avert burr comb development (Figure 1.100) but can provide refuge for wax moth and small hive beetle. The older beekeeping literature makes frequent reference to inner covers whose role is quite different to that of the hive mat. Their function is to insulate bees under telescopic hive lids.

Figure 1.100 Burr comb in migratory lid where a hive mat was not used:
(a) in club apiary 10 October 2020; and
(b) after clearing bees.

Bottom boards

Standard bottom boards

Traditional bottom boards (Figure 1.101) have an impervious floor to protect the hive from soil damp and pests[253]. The entrance, in wild hives, is small enough to enable the bees to defend themselves from pests such as mice, European wasps and other bees and wasps. The hive entrance also serves to provide hive ventilation and is the gateway for incoming honey, pollen, propolis and water and for removal of hive detritus and carbon dioxide.

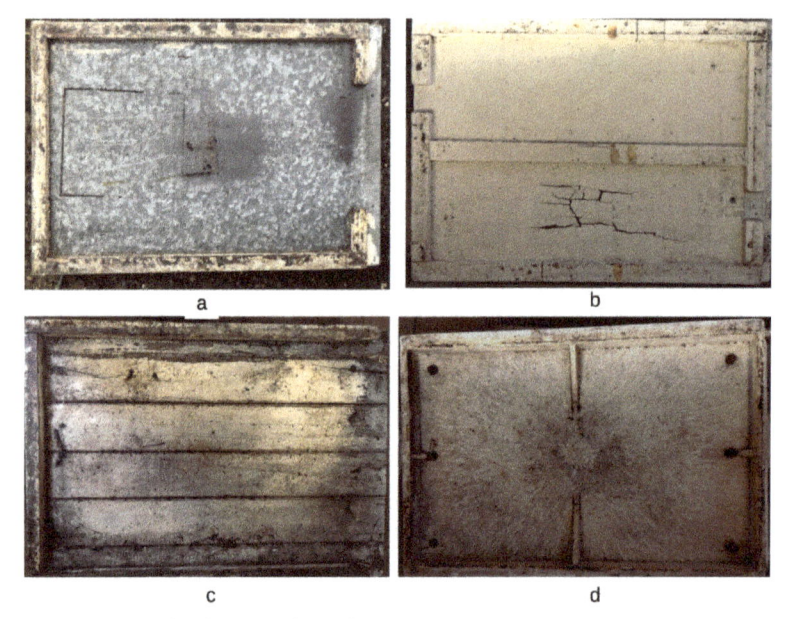

Figure 1.101 Manky bottom boards:
(a) old fashioned 8F galvanised iron bottom board;
(b) two-way wooden 2x4F bottom board;
(c) old fashioned 8F wooden bottom board with wide entrance; and
(d) Parker plastic bottom board with concealed entrances.

This traditional floor arrangement for managed bee homes ignores the fact that a solid barrier retains hive debris, insect frass, disease material and rainwater or condensate that accumulates at the base of the hive.

The introduction of a bee-proof bottom screen has revolutionised bottom board performance at least in cold-to-temperate Australia. Use of screened bottom boards has resulted in the elimination of a breeding ground for bee pests such as small hive beetle and lesser wax moth. Additionally disease-laden debris such as chalkbrood mummies and other fungi are largely eliminated. Screened bottom boards incorporating a catch tray and sticky mat are widely employed to monitor and help control *Varroa* and *Tropilaelaps* mites[254].

Ventilated bottom boards

Screened bottom boards (Figure 1.102) have been found to perform as well thermally as traditional solid bottom board setups[255] as warm air rises. However hives must be located well off the ground to prevent water ingress. The key attributes of ventilated bottom boards are their free draining nature, their capacity to avoid build up of insect fras and comb debris and their provision of free air exchange. Since snow in apiaries in most districts is very rare and temperatures rarely fall below -10 °C (~15 °F) use of screened bottom boards are now the norm year round in Australia.

In a typical tree cavity hive[256] the upper zone of the hive is sealed and well insulated while excess metabolic heat and surplus moisture and carbon dioxide are vented through the entrance normally located near the base of the cavity. This is corroborated by Möbus's finding that bees establish vertical (up-and-down) air currents and that the bees have a large measure of control over nest ventilation.

Figure 1.102   Screened bottom boards:

(a) Paradise 10F bottom board;
(b) sliding screen 8F bottom board with provision for sliding catch-tray and a sticky mat for parasite monitoring;
(c) Bluebees slotted bottom board insert; and
(c) homemade 10F screen bottom board.

Feeders

The requirements for feeding bees are discussed in *Should I feed my dog?*. Beekeepers use supplementary feeding to build bees for the honey flow, especially in early spring, to provision bees in times of dearth and to build good stores to tide them over winter.

However, feeding must be conducted judiciously to avoid the risk of honey adulteration where the practice of adding a food colourant to sugar syrup as a tracing agent has considerable merit.

A typical colony uses about 120 kg of honey and 30 kg of pollen annually for self maintenance. It stores of the order of 60 kg honey surplus – at least in my yard and out-apiary – but this may be diverted in part to the high energy cost process of swarming. Since bees consume the bulk of the nectar they gather to raise bees, to construct comb and to fuel harvesters, the risk of inverted cane sugar ending up in stored honey is minimal. There is one clear proviso: supplementary feeding should cease once the flow and honey storage commence.

Feeders can be classified into three main types, entrance feeders, frame feeders, and top feeders.

Entrance feeders The simplest are external Boardman feeders (Figure 1.103). This style of feeder is still popular amongst hobby beekeepers, although the enthusiasm for their use has waned of late. Entrance feeders:

- are portable and simple to fill;
- can be plugged into most hive entrances without disturbing the hive; and
- can be observed for syrup uptake.

Their main detractions are that:
- they deliver, at best, only small quantities of sugar syrup;
- their use is restricted in cold weather due to:
  the inadvisability of feeding liquid supplements under such conditions; and

the limited ability of bees to process chilled sugar water; and
- they are easily dislodged from the hive entrance.

Figure 1.103   Hive entrance Boardman feeders:
(a) standard feeder delivering minimal sugar syrup;
(b) employed to feed starving swarm 1 November 2022; and
(b) a homemade feeder of reasonable capacity.

Frame feeders These, as the name suggests, are modified frames that temporarily replace one or several regular frames. Their main attributes are that:
- they generally have a moderate sugar syrup holding capacity (2 – 5 litres);
- they are fitted inside the hive and thus are not affected by external weather conditions
- they can be substituted for any standard full-depth frame; and
- they do not encourage robbing.

Their major detractions are that:
- frame space available to the bee colony is reduced; and
- the hive must be opened to refill feeders.

We locate frame feeders immediately under the hive lid and on hive walls to facilitate easy access and to ensure that they are removed quickly once bees commence bringing in nectar.

In an earlier era of moderate austerity, we used old 2 L powdered milk tins – with perforated press shut metal lids inverted over wooden slats housed in an empty super – to feed sugar syrup. We have also very successfully employed home made frame feeders (Figure 1.104), and punctured plastic bags filled with sugar syrup placed under the lid. We continue to use them despite the disdain of 'better educated' beekeepers.

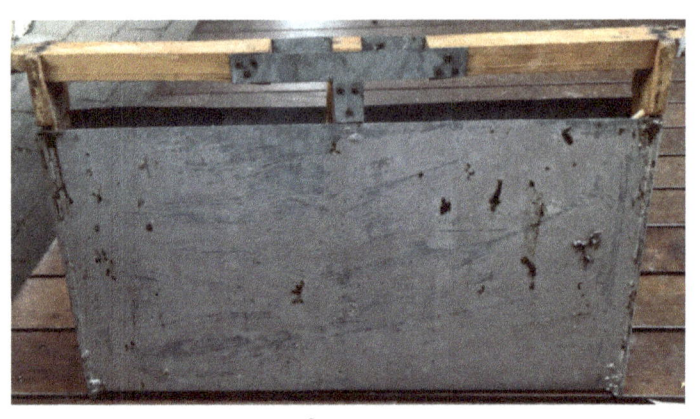

a

b

Figure 1.104   Home made 2 L single-frame feeders:
(a) galvanised iron clad feeder; and
(b) waxed masonite clad feeder.

Frame feeders are now available commercially (Figure 1.105d) though some are wider than the standard Hoffman frame. They are easily refilled but all frame feeders require floating inserts – or dry grass – to prevent bees drowning.

Top feeders With the advent of insulated hive gear, the open tray feeder designs facilitate early season feeding of both pollen and sugar supplements.

Under hive lid top feeders come in a range of styles:
- home made inverted feeder containers – old powdered milk tins and yoghurt tubs with nail punch hole lids suspended over slats of wood in a spare super;
- commercial tray feeders such as Paradise, Hive*iQ*, Technoset-Bee that confine the bees to the hive but provide access to the feed well through guarded ports; and
- plastic tray feeders that closely fit a half depth or ideal super that suit the conventional wooden or plastic Langstroth hive.

The advantages of top feeders come down to the fact that:
- they have a larger capacity than other feeder types;
- they are be readily refilled and avoid disturbance to the colony;
- they eliminate shock heat losses associated with lid removal;
- they are easily accessed to observe feed uptake;
- they, at least the insulated types, are not subject to external weather conditions; and
- they eliminate the risk of robbing.

However some top feeder designs are incorporated into a propriety hive system that make them incompatible with other hive designs. We have made tray insert top feeders for 10-frame wooden hives and for an apiary Warré hive.

However a variety of commercial in-hive feeders (Figure 1.105) are now widely available and have much to commend them. They are easily refilled and have plastic inserts or barriers to limit bee access to sugar syrup wells that prevent bees from drowning. Removal of these inserts allows feeding of dry sugar and pollen or candy made up as a moist fondant.

Figure 1.105 In-hive feeders used to feed sugar syrup or dry feed and candy supplements:
(a) under-lid doughnut feeder with feeder board fitting holding about 2L of sugar water;
(b) two-way feeder with common syrup well atop a 2x3-frame Paradise nucleus box;
(c) Paradise feeder of approximately 10L capacity; and
(d) commercial 3L frame feeder.

Let us spray: Honey bee pheromone chemistry[257]

Part I   Honey bee nest chemistry

As in other animals, bees employ two types of glandular chemicals regulating behaviour, primer pheromones and releaser pheromones[258].

Primer pheromones act at a physiological level, triggering complex and long-term responses in the receiver and generating both developmental and behavioural changes. Releaser pheromones have a weaker effect, generating a simple and transitory response that influence the receiver only at the behavioural level. As well there are a range of natural chemical volatiles, such as those produced by flowers, that bees quickly learn to recognise.

Let us start with the influence of these chemical agents inside the hive. There we can peruse the range of bee glands that release these pheromones, noting that they control activities as disparate as comb construction and brood rearing.

Queen mandibular pheromone

When the principle queen mandibular pheromone 9-ODA was first isolated and synthesised by Colin Butler and coworkers[259] at the Rothamsted Agricultural Experiment Station in 1962, our understanding of chemical communication in honey bee colonies was revolutionised. However the finding was woefully lacking in detail: much more is now known and other mandibular pheromones were soon discovered (Figure 1.106)[260].

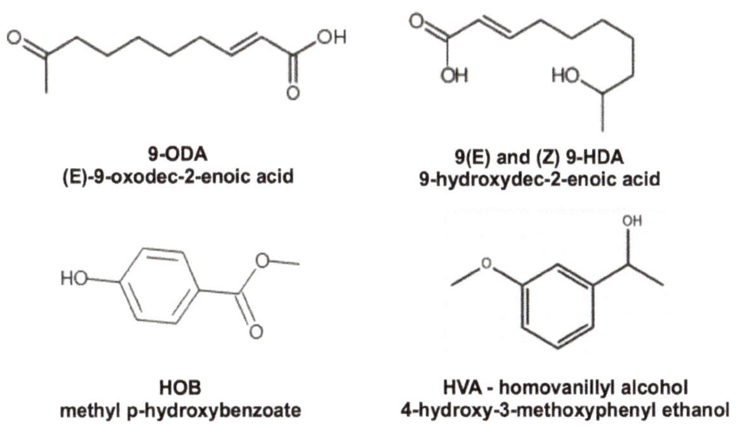

Figure 1.106 Queen mandibular pheromone complex, 9-ODA, the trans and cis enantiomers of 9-HDA, HOB and HVA.

9-ODA
(E)-9-oxodec-2-enoic acid

9(E) and (Z) 9-HDA
9-hydroxydec-2-enoic acid

HOB
methyl p-hydroxybenzoate

HVA - homovanillyl alcohol
4-hydroxy-3-methoxyphenyl ethanol

Slessor, Kaminski, King, Borden, and Winston outline the very specific chemical basis for retinue response of worker bees to the presence of a queen[261]:

> The substance produced by the mandibular gland complex of queen honey bees is considered to be responsible for the retinue formation in honey bees, Apis mellifera L. The retinue response includes the licking and atennation behaviour which signals the presence of a dominant reproductive queen and thereby establishes and stabilizes the social fabric of the colony…

They then go on to describe how that each of the constituents found in mandibular extracts elicits very weak retinue behaviour but that, together with 9-ODA, they perform the full function even though some of these chemicals are present at vanishingly low concentrations. There are a plethora of other controls effected by queen mandibular pheromone, for example the regulation of construction of drone comb and prevention of raising of new queens. Sufficient to say is that a young and well mated queen runs a very tight ship.

Interestingly Paul Hurd at Queen Mary University of London has outlined the importance of the epigenetic effect of the pheromone HDA (9-hydroxydecenoic acid) in royal jelly in influencing caste development (queen versus worker) in honey bee larvae. By day four after eggs hatch, the pathway to queen development is set while that of the worker is far from complete.

The obvious outcome is that the queen is well fed and that cells, cleaned and polished by worker bees, are laid into.

Bortolotti and Costa[262] provide a wider overview of the influence of this chemical cocktail:

> The honey bee queen represents the main regulating factor of the colony functions. This regulation is largely achieved by means of pheromones, which are produced by different glands and emitted as a complex chemical blend, known as the queen signal.
>
> The queen signal acts principally as a primer pheromone, inducing several physiological and behavioral modifications in the worker bees of the colony that result in maintenance of colony homeostasis through establishment of social hierarchy and preservation of the queen's reproductive supremacy. More specifically, the effects of the queen signal are maintenance of worker

cohesion, suppression of queen rearing, inhibition of worker reproduction, and stimulation of worker activities: cleaning, building, guarding, foraging, and brood feeding.

On the topic of bee larvae feeding stimulus, see the notes on brood pheromone below. Maissonaise and coworkers[263] go further suggesting that queen mandibular pheromone alone:

...does not trigger the full behavioral and physiological response observed in the presence of the queen, suggesting the presence of additional compounds.

Using other queen body part extracts they found that additional chemicals were able to exert further influence on worker bee ovary development, comb construction and worker bee queen attendance (retinue behaviour). Keeling and coworkers[264] identified yet more chemicals produced by glands other than those present in queen mandibles (Figure 1.107) that further enhance queen bee retinue formation.

While each of these chemicals alone elicited no response they could together still attract workers even in the absence of any queen influence[265]. We can conclude that chemical signalling is a complex process not always limited by the presence or absence of a specific pheromone.

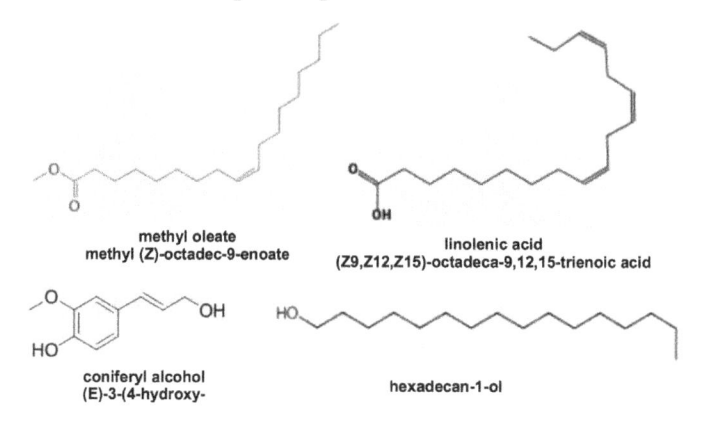

Figure 1.107
Queen retinue pheromones produced by glands other than queen mandibles.

methyl oleate
methyl (Z)-octadec-9-enoate

linolenic acid
(Z9,Z12,Z15)-octadeca-9,12,15-trienoic acid

coniferyl alcohol
(E)-3-(4-hydroxy-

hexadecan-1-ol

Devolved decision making by worker and drone bees

Despite the common beekeeper perception of the supremacy of the queen in the honey bee colony, the colony itself – and largely the worker bees and to a limited extent even drones – exercises an enormous influence over other facets of colony functioning. One may think of the colony as a large organism with facility for workers to regulate and direct such complex tasks as foraging, comb building in response to nectar inflow, brood raising and queen replacement as well as swarm initiation largely independent of any queen signalling.

It is helpful to restrict the discussion to specific instances where chemical signalling can be traced to bees other than the queen beginning with the brood itself and then exploring the realm of both workers and drones in controlling behaviours both at the colony and individual bee level both within and outside the hive.

Brood pheromone

Brood pheromones comprising straight chain methyl and ethyl esters of saturated and unsaturated fatty acids (Figure 1.108) are well known[266] to inhibit worker bee ovarian development and to enhance worker foraging behaviour. Queen mandibular pheromones play a similar role in controlling the laying worker ovary development, with the unsaturated terpene (E)-β-ocimene[267] being only added to the list of active constituents produced by brood early in the last decade.

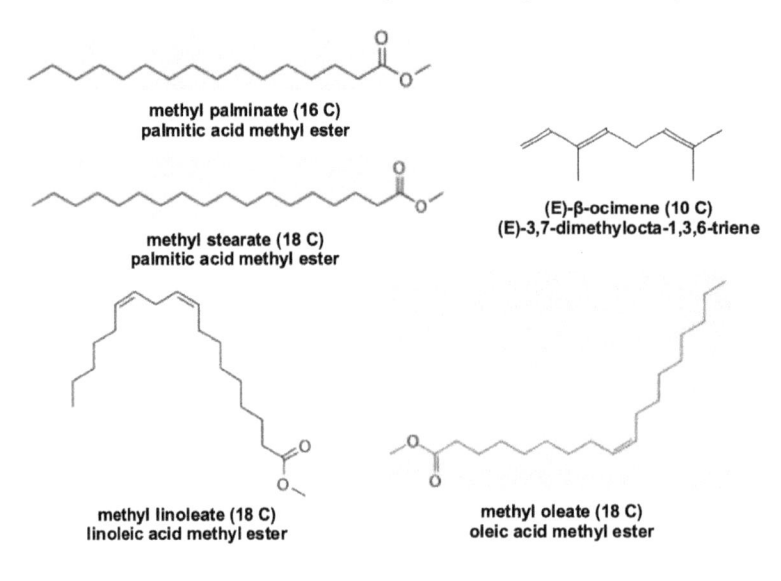

Figure 1.108 Brood pheromones.

methyl palminate (16 C)
palmitic acid methyl ester

methyl stearate (18 C)
palmitic acid methyl ester

(E)-β-ocimene (10 C)
(E)-3,7-dimethylocta-1,3,6-triene

methyl linoleate (18 C)
linoleic acid methyl ester

methyl oleate (18 C)
oleic acid methyl ester

Worker footprint pheromone

This footprint (or trail) pheromone, which is perceived olfactorily and possibly also chemotactically, is persistent but is probably not colony specific[268].

Lensky and coworkers have studied the biological effect of the secretion of queen, worker and drone honey bee (*Apis mellifera*)[269] tarsal (Arnhart) glands located on the 5th segment (tarsomere) of each bee leg. They comprise a mixture of alkanes, alkenes, alcohols, organic acids, ethers, esters and aldehydes – twelve queen compounds, eleven for workers and one for drones. For example the appearance of queen cell cups is inhibited by queen footprint pheromone, so that crowding rather than imminent swarming may give rise to their formation. In practice I always check out lone swarm cell cups but they rarely contain eggs or larvae. The role of footprint pheromones amongst workers extends – at least in part – to marking hive entrances and flower marking.

Beyond the honey bee nest

Many honey bee pheromones appear to operate mainly within the confines of the nest, while others – augmented by the likes flower scents – have distinctive roles outside the nest. While the notion of chemical signalling being different inside as apposed to outside the nest is artificial it is nevertheless useful conceptually. Many behaviours of honey bees outside the confines of the hive are important in maintaining honey bee cohesion, notably in achieving successful matings in drone congregation areas, in maintaining swarm cohesion and in allowing bees to reassemble when dislodged from combs. As we shall see there are a dazzling array of controls exercised beyond the honey bee nest as disparate as queen-drone mating behaviour and swarm behaviour[270], flower recognition and bee aggregation.

Part II Honey bee chemistry beyond the nest

What do bees use to reorganise themselves when they are dropped off a frame outside the hive or the hive topples over? How do bees maintain cohesion when migrating as a swarm? Why do bees raise their abdomens and fan vigorously when they enter a swarm box? And how do drones communicate with each other and attract virgin queens to congregation areas well away from their natal domicile?

And why do bees flit from one dandelion bloom to another with pinpoint accuracy never returning to flowers just visited? How to foragers recognise and remain faithful to a particular flower type?

We might classify these responses into two groups: those pheromones produced by bees that allow them to recognise other bees and chemical scents produced by flowers and water sources.

Here we examine the chemical sensing mechanisms that bees employ beyond the hive. Some of these, such as those produced by worker Nasonov glands are used to help bees congregate when stranded away from the hive and that fanning bees employ to attract their fellow workers when entering a new hollow. The same Nasonov gland in the giant honey bee (*Apis dorsata*) produces alarm pheromones employed in aggressive defence.

Smelling the roses

Spray lavender water into a hive and bees will emerge en masse, fly across wind and head up wind to the flower patch all the time sensing an increasing gradient of lavender oil comprising primarily linalool, linalyl acetate, lavandulyl acetate, terpinen-4-ol, lavandulol, 1,8-cineole and camphor[271]. It seems likely that bees can discriminate different species and potentially different patches of lavender (*Lavendula* spp) as each would have a different reward and different mix of attractant chemicals.

In examining the extrinsic sensory world of the bee outside the hive we learn that bees rely on a whole more on what they smell than what they see to gain orientation. As flying insects, they of course need to see but there is far more to the bee senses than the waggle dance messenging that text books religiously regurgitate[272]. This world is entirely different to that of the darkened confines of the nest. While visual signalling would appear to dominate successful navigation of the fifty or so square kilometres (this assumes a foraging radius of four kilometres) bees patrol, when it comes to targeting sources of nectar, pollen, propolis and water they are likely almost entirely reliant on their sense of smell?

Sniffer radar

A good starting point to examine these mysteries is to examine the general morphology of the bee. Insects such as bees have extraordinarily sensitive 'sniffer organs', things more accurately termed antennae, more colloquially feelers. Each bee antennae has 170 individual chemoreceptor cells: each locus can detect a single molecule of a volatile substance. Large mushroom bodies in the forebrains of invertebrates that process olfactory signals certainly suggests that bees have an acute sense of smell.

There are several categories of chemicals that invertebrates such as insects can recognise with remarkable ease. We know that dung beetles can detect a cow dropping like a homing pigeon and that dipterans such as mosquitoes and flies — the filth insects — know where their next meal is. Mosquitoes can detect thermal heat and carbon dioxide gradients, flies excrement within seconds of deposition. And for bees?

Inside the hive, trophallaxis — exchange of food — is the primary means of transmission of pheromones that control collective behaviour. Volatiles, both those generated by flowers and those manipulated or manufactured by bees themselves are also detected by antennation. Prominent amongst the natural volatiles are the cyclic and acyclic terpenes (ten and fifteen carbon units built on the basic five-carbon building block isoprene) such as isomers of limonene (oranges), camphor (from camphor laurel) and pinene (pine tree oil) and simple aromatics such as eugenol (oil of cloves), cinnamaldehyde (cinnamon) and coumarin (principal constituent of new mown hay) (Figure 1.109).

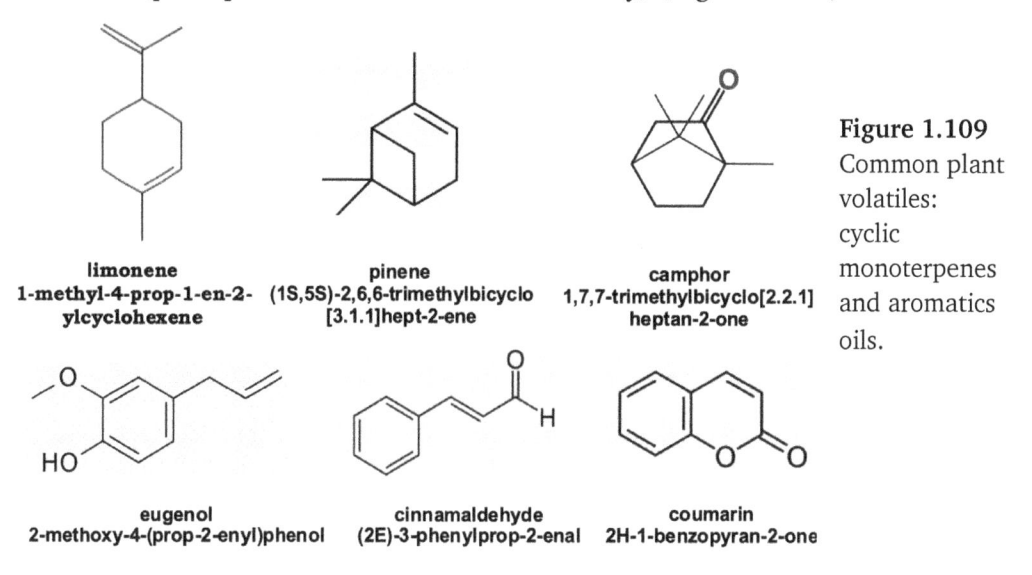

Figure 1.109 Common plant volatiles: cyclic monoterpenes and aromatics oils.

limonene
1-methyl-4-prop-1-en-2-ylcyclohexene

pinene
(1S,5S)-2,6,6-trimethylbicyclo[3.1.1]hept-2-ene

camphor
1,7,7-trimethylbicyclo[2.2.1]heptan-2-one

eugenol
2-methoxy-4-(prop-2-enyl)phenol

cinnamaldehyde
(2E)-3-phenylprop-2-enal

coumarin
2H-1-benzopyran-2-one

The Nasonov gland

Bees also spray a natural array of terpenes (Figure 1.110) through their Nasonov gland located under the seventh dorsal abdominal tergite. The Nasonov gland was first described in 1882 by the Russian zoologist Nikolai Viktorovich Nasonov (1855 – 1939)[273].

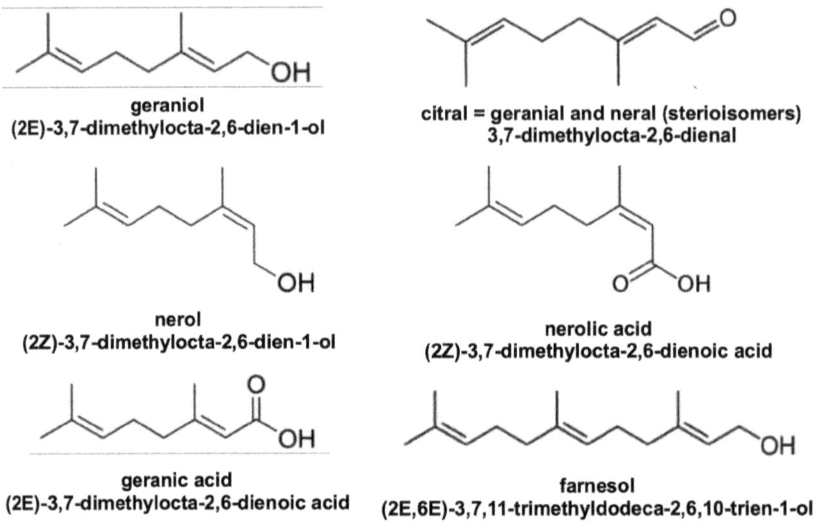

geraniol
(2E)-3,7-dimethylocta-2,6-dien-1-ol

citral = geranial and neral (sterioisomers)
3,7-dimethylocta-2,6-dienal

nerol
(2Z)-3,7-dimethylocta-2,6-dien-1-ol

nerolic acid
(2Z)-3,7-dimethylocta-2,6-dienoic acid

geranic acid
(2E)-3,7-dimethylocta-2,6-dienoic acid

farnesol
(2E,6E)-3,7,11-trimethyldodeca-2,6,10-trien-1-ol

Figure 1.110 Nasonov gland volatiles [essential oils] found with *Apis mellifera*: geraniol 100 parts, geranial 1 part, neral 1 part, nerol 1 part, nerolic acid 75 parts, geranic acid 12 parts and farnesol 50 parts[274].

In a follow up study Williams and coworkers were able to show that honey bees were attracted to each of these terpenes.

Interestingly geraniol is the base monoterpene natural product from which many terpenes are synthesised. As I am not an invertebrate neurobiologist I will not hazard a guess as to how bees detect and process signalling from these chemical mixtures. I can really only make sense of these chemicals from having handled terpenes in the laboratory and, of course, by sniffing geraniums.

However bees recognise many chemicals innately – they employ individual olfactory receptors on their antennae – including learning how to detect a wide range of naturally occurring volatiles. Chemical signalling is not only of importance to bees themselves: proprietary formulations are being increasingly employed to act to lure bees[275] and to detect and attract exotic species of honey bee such as the Javanese strain of the Asian Honey Bee[276] threatening Australian and global beekeeping enterprises.

Nasonov terpene signatures differ across the honey bee genus. Only neral has been detected in *Apis indica* (syn *Apis cerana indica*)[277] though a mixture of linalool, linalool oxide and citral (geranial and neral) has been found in the Japanese race of *Apis cerana*[278] (Figure 1.111).

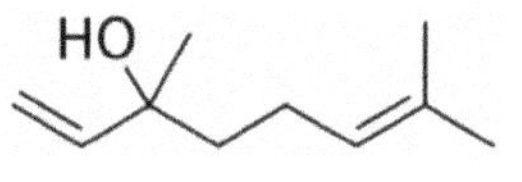

linalool
3,7-dimethylocta-1,6-dien-3-ol

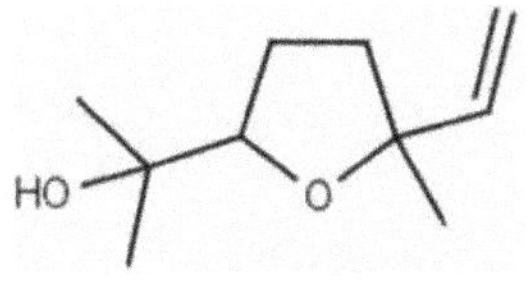

linalool oxide
2-(5-ethenyl-5-methyloxolan-2-yl)propan-2-ol

Figure 1.111   Nasonov pheromones generated by *Apis cerana japonica* Asian honey bees.

There are a number of signalling chemical groups impacting on behavior of bees outside the hive, alarm pheromones used in defence, footprint pheromones used to mark hive entrances[279] and flower marker pheromones. Not surprisingly different species of honey bee employ their pheromones in various ways to achieve different ends. For example the Nasonov pheromones in in *Apis dorsata* are less associated with aggregation behaviour and more with pheromone release and abdomen waggling behaviour associated with defence[280].

Alarm pheromone

In *Apis mellifera* alarm pheromones (Figure 1.112) are produced by both mandibular and sting (Koschevnikov glands[281, 282]): the mandibles release 2-heptanone while stinging bees release a range of alcohols and esters. One of these esters, isoamyl acetate, a key constituent of ripe bananas, is well known to initiate stinging behaviour and is the familiar smell of a satchel in the school locker room. As well as that you should be able to detect when a hive is overly disturbed. Good rules of thumb are to avoid packing bananas for that visit to the apiary and to work gently and smoothly when opening hives.

Cassier, Tel-Zur and Lensky[283] discuss the role when both sting and associated Koschevnikov[284] glands act in synergy and the vast array of aliphatic and aromatic hydrocarbons, alcohols, aldehydes and ketones, esters and carboxylic acids – some forty six or so – that have been identified and but whose individual functions are not understood.

Camargos and coworkers found twenty three Koschevnikov gland (the gland associated with the worker bee sting apparatus) compounds, identifying twenty one: hexan-1-ol, isoamyl acetate [isopentyl acetate], hexyl acetate, octan-2-one [n-hexyl methyl ketone], nonan-2-one [methyl heptyl ketone], nonan-2-yl acetate [1-methyloctyl acetate], 2-decen-1-ol [(E)-dec-2-en-1-ol], 2-dodecanyl acetate, pentadecyl acetate, 9-octadecanal [(E)-octadec-9-enal], oleic acid [(Z)-octadec-9-enoic acid], nonadecan-1-ol [nonadecyl alcohol], (Z)-9-icosen-1-ol [gadoleyl alcohol], (Z)-9-tricosene, tricosane, 2-methyltetracosane, 7-hexylicosane, heptacosane, pentatriacontane and hexatriacontane.

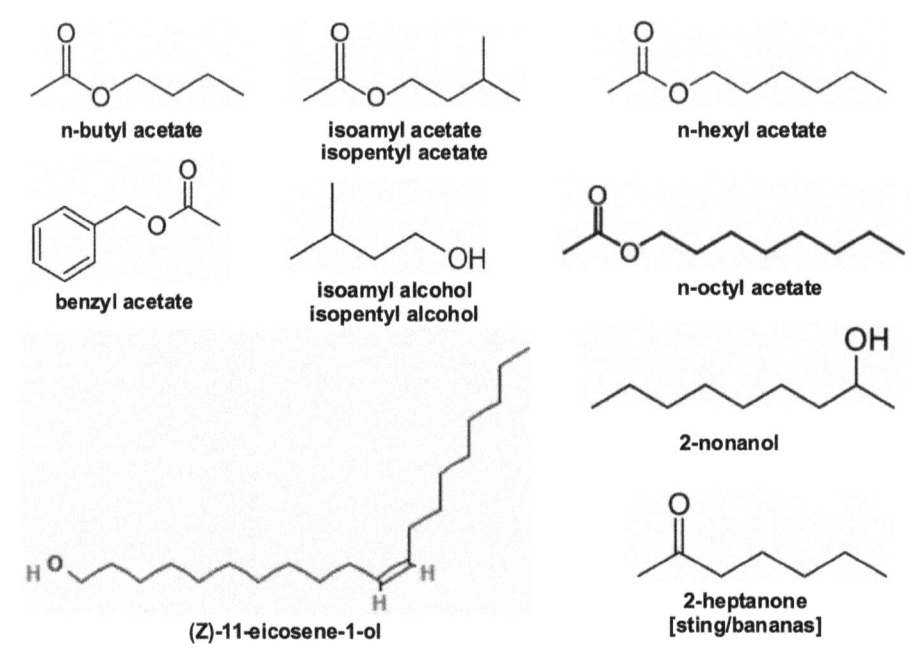

Figure 1.112   Alarm pheromones: a sample of bee sting gland volatiles.

Collinson[285] explains that there are many subtle and age differences between foragers (gatherers), guard (alarm pheromone emitting) and soldier (harassers of predators) castes of worker bees so clearly defense behaviour requires nuanced interpretation by both bee and the observant beekeeper.

Drone congregation pheromones

Drone pheromones comprise a mixture of forty seven lipids, saturated, unsaturated and methyl branched fatty acids and several long chain alcohols[286] (sample depicted in Figure 1.113) identified from extracts of drone mandibular glands.

Again this looks like an unholy chemical concoction though some are common constituents of animal fats. Think of them particularly the saturated type – the ones without the double bonds – as a fairly exotic mix of unhealthy long chain fatty acids. It is hard to conceive how these essentially non volatile acids might act on the wing but they appear to affect the behaviour of other drone, but not worker, bees inside the hive. Interestingly queen mandibular pheromone appears in a reciprocal relationship in attracting drones on mating flights[287].

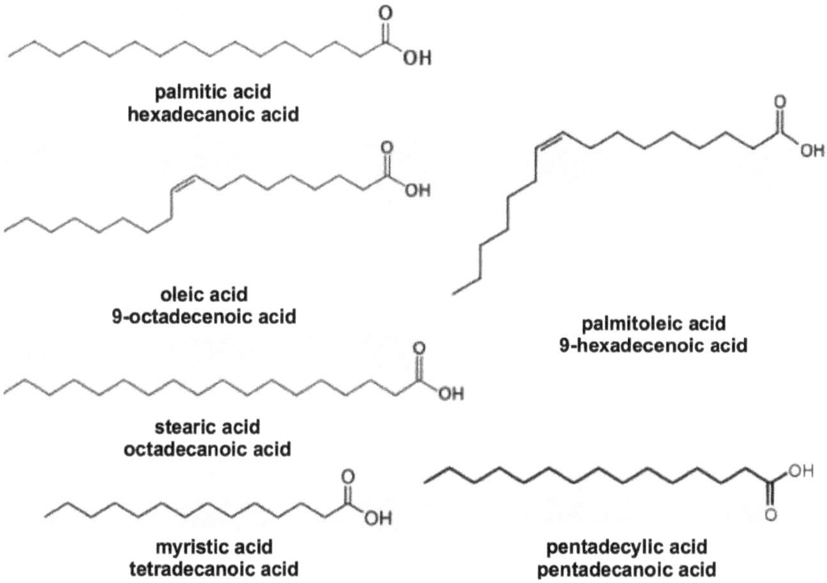

Figure 1.113 Drone congregation pheromones: palmitic acid 10.5 parts, oleic acid 9.4 parts, stearic acid 3.6 parts, myristic acid 1.7 parts, palmitoleic acid 3 parts, and pentadecylic acid 1 part.

Forager pheromone

Ethyl oleate (Figure 1.114), released by older forager bees, regulates the maturation of nurse bees to avoid too many house bees becoming foragers[288].

We can expect to learn a whole lot more about chemical communication in bees, not withstanding this limited overview. From a practical perspective understanding the role of pheromones and plant volatiles gives us some cues about how we as beekeepers might more sensibly manage bees. Not lunching on bananas, employing smoke to mask chemical signalling amongst bees, recognising that robbing bees are best left alone and experimenting with perfume sprays – as an alternative to employing smoke – on total fire ban days may all be ways to more effective control of bees. And replacing an ailing queen, neither capable of laying well nor communicating well with her offspring, does make a lot of common sense.

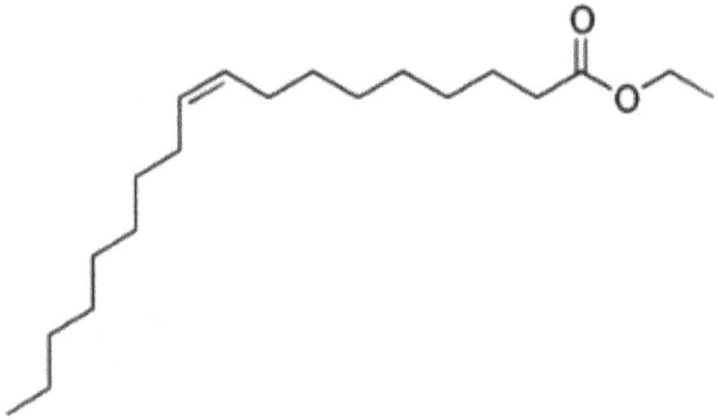

ethyl oleate
ethyl (9Z)-octadec-9-enoate

Figure 1.114   Forager pheromone, a simple ethyl ester of oleic acid.

Keeping bees in good fettle

Many owners of bees keep their bees on a wing and a prayer. They check to see that the queen is laying and has a decent brood pattern and harvest a few frames of honey at the end of summer, but otherwise they leave their bees alone. And of course some overly worry them by opening them too regularly for no express intent or purpose.

These days – I live in a cold climate by Australian standards – I close my bees down somewhere between late summer and mid autumn, checking them for stores with the heft test in late winter. I only ever venture to open them in the first days of spring. I take a bee holiday. But I regularly requeen and where possible run up to a handful of spare nucleus colonies. I also feed them to the hilt with heavy sugar syrup well before the weather cools down to ensure I that have a full box of stores for overwintering. To that I will make up an icing sugar (or finely grained sugar) mixed with irradiated pollen to make up a fondant if I think the pollen stores are a little short. Recent years have been wetter than average and pollen stores have not been a problem.

Disease problems aside and some tweaking of hive gear, I believe my bees have been always well able to look after themselves. Where does this leave the regular punter?

When I thought about this a little more I concluded that I had received some sage advice long ago. Never string bees along with a queen that has patently poor workers of poor hygienic disposition or lays a poor brood pattern. Always make sure there are spare stores in the hive. The latter stores issue has little to do with bees starving, though they may well do so if you harvest all their summer bounty and do not lavish them with supplements. Leaving bees short at any time of the year, notably in the build up and in the autumn bee replacement phases, will put bees back. Low reserves check colony development and its potential to survive and harvest as much nectar and pollen as possible.

Part I Do you carry a spare tyre?[289]

No, I am not trying to cast aspersions on your midriff. But everyone who drives a car carries a spare in their boot. That tyre is ready and inflated when your tyre picks up a nail while you are doing some retail therapy in Fyshwick, an industrialised suburb of Australia's Capital City. And you might want to carry at least two spare fuel drums as well as a repair kit when you head out along the Oodnadatta Track, an Antipodean destination, or as we say in Australia to go 'On the Wallaby'.

A spare nuc (a baby or nucleus hive) in any apiary can also be liquid gold when you consider either trading in the monetary value or the enjoyment brought to you by your bees when all does not go according to Hoyle (Figure 1.115)[290]. There should be a proper way to change a tyre and, no doubt, a manifesto on how to go about feeding bees. We shall revisit Edmund Hoyle in another guise when we discuss the potential of bees to produce rivers of honey in hives run with two queens.

Figure 1.115 Edmond Hoyle 1672-1769.

In times past I just ran a handful, and occasionally too many, full strength hives with no such backup. Somewhere way back I then changed tack. I built a Swathmore swarm box, rattled up about thirty nucleus boxes and finally purchased a breeder queen and raised more queens than most sideline beekeepers would poke a stick at. A little older and now somewhat wiser I seek out commercial queen breeders with a good reputation and buy a few more queens than I need. Much simpler. Not only do I get gentle, productive queens, ones that have a good measure of disease resistance, but I get the spares I really need.

But queens still fail and I've had more than my fair share of failed introductions, something I put down to a poor understanding of the queen (or gyne) status of the colonies I have tried to requeen. So I keep two or three spare nucs and rarely directly requeen any large colony. These days I haul out the few nuc boxes I still posses, dust off the cobwebs and introduce queens that I purchase to them – the nucs, not the spiders. Those nucs I make up a day or two before they arrive in the post to make sure the new girls are made instantly welcome, a practice a little reminiscent of preparing for that first date. Easier and less stressful than trying to find twenty queens in late season buildup conditions when the bees are spilling out of three supers.

The wisdom of having a spare

Often enough any seasoned beekeeper is asked by a beginner: 'How many beehives should I keep in my back yard?' Many beekeepers only want to dabble and enjoy the acumen of actually owning bees. Yet the advice proffered is almost invariably to get a second hive. However be extremely wary of owning three or more. Why so?

Well firstly honey bee colonies are about as variable as people and about as unreliable. Some are interesting, only some are productive and there is the ever constant risk of losing a good queen (or hive) and being in a quandary about how to find a replacement. The same rules apply to bees and people. Being left on your own is always a risky business.

The real value of having two hives is being able to compare their performance, one against the other. For the beginner, this can be of inestimable value. It teaches you that a hive can perform beyond all expectations and it will teach the skill of reading hives, knowing exactly when bees are failing or below par. But owning 'two instead of one' has unparalleled merit when – as any beekeeper soon enough discovers – one queen falls over. Not surprising when you consider the risk a fly takes when that fly swat comes out and she knows there are no replacements to raise maggots in that meat left on the sink. Most commercial beekeepers I know run a nucleus apiary to keep up a good supply of spare queens. They work on a one-in-ten rule to make sure they are rarely short of spares. This will change with the mite arriving in Australia in late 2021. As naïve Australian beekeepers seem likely to discover, high annual hive mortality and reduced queen longevity, associated with the arrival of *Varroa*, will force a new thinking. It will be more a case of 'How many spare nucs can I run?'

Instead of being (s)wiped out you can quickly establish a new colony by simply splitting the remaining healthy colony once it reaches double box strength. Or, if you just lose or squash a queen, the simple measure of swapping over a comb of young brood to the seemingly queenless colony will rectify the problem instantly. In both cases, the queenless split and the colony to which young brood has been added will raise an emergency queen. From an egg, she will take a bit over three weeks to first hatch, then grow, pupate, emerge, mate and start laying. So the bees will miss a brood cycle but, in bee terms, the problem is solved.

In the back yard – assuming you regularly inspect your bees to check their disease status – the second hive is almost the best guarantee that you won't be wiped off the beekeeping slate. That second colony, so to speak, is your spare tyre.

At a larger scale you want all your colonies to be productive. Continually removing frames of brood to improve the condition of underperforming colonies and the potential for transferring disease from one hive are not the best strategies. Ask me? I once lost a whole backyard apiary to American Foulbrood and Small Hive Beetle. That was an occasion when my surveillance became inexplicably lax and I spread the problem around by interchanging gear.

A simple solution to keep all your full size gear in active service is to run spare nucs (one for around five hives, and an additional nuc for every ten colonies thereafter). If you make them up from hives to be requeened (leaving the queen behind) the risk of transferring disease is minimised and the chance of successful queen acceptance is greatly increased. And you buy time – the old queen does not need to be found – and you eliminate the chance of losing two queens – by first killing the old queen and then failing to establish the one you have just opened your wallet to make a replacement.

The spares (tyres) are a great investment but there is the additional work of maintaining the nucs, giving the queen enough room to lay (by exchanging frames of sealed brood for empty combs in hives that need strengthening). However by owning (Figure 1.116) and running spares you will always be in a position to resolve hive queen problems quickly without having to get on a long waiting list from an oversubscribed queen breeder.

Figure 1.116
Canberra Region Beekeepers nucleus assembly workshop July 2017.

And of owning three or more hives? Beekeeping can be infectious but beware of keeping more bees than you can sensibly manage. Too many hives will also test domestic alliances, that is unless you come from a beekeeping family where owning bees is part and parcel of the way of life.

Nucs and swarming

Whenever things go awry, say when you find a poorly laying queen or bees are preparing to swarm, one of the resources you can always turn to is that spare tyre you have saved up for the rainy day. And if you think the extra queen is money down the drain, it's worth remembering that a queen in a small colony won't lay nearly as fast as the same queen in a full strength colony. She will have only a few thousand workers at most and they can only nurse a frame of two of bees. The queen ages slowly just as diutinous (long lived overwintering) bees live much longer and only die when they have the stress of raising bees and foraging in early spring.

So when your bees are roaring and spilling out the front of the hive in spring, pulling out that nuc to requeen a colony headed up by an aging mum (Figure 1.117) is one of the best ways – apart from providing bees more space – to head off swarming.

Figure 1.117   Nucleus hive setup up to requeen Paradise colony J04, Jerrabomberra Wetlands, January 2019.

Nucs add a dimension to your apiary, are easy to inspect and manage and provide an entree into the skills of checking for disease and finding queens so much more challenging in a full strength hive.

My only recommendation is that you acquire an insulated high density polystyrene six-frame nucs (Figure 1.118) if you want to successfully overwinter such small colonies.

Figure 1.118 Very late season call, moving a strong nuc from a wooden 6-frame nuc box to an insulated box May 2019:
(a) strong six-frame nucleus colony in wooden gear with makeshift heavy sugar syrup bags positioned over frames; and
(b) frames and bees rehoused in six-frame highly insulated hive for overwintering.

Do you really need that extra nuc gear?

You really don't have to buy nucs or nuc gear. A regular full depth box will do. Just slip in a piece of corflute or heavy duty cardboard and move it across the box as the colony expands. And we really haven't explored the many uses of nucs or some of their weird and wonderful designs. However there is nothing wrong with using standard gear to make up nucs. Alternatively you can make nuclei from broccoli boxes by inserting dowel rods or bamboo sticks to rest frames on. Locally available polystyrene broccoli boxes readily accommodate five full depth frames. However they need improvised support rods to rest frame lugs on and bee space is not perfect on walls or frame ends so they are a temporary solution for housing bees, not for keeping bees for more than a few months. The central function of nucs is to maintain spare queens that will greatly increase the flexibility in running any apiary.

Part II Should I feed my dog?[291]

Bees have gone to the dogs

In an ideal world our dogs – and bees – could rely entirely on nature's bounty. Sometimes, as evidenced by the number of apple cores and discarded sandwiches my Scottish terrier Boots finds on regular walks, I'm inclined to the view that this may indeed be possible. I'd be less certain that, had we failed to feed, educate or immunise our children, they would have grown up to be healthy, wealthy and wise and be there ready to meet our every need in our dotage.

So what to expect of our bees? It would be splendid if they were never to fall prey to disease, if they always gathered all they needed to sustain themselves and for them to happily share their summer bounty. The Land of Dairy Flat Road and Club Honey – the Canberra Region Beekeepers' apiary – returned a seemingly modest 180 kg of liquid gold in the very dry year of 2018. However that trickle of gold should be reckoned against the fact that it took a bout 60 kg of sugar in early spring to kick start near starving bees and rather more than 60 kg after we had extracted in early March to tide them over the winter to follow. That drought experience put pay to our Garden of Eden thesis: we fed almost as much as we got so almost a zero sum game but we won if only just. Our bees were healthy and we got enough honey to knock up a batch of mead.

In good seasons, and with good disease surveillance and young queens, bees exceed all their needs and will invest by establishing new colonies through the process we know as swarming. So what if, our pets protest, they find they cannot feed themselves[292]. Not feeding animals is famous for attracting press attention and for having offenders reported to the Royal Society for the Protection of Cruelty to Animals, better known colloquially as the RSPCA:

> ...A Port Noarlunga South man convicted for failing to feed three dogs.... The magistrate decided not to record a conviction, and banned the defendant indefinitely from owning...

In most cases, simply leaving enough honey with the bees makes great sense. On the other hand many commercial beekeepers will extract all but brood comb honey near seasons' end and feed them heavily with sugar and pollen supplements to make up the shortfall. One might reconcile countervailing arguments about leaving ample stores or feeding sugar – maybe just a top-up – to meet overwintering requirements. However I find it extremely difficult to accept that bees can be left in the lurch simply because they might survive.

Common practice everywhere is to leave a super of honey and is a no fuss solution to meeting honey bee colony needs for overwintering. On the other hand, an apiarist making a living will do much better to extract and sell that honey at say. A $15 - $20 per kilogram and putting the strong end-of-season bees to work processing sugar syrup. Sugar currently retails at about $1.20 per kilogram. The choice is yours but leaving enough honey accords well with Thomas Seeley's sustainability notion of Darwinian Beekeeping[293]. In this context it is interesting to stop and reflect on how bees operate in the wild. In many instances colonies do not survive winter especially if this follows drought. Drought sorts bees out and is a factor – apart from available nesting hollows – that limits the number of bee colonies the landscape can support.

Thus stated, it does say a lot about overstocking an apiary. For the back-yarder not overly concerned with getting any return on their bees, and willing to sacrifice their bees to nature's call, there may be a case for not feeding bees and allowing them to perish. However, and this is a personal take, I'm a little challenged to accept that valuable bees can be allowed to starve as I believe that allowing bees to fail presents two problems apart from colony loss. Firstly dead outs are easy prey to wax moth and, if nothing else, the integrity of your equipment. Who wants to deal with tangled wax moth web and pupated wax moth larvae buried in wooden hive gear? Rather more seriously, weakened hives are easy prey to robber bees and that increases the risk of spread of disease, notably American Foulbrood and increasingly to *Varroa* mites, presenting a serious problem for other recreational beekeepers and increased risk of disaster for the commercial operator. Poor winter nutrition also risks the appearance of stress-related diseases especially European Foulbrood, chalkbrood and, very importantly, nosema disease.

Like my dog, I pay close attention to my bees, and feed them but only if and as needed. As already noted elsewhere healthy bee colonies consume large amounts of honey and pollen just to keep the engine running. We are only guaranteed a surplus, a resource we can sacrifice by allowing our bees to swarm, if the bees are healthy and there is ample forage. Feeding them more then they need invites the 'Chinese honey syndrome' making invert glucose and fructose honey from cane sugar. Only feed your bees in spring if they are running low on stores or in early autumn if they have not stored a surplus to get them through winter!

Bees nutrition in the wild

What can we learn from bee survival in the bush or on the African savannah? Honey bees, like ourselves, have multiple strategies for survival. In the normal course of events colonies of all twelve species of honey bees *(see Global distribution of honey bees)* swarm to replace their losses. This is necessary to reestablish bees lost to predation (think bears, elephants, toads, bee eaters and traditional bee hunters) and disease as well as to opportunistically colonise landscapes (think New World landscapes including The Americas, New Zealand and Australia). As we have noted honey bees also lose colonies to starvation, a condition that destroyed Kenyan Top Bar colonies caught short in our Jerrabomberra Wetlands apiary over the past handful of years. African and many races of the Asian Honey Bee swarm very prolifically and – unlike the northern European races of the Western Honey Bee – frequently swarm, abscond or migrate. They do so to escape predation or abandon diseased comb but very often their chance of survival is increased by moving to regionally available flower resources. For a discourse on bees and plant phenology see Hepburn and Radloff's Honeybees of Africa[294].

European races of the Western Honey bees do not have this choice. Abandoning the nest in winter, or even in times of summer dearth only decreases, not increases, their chance of survival. They must store all the reserves they need for a full year cycle and may do so (for example in parts of Canada and equivalent northern European climes) in a period a short as several weeks. So in a severe winter or following a poor honey flow the chances of the individual wild colony surviving can be quite low. But a least a few colonies survive and those that do repopulate the landscape in good years.

How and when to feed bees

Traditional wisdom has it that you leave a complete box of honey so that bees don't run out during winter and more likely run low in August and September – Australia's late winter and early spring – when they are trying to build their numbers up quickly to take advantage of an early honey flow. Starting early also allows them to maximise their honey gathering on the main flow, usually in box woodland from November through to January.

Bees eat a power of honey – and have insufficient means to collect more – in the late-winter, early-to-mid spring phase of colony build up. Think of it, a queen can lay out a full frame of brood in about three days. Three weeks later that frame of brood will emerge, consume a frame of honey, and cover three frames of comb. The colony soon exhausts its supply so supplementary feeding of bees (Figure 1.119) is both wise, if supplies are short, and necessary.

Figure 1.119 Large volume top feeder on a Paradise nuc.
Note bee-smart window at one end. By mid spring they can easily process about 5 L of 2:1 sugar syrup in less than two days.

So assuming your bees have not put away sufficient winter stores the questions are of how much and when to feed. It's always a good idea to keep feeding bees until they have that full super of stores. That may be up to say 15-20 kg, a lot of sugar best fed as heavy (2 parts sugar to 1 part water) syrup. Frame feeders (Figure 1.120), inverted containers and 2L kidney shaped feeders in a spare top super as well as large volume top well feeders – such as found on Paradise and HiveiQ Hives– may all need constant replenishment so don't expect that 100 mL Boardman entrance push in feeder to go far. I like to complete feeding in February and March after the end of the honey flow – a good time to assess whether the bees will make it themselves and then not open or disturb the bees from Anzac

Day (25 April) until the first week in September. It's relatively easy to stock bees up with heavy sugar syrup in warm weather but never feed sugar as syrup in winter – it is almost certain to trigger serious nosema and induce outbreak of other stress-related disease problems. Commercial apiarists chasing pollination contracts are forced to stimulate their bees early but are wary of leaving hives open or stressing brood.

Figure 1.120 Author's galvanised zinc feeder made thirty years ago.

Other home-made durable feeders, easy to construct, can be made from a trimmed frame covered with masonite and dipped in paraffin wax. You have to open the hive but frame feeders are readily filled with added straw or grass to prevent drowning.

This 'be prepared' strategy allows me to forget the bees and take a long break. That never dissuades Beginners' Corner questioning of how to feed bees in winter. If you tilt a hive, it is pretty light weight and you didn't make that preparation, feeding is of course necessary. In an emergency place granular white sugar on top of the hive mat under the lid – pick the warmest spell possible and don't leave the hive open for more than a minute or so – or, much better, place dry sugar in a top feeder that neatly avoids opening the hive.

There are two risks in adopting this must-feed approach apart from risking upsetting the warm bee cluster. Firstly under cold conditions bees may not be able to reach the sugar (or outside frames of a top bar hive) and may starve in a seeming tide of plenty. Secondly and perversely bees may remove sugar and drop it outside the hive front entrance. This happens most often in spring when there may be a light nectar flow and the bees are not in immediate need of feeding. I've given up on dry sugar feeding, largely because the bees can be slow to use the resource but more particularly as the bees can quickly consume heavy sugar water, typically 2-5 L per day under good conditions, and because I prefer not to risk colony wellbeing coming into winter. Never make the mistake of thinking your bees can cope with cold wet sugar syrup in the depths of southern highlands winters.

Do I feed my dog? Yes, it does need to be fed daily, but never overfed and never at the point of starvation.

Keeping bee records

You wanna be a good beekeeper
You betta keepa good records

One skill that helps beekeepers gain insights into honey bee behaviour comes from keeping detailed bee inspection records. Ever tried to work out if the queen you last saw in early autumn had been replaced or that the swarm control measures you practiced actually worked? Grandpa can probably tell you the day he collected his first swarm and tell you how much honey he got off his two hives when he started keeping bees. But that was when Noah came out of the ark. Quite often I find I'm hard put to tell you which day I checked my bees last week or whether I requeened last autumn or in spring last year.

Keeping good records is part and parcel of gaining experience in handling bees and learning to observe the unexpected, say quickly picking up an American Foulbrood infection when doing a cursory brood check or observing the eucalypts coming into flower as you enter the farm gate.

That brings us to the type of records one needs to keep, ones that reflect the condition of all the colonies you are opening, not just the individual frame or super of bees, and to relate what you see in one hive to the other hives in the apiary. Bees compete vigorously and I never assume that an apiary is a collection of colonies living in harmony.

The ability of the keeper to pick up whether his or her bees are beginning to seriously rob is a signal to close down any hives, pack up and go home. Come back another day determined to work quickly and to cover all exposed frames and honey as best possible. For the same reason you might check the tread on your tyres and glance at the fuel gauge before setting out to the apiary: a break down with a trailer load of bees will land you in all sorts of strife.

A nagging question about keeping good hive and apiary notes centres around what it is important to record. First and foremost you will probably want to know which hives are weak or smitten with chalkbrood so that any problem, one that can't be fixed immediately, can be rectified at the next visit. In writing the shopping list and clearing out the larder, I am reminded of an old family saying:

Eat what you can
And can what you can't.

But there is another reason for keeping sound records. One of the conditions of being allowed to keep bees is that you must register them, you must check them and you must promptly report any adverse finding. So, assuming you are a law abiding citizen, you must keep records to demonstrate that you have been diligent in conducting those checks. For anyone keeping bees in the Australian Capital Territory such details can be found in the *ACT code of beekeeping practice*[295] or better, with the recent *Varroa destructor* mite incursion, the Australian Honey Bee Industry *Biosecurity code of practice*[296].

The penalties in most jurisdictions include being fined, having your bees confiscated and – for the worst offenders – being banned from keeping bees. In practice we have all come across the lackadaisical bee owner, someone who is convinced that he or she is improving the environment by having bees in the backyard behind the shed. That those bees are likely disease ridden or inclined to sting the neighbours should make us wary of bestowing them the title of bee keeper.

There are many practical things you can do aside from formal record keeping. The simplest among them are to mark your queens, brand your hives – a legal requirement and always anti theft insurance. Many beekeepers I know record brief details of inspections on the hive lid. These can be helpful especially in a large apiary where you have identified diseased hives or hives in need of urgent requeening. These extra measures can be useful but are an inadequate record of how bees overall are faring.

What type of records

The best option for a keeper of a handful of hives is to keep a notebook in a protective plastic bag or on a clipboard. Record essentials include colony number, date, location and any actions taken and observations made. I once kept records on loose sheets of paper, either as simple notes or on a pro forma sheet, but these get lost or hopelessly mixed and end up in the bin.

The alternative for keepers who want really good records and willing to make full hive assessment is to subscribe to a good quality online recording system. Some applications have an online portal that allows summary reports to be written automatically though I find these fall a long way short of any nuanced interpretation of bee condition. Artificial intelligence and skilled appraisal of hive condition are distant cousins.

Virtual record keeping does however have its limitations. In practice sticky fingers, internet access and software problems have mitigated against any effort on my part to fall for direct recording on an electronic device, photographs apart. Paper copy on a clipboard is easier in the paddock and sure beats a flat iPad battery.

Routine inspection checks versus full seasonal inspections

Recording date, hive number and location aside, brief notes, the numbers of supers added or harvested, the amount of supplement fed, and a few notes on brood status, is all that is needed. Maintaining barriers, not swapping gear around and banning import of gear by other beekeepers and from other apiaries, is hard to practice but has kept American Foulbrood out of the club apiary for twenty years and facilitated multi-hive recovery from serious chalkbrood outbreaks.

In both spring and autumn I fully inspect all hives recording full details of things such as number of cells with chalkbrood and sacbrood, number of frames of bees, brood, honey and pollen knowing well that our newly acquired mites will make this recording seem like a piece of cake. I take inspections seriously shaking bees off all brood frames to make sure I do not miss that occasional American Foulbrood or European Foulbrood cell.

The Canberra Region Beekeepers hive inspection sheet is appended. It picks up details that you may want to record in undertaking a full seasonal inspection cuing for important details you may later regret not having. For the routine inspection however, seeing if bees are building as well as might be expected, anticipating swarming or natural queen replacement and assessing the optimal amount of space needed and the progress of a honey flow is all that is required.

It involves as little as lifting a lid, inspecting a frame of two and checking that there are no swarm cells clinging to the bottom of brood frames. Too much inspection too often can be counter productive as it takes a day or two for a colony to resettle after a prolonged hive opening.

Here are records I keep:
- disease records and follow up action especially for AFB (and also for EFB and serious chalkbrood outcomes);
- details of mite inspections in Australia at present limited to basic checks for absence-presence of *Varroa* mite;
- queen records such as date of hive requeening, colour ID (if you mark queens), poor brood pattern and hive weakness that might be attributed to a failing queen;
- stores levels especially in early to mid spring and in autumn where supplementary feeding may be needed;
- abnormal hive behaviour such as evidence of queen replacement; and
- crowded brood nest or supers in spring and early summer signalling likely onset of swarming.

Sadly two main killers, *Nosema apis* and *Nosema ceranae* cannot be recorded directly, the existential mite equation aside. However if you have good queens and some hives are dwindling quickly send off a sample, as you do for AFB and EFB, to get a diagnosis. Of course take immediate action such as splitting and supering hives as needed to keep idle bees fully occupied and building if there is a likelihood of their switching to swarm preparation.

Unmanaged bees

Of course there are unmanaged bees everywhere, not just those of the recalcitrant bee owner. Such bees are well capable of looking after themselves, but keep in mind that bees do not exist to store honey. They collect just sufficient to get themselves through most lean times, droughts and long winters and to achieve new colony establishment, that is swarming amongst wild bees.

From a beekeeper perspective the very existence of wild hives may be of little consequence. However being blamed for nuisance problems such as bees in swimming pools, their potential to produce drones that will affect the calibre of hives requeening themselves and their being a reservoir for pests such as small hive beetle and mites and other diseases is another matter. There the need or potential to remove such hives will be dependent on circumstance.

A side note: as honey bees are not native to Australia, wild honey bee nests are widely regarded as a feral animals. While their importance to agriculture is generally accepted by the wider community their presence in many natural settings, such as Australian National Parks, is openly decried.

Keeping hive records is very much like keeping bees. Doing both well is hard work, especially making sure your records are in good order after you have cleaned up, well stung, following a day among the bees. I rarely pour over beekeeping records but when you are asked to produce them or are planning things like requeening or disease control, those records are liquid gold.

Hive Inspection Sheet

Colony ID: ....
Date: ....
Inspected by: ....
Number of frames: ....

Colony resources	Notes	Observations
Frames of bees	☐	Note over-crowding or dwindling
Frames of brood	☐	Note anomalies, e.g. high drone numbers
Frames of honey	☐	Note fresh nectar or stores depletion
Frames of pollen	☐	Note fresh pollen storage
Colony condition	**Notes**	**Observations**
Queen seen (Y/N)	☐	Note if unmarked or marked colour
Eggs & larvae seen (Y/N)	☐	Note absence of any brood stage
Brood pattern solid (Y/N)	☐	Note if spotty or poor
Poor frames (Y/N)	☐	Note number, then mark or remove
Swarm/supersedure cells (Y/N)	☐	Indicate if raising queens and type
Hive entrance activity	☐	Make notes, e.g. pollen colour/flow
Colony health	**Notes**	**Observations**
Small hive beetle (Y/N)	☐	Estimate numbers, note damage, trap use
Greater and lesser wax moth	☐	Note tunnelling, webbing, damage
American or European foulbrood (Y/N)	☐	Sample and submit for lab testing
Varroa mite count (numbers)	☐	Treat or cull bees pending count
Chalk brood (number of cells)	☐	Note any mummies on floor
Sac brood (number of cells)	☐	Note crawling/deformed bees
Varroa mite (numbers)	☐	Note method, e.g. sugar shake
Nosema – (any symptoms Y/N)	☐	Note dwindling, crawling bees
Predators (Y/N)	☐	Note type, wasps, birds, ants
Colony manipulations	**Notes**	**Action**
Honey removal (Y/N)	☐	Note number frames removed
Super addition/removal (Y/N)	☐	Note details: super reversal, frame lifting
Nucleus colony use (Y/N)	☐	Note use to requeen or manipulations
Frame rearrangements (Y/N)	☐	Note details, e.g. brood splitting
Swarm control measures (Y/N)	☐	Note if spitting, Demareeing
Requeening (Y/N)	☐	Note queen type and marking
Colony nutritional feeding	**Notes**	**Purpose (building, winter stores...)**
Dry sugar or fondant (g)	☐	Note how fed
Sugar water 2:1 (L)	☐	Note feeder type
Sugar water 1:1 (L)	☐	Note if stimulatory
Pollen or substitute (g)	☐	Note brand, whether irradiated...

Rivers of honey[297]

Edmond Hoyle[298] (1672 – 1769) is famous for establishing rules for playing the game of Whist and more generally for card playing etiquette (Figure 1.121). If we were to run bees 'According to Hoyle', we could ever only do so in hives headed by a single queen.

Figure 1.121   The gamble of running multi-queen colonies 'According to Hoyle'.

Only one queen rules

Honey bees have a clear preference for running with a single queen! But could this rule be broken?

Some 30 million years ago:
- ancestral honey bees such as *Apis henshawi* (Figure 1.122) had the same worker morphology[299] as extant honey bees signalling that they likely ran best with a single queen;

- a queen in such a social group would have initially mated with one drone but colonies evolved to better survive when the queen mated with multiple drones;
- honey bees stuck to the single-queen formula; and more recently
- the first beekeepers arriving on the scene (from ~4000 ybp) soon learnt that trying to introduce an extra queen was more or less impossible.

Exception proves the rule

Surprising as it may seem honey bee colonies can sport an extra queen, the most notable examples of which include:
- the fleeting multi-gyne (a gyne is a potential as opposed to a laying queen) condition when bees are preparing to swarm;
- the occasional persistence of two or three laying queens when bees are replacing their ailing queen (supersedure);
- the recombinant prime swarm condition where two laying queens may both settle and lay in the one colony; and
- the Cape Honey Bee that parasitises West African Honey Bee colonies and that survives briefly as a two-queen hive.

Figure 1.122   Fossil *Apis henshawi* ~ 23 my bp.

Should you try to run hives with extra queens?

The benefits of running hives with extra queens are that more bees equates to more honey, more queen pheromone means less swarming and that colonies are unlikely to become queenless.

These advantages must be traded off against the fact that there are serious downsides.

Firstly any drought period, case of poor queen performance, or any bumper flow condition – the latter sometimes leading to gargantuan swarms – results in colonies reverting to the single-queen hive condition. Throwing in the fact that giant colonies are physically difficult to manage, and that timing of build up and extraction is doubly challenging, has made us wonder whether all the extra effort in running two-queeners is worthwhile. However beekeepers like us, bored with the routine of just trying to keep bees in order, have started the whole learning process again and given extra-queen beekeeping a shot.

The single-queen hive

The classic Seeley and Morse 1976 study of wild hives (Figure 1.123)[300] is a good place for us to start. Here we learn about the dynamic nature of the superorganism we call the honey bee colony.

We note first that while the colony persists all its members are regularly replaced. We once thought the honey bee colony were potentially immortal. Tom Seeley soon put us right with another study of wild hives suggesting that they have a mean lifespan of around six years as queen replacement occasionally fails. Most unmanaged and wild hives swarm annually so that means a new colony queen every twelve or so months. Where such swarming is controlled we've observed that most queens are replaced every 12-18 months anyway.

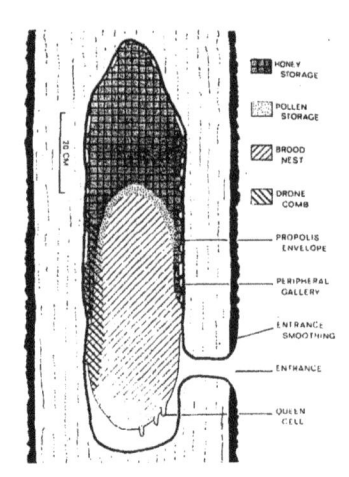

Figure 1.123   Classic Seeley-Morse wild hive.

So we learn that:
- to maintain colony vigour, honey bees replace their queens regularly;
- honey bees can, in theory, support more than one queen;
- Demareeing[301] (artificial swarming) – where the queen is isolated from most of her brood using an excluder – can give rise to a two-queen colony; and
- honey bee colonies with an extra queen have a strong tendency to revert to the single-queen condition.

Farrar showed very elegantly that honey yield correlates closely with bee numbers[302] (see Figure 1.6 in *Long live the queen*) demonstrating that the slope of yield to bee numbers turns out to be rather greater than one. This equates to doubling the number of bees more than doubling the yield of surplus honey all other things being equal.

How we came to two-queen beekeeping

The lure of an extra queen producing more bees and thus more honey has long fascinated beekeepers but, in execution, had eluded all but the most capable operators.

First though we relate tales of:
- Dannielle Harden's narrative of how a small back yard and not enough room to run all the hives she wanted motivated her to run two-queen hives from the outset; and

- Alan's early foray into running about eight two-queen hives on a giant honey flow back in the early 90s and getting a tonne of honey[303].

There is much more information on doubled and two-queen hives in Alan's book published by *Northern Bee Books*.

Rivers of honey aside, we have been severely chastened by the natural proclivity of bees to revert to their ancestral single-queen condition. An extraordinary once in a lifetime Southern Hemisphere 2021-2022 bumper season has taught us ever new lessons. Bees really intent on swarming – even with two-queen colonies – did so irrespective of the beekeeping practices we adopted.

The doubled versus the two-queen hive

Two distinctive types of multiple queen hives have been discovered:
- hives where single-queen brood chambers, any number, are bridged with common honey storage supers; and
- hives where there are one or more extra queens in the same colony.

Until towards the end of the 19th Century, the conventional wisdom was that hives could only ever be operated with the one queen. We hark back to the dawn of their discovery 140 years ago.

Firstly in April 1892 George Wells[304] from Aylesford, Kent reported successfully operating 'doubled hives'. We can describe these sorts of hives as a line of (mostly two) single-queen hives capped with bridging honey supers (sometimes large single 'coffin' supers) over 'excluder zinc'. These captured the early fruit tree blossom flow and powered into summer flora – presumably clover, brambles and hedgerow – producing 'astounding amounts of honey'. They are predicated on splitting and pairing of single queen hives in autumn.

Then twenty five years later in 1907 E.W. Alexander[305] from New York State and Craudh[306] from Ballyvarra, Ireland independently reported hives with 'a plurality of queens', the early progenitors of the two-queen hive. This is a more complex affair (two or more queens in the one hive) but they were variously designed to build bees to giant colonies by the commencement of the main flow (e.g. aster and goldenrod in the US) and in the UK and Southern Ireland (clover and the late heather flow).

We need not dwell on the detail or the long history of development of the seemingly impossible two-queen hive except to recall Alexander's remarkable achievement of introducing 'any number of queens' to an already queenright hive. Ellis and Medicus[307] accounts of the late heather flow demonstrated that even the standard slow development of two-queen hives could greatly increase honey crops. These schemes reached an apogee in the latter half of the 20th Century with the development of the Consolidated Brood Nest two-queen hive – the CBN hive – formulated by Floyd Moeller and John Hogg. Their schemes put all the brood together, all honey supers atop, mimicking single-queen hive operation.

The essential message is that these pioneer apiarists were able to generate colonies of extraordinary strength. Their hives produced rivers of honey.

The chastening reality was that maintaining the two queen condition was well beyond the skills of all but the most capable and dedicated beekeepers. The exceptional demand on attending to the needs of such colonies – building tower colonies or employing non standard gear, timing operations, taking off the crop promptly and adapting to end of flow conditions – proved to be too much effort.

These requirements have been somewhat ameliorated. Any competent beekeeper can now run hives with an extra queen using standard gear, but it is important to realise that running such hives is, and will always be a challenge of the first order and require some beekeeping acumen.

Setting up and running hives with two queens

The doubled hive

The essential element of this scheme (using regular gear) is autumn preparation of contiguous single queen hives and the set up of straddled supers on or about the first day of spring, that is 1 September in the Antipodes and 21-22 March in much of the Northern Hemisphere. Set up of doubled hives is like falling off a log. The provisos are autumn requeening, disease checks and close attention to optimising early spring nutrition. Two queens require double stores and double pollen input to build for the flow.

Here is a traditional scheme (Figure 1.124) for running a doubled hive year round, good in principle but rather harder to achieve in practice.

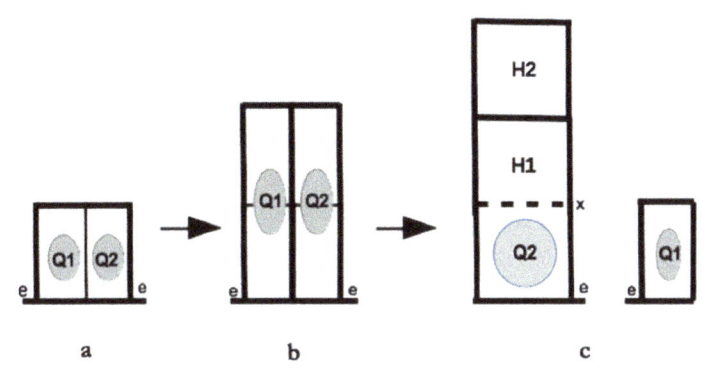

Figure 1.124 Traditional doubled hive scheme for honey production:
e = entrance; x = excluder; Q1, Q2 = queens; H = honey supers:
(a) overwintered doubled – centrally divided – colony in early spring;
(b) colonies are built to full ten or twelve-frame condition prior to the main flow; and
(c) colonies are combined as a single-queen hive with sealed brood, and most of the bees are supered for the flow: the spare queen is offset to a nucleus hive.

A simpler and modern start up option is to run pairs of closely juxtaposed single-queen hives – prepared in autumn – rearranging them at the first opportunity in late winter or early spring here shown (Figure 1.125) for an eight-frame set up. Larger nine, ten or eleven frame hives (such as the British National) can be overwintered as single story hive pairs.

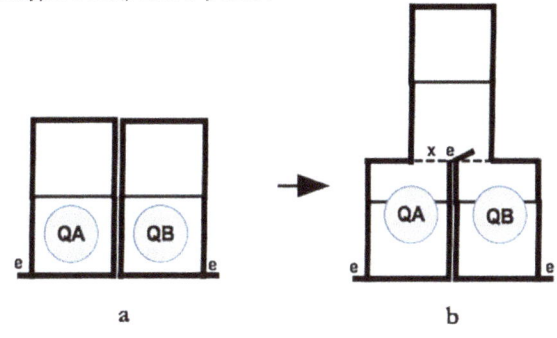

Figure 1.125 Modern doubled hive scheme for honey production:
e = entrance; x = excluder; Q1, Q2 = queens; H = honey supers:
(a) overwintered pair of juxtaposed eight-frame hives prepared in autumn; and
(b) double hive colonies united and reorganised in early spring using additional shallow brood boxes to allow queens to lay to maximum capacity.

The two-queen hive

Setting up and running two-queen hives is much akin to herding cats. Where there is a long lead-in time to the main flow – say an autumn heather flow – a second queen can be established as a nucleus above a division board (preferably a double screen) over a strong colony in spring. Once well established the two colonies can be united by simply replacing the double screen with a queen excluder (or using a queen excluder and newspaper if a nuc/division board is used).

We often think back to the late 1960s and Robert Banker[308] running 1500-1600 two-queen hives and having another 1000 to 1200 nucs and packages to back up his operation as some measure of the challenge of running two-queen hives.

Here is a much simpler scheme (Figure 1.126) we use for starting a two-queen hive, one that – like the doubled hive setup – gives bees an early start to achieve a crop on spring blossom flows. Note we run the two brood nests together so that – with brood below and honey supers atop – the hives can be run in the same way as single-queen hives are operated.

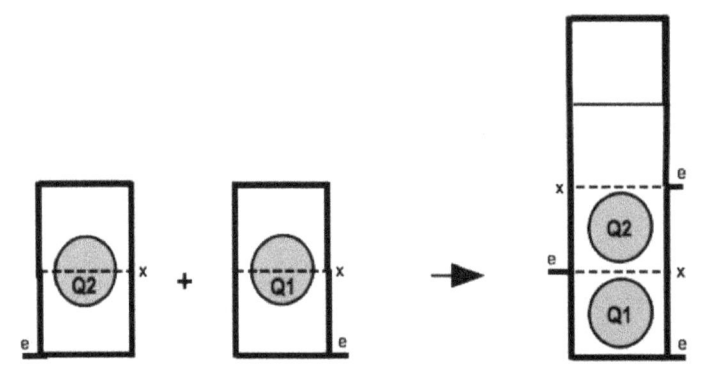

Figure 1.126 Two-queen hive system formed by uniting juxtaposed overwintered pair of hives to form an advanced Consolidated Brood Nest two-queen hive:
e = entrance; x = excluder; Q1 and Q2 either an old queen and an introduced queen or two autumn-introduced queens: e = entrance; x = excluder; Q1, Q2 = queens:
(a) frames in each colony are sorted so that all unsealed brood is located in bottom brood chambers; and
(b) the colonies are united by piggybacking these overwintered doubles using newspaper to unite the colonies and an excluder to keep the queens apart.

Here are a couple of two-queen hives (Figure 1.127) then working (summer 2021-20220 Yellow Box (*Eucalyptus melliodora*) and Blakely's Red Gum (*Eucalyptus blakelyi*).

Figure 1.127 Two of three hives working eucalypts in full cry at Hillside Station, Symonston, Australian Capital Territory 23rd of December 2021.
Note offset support nuc

Where to for multiple queen hive operation

While you may claim that operating colonies with extra queens on the prairie in Canada or in the best beekeeping locations in southeastern and southwestern Australia is a given, this is emphatically not the case. We have had some success and notable failures and we think we have a modicum of understanding of how such hives operate and why – in real time – they have sometimes failed. Commercial beekeepers we know ever only use them for seamless requeening operations.

Here is a notional hive condition trend predictor chart (Table 1.6) based on colony queen status. Note that outcomes such as nectar inflow can be temporally and locally condition dependent – a strong colony in a period of dearth may result in stores being consumed quite quickly when queens may also stop laying.

Hive category	Number of laying queens	Population trend	Honey production potential	Swarming potential	Requeening trend
Non laying Queen	0	↓	-	0	0 → 0
Single-Queen	1	↑	+	↑↑	1 → 1
Doubled Hive	1 + 1	↑ ↑	+ +	↑	2 → 1
Two-Queen	2	↑↑	+ + +	↑	2 → 1

Table 1.6 Relative colony performance of colonies of different queen status. Note: A colony just made queenless may abscond.

Getting a good honey crop off any hive is always a gamble, but good practice – and being able to build bees before the flow – makes some beekeepers using traditional single-queen hives more successful than others.

While doubled hives are easy to set up and run – as well as fix when they go awry – running two-queen hives is more akin to Russian roulette. Keeping control of queens and getting bees to super strength – especially for early flows is like starting all over again in beekeeping. The schemes we have outlined are at times nuanced, but we do owe it to British and American beekeepers for inventing the multi-queen hive.

We leave you with a famous early discussion[309] between *American Bee Journal* editor Bud Cale and two-queen hive aficionado Clayton Farrar:

> Cale: *Isn't your two-queen system of management too much work?*
>
> Farrar: *If you are interested in getting those kind of crops, you will find a way to adapt yourself to the management.*

Part II
Byways

Most scientific avenues of bee enquiry are more about bees than keeping them. *Byways* tells us about different types of honey bees – including the ones we do not keep – as well as some of the many pests, predators and diseases of kept bees including native stingless bees. It also tells us about the lives of a handful of keepers of bees whom I have encountered whose insights might more greatly influence the way we keep bees.

I   Bee Back Lanes

Most people keep bees to produce honey, pollinate their vegetables and provide an income. Others keep bees for very different reasons, but mainly because they are curious and want to know more about how bees work their magic. Maybe bees are kept because their social ordering tells us a whole lot about ourselves.

Bee back lanes is a serendipitous collection of articles about bees, definitively not a standard text on how to manage them.

Initially I was drawn to the topic of where honey and other social bees fit into the large class of insects, delving briefly into one arm of the insect order Hymenoptera. These, the Apocrita, the ants, bees and wasps, comprise the bulk of wasp-like insects. That naturally led me to a review of the distribution of honey bees – a study extended to an overview of the giant honey bees.

It seemed equally natural to then turn my attention to the many things that pillage social bee stores, their predators, and the parasites and the disease organisms they host. These comprise the obvious pests, the mites, the beetles and the moths that make a free lunch of bees and bee stores. To these are added lists of pests and diseases of honey bees and of native bees, an extensive smorgasbord of free-loaders.

II   Beekeeper Sketches

Amongst myriad beekeepers of yore have been few heroic souls whom we all remember: Langstroth, Huber, Demaree, Farrar, Butler, Morse, Seeley... Their style of keeping bees always defied convention but their stories have been told. Of myriad others who contributed to accepted beekeeping practice, we have no record: their tales have been lost if not all their methods.

> *Some village-Hampden, that with dauntless breast*
> *The little tyrant of his fields withstood;*
> *Some mute inglorious Milton here may rest,*
> *Some Cromwell guiltless of his country's blood.*

Elegy Written in a Country Churchyard
Thomas Gray

Character sketches paints images of a handful of near contemporary beekeepers, inventive but lesser known souls, who have most strongly influenced my idiosyncratic approach to keeping bees. All took the road not taken.

I
Bee Back Lanes

Bee Back Lanes introduces us to the honey bees, their natural ranges, to their keep and to their many opportunistic animals, pests, parasites and diseases, scoping those that bees and beekeepers alike encounter.

Social organisation of ants, bees and wasps[310]

The origins of eusociality

We often take for granted that bees work cooperatively and then of their own volition? We would be quite surprised were we to hive a swarm only to discover they failed to stay as a group and get on with the serious business of squirrelling away resources, honey and pollen, for leaner times.

In this beyond-social-cooperation scenario, so-called eusociality[311], the altruistic worker bees stay at home to provision their parents' offspring or venture out to provision the colony and give up the opportunity to reproduce themselves. They stay ever faithful to their mother when she departs with her swarm.

Understanding their community dynamic is instructive, if only to give us advance warning of global honey bee disruptors such as *Varroa* and *Tropilaelaps* and the viruses they carry. It is sobering to realise that, before jumping ship to our bees, these parasites evolved with their also socially sophisticated cousins, species comprising the Asian Honey Bee and the Giant Honey Bee.

The introduction of the Western Honey Bee from the Old to the New World is the reason there are beekeeping clubs across the globe and that the world at large is provided with scuds of honey and beneficial crop pollination. As collateral they affect the way our forests are structured, they compete with other insects and birds and, as efficient pollinators, they are an unwitting weed promoter.

But bees are the just the tip or, rather, right at the top of the social iceberg. Out there, there are a plethora of social lookalikes both amongst other bees and also the ants and the wasps. The evolution of social castes, where offspring are partly or entirely subordinate to the sexually dominant parents is an extremely rare phenomenon. So much so that the origin of true social behaviour has been tracked back to just nine separate ancestors amongst the wasp-bee-ant community, seventeen amongst thrips and to only eight other lineages in the remainder of the animal kingdom.

The common supposition is that social behaviour arose from the condition where food was plentiful and where getting together meant that the group could fend off predators. In other words the group could make a better fist of surviving than going it alone. But this was not enough. Like a bunch of precocious children, the group had to learn to share resources and to help with the housework. The chances of this stay at home, never-ask-for-a-date strategy working were so remote that by and large insects have retained the very successful strategy of one girl meets one boy to pass on their story to the next generation. Most insects and indeed bees are solitary.

However once rare sociality was initiated, the tide turned (Figure 2.1). To mix two escape metaphors either the 'cat had got out of the bag' or 'the horse had bolted'. Amongst the most socially advanced insects, the division of labour finally evolved to the point where the separate caste of worker (and in the case of some ants also soldiers) is sterile or at least functionally asexual. As Wilson and Hölldobler have put it, workers reached a point of no return where they could no longer pass on their genes. The workers, albeit closely related to their parents, started to care cooperatively for the young and became a non reproductive (or at least less-reproductive) caste incapable of founding new colonies. In this they were not alone: the fertilised queen, except in the case of the likes of the less social bumblebees, could now never found a colony on her own.

Figure 2.1   A social occasion, a 1912 wedding interrupted by bees that had long learnt how to organise their own nuptials[312].

A quick rain check

Social ants, social bees and social wasps dominate the global invertebrate community both ecologically and in terms of other simple metrics such as total biomass. Once started, the evolutionary pressures resulted in the development of menagerie of social lifestyles and an explosion in the number of their species. Collectively they out-compete their much much more diverse and numerous solitary relatives.

The ants, bees and wasps together

Apart from one beetle, a bevvy of thrips, aphids and termites, and a few snapping shrimps, not to mention a couple of African mole rats where the youngsters all stay at home to look after their siblings, the really advanced social animals all belong to a single insect suborder of the Hymenoptera, the Apocrita. These are the 'lacey-winged, narrow waisted, ovipositor-specialist insects' made up of ants, wasps and bees. While ants are all social in varying measure, most bees and wasps are solitary. Social species are conspicuous by dint of their sheer numbers and gregarious habit.

The advanced eusocial honey bees (Apini) are unusual in that they have a single genus, *Apis*. The equally advanced neotropical stingless bees (Meliponini) have of the order of 57 genera while the also diverse South American orchid bees (Euglossini) are mainly solitary, a few species of which are semi-social.

As noted, eusociality has evolved nine times amongst hymenoptera (Table 2.1). Here we can go further and look to the breakdown: three times amongst halictid bees, once amongst the pollen basket (corbiculate) bees: the honey bees, the stingless bees, the primitively eusocial bumble bees and the now solitary or weakly social orchid bees, three times amongst the wasps and only once amongst the ants.

Social bee, wasp and ant progenitors

Social Taxa	Progenitors	Origin date (mya)	Number of queen mates	Number of social species
Halictid bees (Halictidae) sweat bees	3	15-28	Usually 1 1-3 in *Halictus*	140 + 217 + 544
Allodapine bees (Allodapinae)	1	41-65	1?	?
Corbiculate bees (Apidae)	1	78-95 Apini (56-61) Meliponini (12-21) Bombini (28-35) Euglossini (65)	8-44 in *Apis* 1 (e.g.*Tetragonula-Austroplebeia*) 1-3 in *Bombus* 1 (rarely social)	1000
Sphecid wasps (Sphecidae)	1	~140	1	1
Stenogastrine wasps (Stenogastrinae)	1	~80	1	50
Polistine and vespine wasps (Vespidae)	1	~80	Usually 1 2 in *Vespula* spp.	860
Ants (Formicidae)	1	115-135	1	12,000

Table 2.1 Simplified record of ancestral ants and bees[313].

We now turn to look at how these insects are related to one another and touch on some common Australian social ant, wasp and bee taxa.

Eusocial hymenoptera fauna

How can we categorise the roughly 13,000 social ants, wasps and bees? Firstly we can examine their relationship or phylogeny, that is where they came from, in time and space. The chart (Figure 2.2) shows the overall relationships of social members of this large insect group. There are many other families that contain only solitary and parasitic species of which few are social. For those you will need to resort to specialist texts such as CSIRO's *Insects of Australia* and Charles Michener's *The Bees of the World*.

The Hymenoptera are divided into two broad suborders, the Apocrita containing all the ants, bees and wasps and the Symphyta containing the phytophagous (herbivorous) sawflies (Figure 2.2). While gregarious as larvae, the sawflies contain no social species and are a small and poorly diversified group of solitary insects.

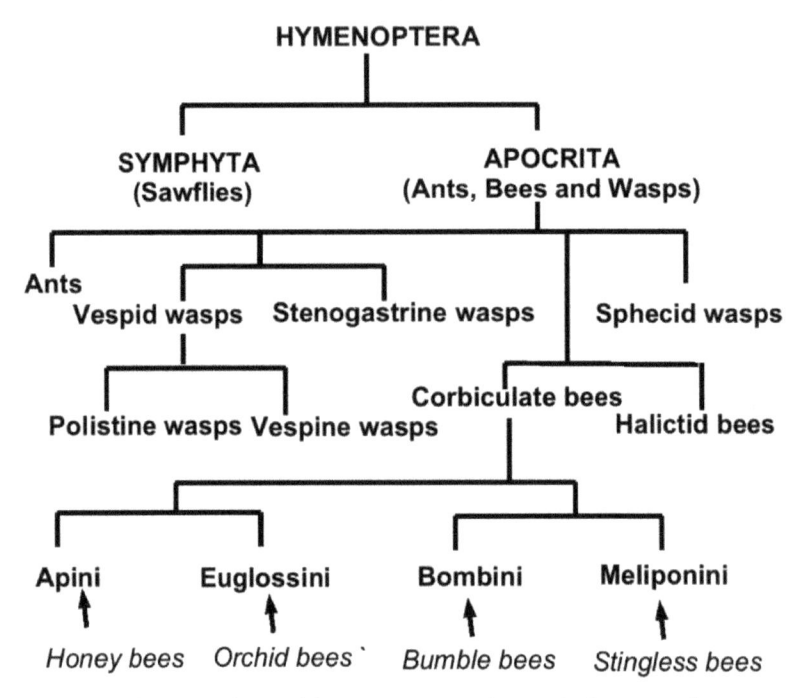

Figure 2.2 Phylogeny of social bees, wasps and ants in insect order Hymenoptera.

The social ants

Ants are the dominant land invertebrate. And they are more than just the nuisance to be contended with in your honey pot or sugar bowl. Think of the ubiquitous Green Tree Ant (*Oecophylla smaragdina*) of the Australian and Asian monsoonal and wet tropics, or of the South American Red Fire Ant (*Solenopsis invicta*) from the Pantanal region of the Paraguay River and introduced to Alabama in the 1930s and to Brisbane in 2001. These are the quintessential barbeque stoppers:

I wake in the morning as soon as 'tis light,
And go to the nosebag to see it's all right'
That the ants on the sugar no mortgage have got,
And immediately sling my old black billy-pot...

The Billy of Tea Anon

Ants are great scavengers and serve a range of important ecological functions. Take the two most conspicuous mound-building ants found in the Australian Capital Territory, the Banded Sugar Ant (*Camponotus consobrinus*) (Figure 2.3a) and the Meat Ant (*Iridomyrmex pupureus*) (Figure 2.3b). Their nests are often co-located.

a   b

Figure 2.3   Social ants found in nature reserves on Mt Taylor near the author's home[314]:
(a) Banded Sugar Ant (*Camponotus consobrinus*); and
(b) Meat Ant (*Iridomyrmex pupureus*).

The Sugar Ant taps aphids and harbours larvae of the Azure Purple Butterfly (Figure 2.4a: iridescent tree-topping male left and female right) in their nests and predate on the Meat Ant. Meanwhile the also omnivorous Meat Ant protects and taps the larvae of the brilliant Imperial Hairstreak Butterfly (Figure 2.4b). The giveaway for this spectacular butterfly is a largely defoliated black wattle such as *Acacia mearnsii*. I've often seen adult butterflies, their pupae and their larvae attended by these ants festooning a largely defoliated wattle.

a   b

Figure 2.4   Butterflies of the Australian Capital Territory tended to by social ants[315]:
(a)   Azure Purple Butterfly (*Ogyris genoveva*); and
(b)   Imperial Hairstreak Butterfly (*Jalmenus evagoras*).

I was once party to a published article on the Meat Ant. We were at Kinchega National Park studying the trophic space – where these scavenging ants fitted into the food web – of that arid landscape. Our ants were bringing home a veritable smorgasbord of delectables. Meat ants scavenge like teenagers, and what they eat was as easy to collect and recognise as the wrappers from muesli bars and icecream cones left in the dustbin. And like adolescents, meat ants are not picky eaters. We found that they harvested the remains of nearly a dozen insect orders including live termites. Never admit to owning dying livestock or suffer the misfortune of becoming a lost shearer out there if you expect to come out alive.

We were happily collecting these tasty tit-bits, that was until a thunderstorm rolled in an hour before dusk. Like bees facing a sudden change of weather, our ants stopped partying and headed home. However, it was soon apparent that some ants were heading away from, and not towards, the safety of their red-coloured subterrainean sand nest. Where were these lost souls off to? A small amount of bush tracking showed they were returning to a small nest over a dune and about 50 m away.

We had first hand evidence that the parent colony had budded off a daughter nest, swarm-like, and were gradually establishing a new colony. This is but one way that social insects start again. Newly mated winged-queens of the Sugar Ant, on the other hand, fly away in droves conferring the advantage of being able to establish new nests far distant from the parent nest.

The social wasps

Social members of the complex assemblage of wasps comprise the bee-like sphecid (thread-waisted) and vespid (stinging) wasps. The vespids are further divided into the vespine, polistine (paper) and stenogastrine wasps all with an independently evolved social ancestor. Common exemplars of the vespine and polestine group are the exotic ground nesting European Wasp (*Vespula germanica*) and the overhang nesting Chinese Paper Wasp (*Polistes chinensis*)[316], both significant honey bee competitors locally. The local native relative *Polistes humilis* has been largely displaced by the Chinese Paper Wasp.

The social bees

Although many wasps and ants are collectors of nectar, honeydew and pollen, bees have abandoned their mainly predator cousins and adapted to collecting nectar and pollen alone or predating on the stores of other bees. By way of example, sphecid wasps, closely related to the corbiculate bees provision their larvae with prey but as adults are nectar and pollen collectors. Of the halictid (sweat) bees that are social, all are at best primitively eusocial. Perhaps the most spectacular Australian halictid species is the solitary blue-banded bee, *Amegilla pulchra*. The genus has no social representatives in Australia but the solitary *Amegilla* species are widespread and common.

The corbiculate (pollen basket) bees contain all of the most advanced eusocial species. They have an ancient lineage with a single progenitor dating back about 90 million years. From this social grouping came the modern tribes, the primitively eusocial bumble bees, the highly developed and very successful stingless bees (Australian aborigines call them sugar-bags[317]) and the similarly socially advanced honey bees. The remaining orchid (long-tongued or buzz pollinator) bees are nearly all solitary and are at best weakly social. Despite this most corbiculates are solitary bees.

One distinguishing feature of the tribes of social corbiculate bees is the variation in their reproductive strategies. The bumble bees produce queens and males over summer while the fertilised queens aestivate over winter and establish new colonies each year. Both the stingless and honey bees reproduce by swarming and both have a single queen, giving new colonies a jump start. However stingless bees continue to provision the new nest – as do meat ants – giving some surety to establishment of the new colony but curtailing the survival and development of the parent colony. Of the approximately fifty seven genera of stingless bees distributed globally, Australia has only two, *Tetragonula* and *Austroplebeia*. Only one species, *Tetragonula carbonaria,* has a range extended to the temperate east coast. On a global scale most species of stingless bees are restricted to the tropics, their greatest diversity being found in the Central Americas.

Honey bees have a special reproductive advantage. Their swarms, once separated from the parent colony, are entirely independent. This has conferred a capacity for rapid geographic dispersal. Resource constraints taken into account, a few species have quickly colonised whole continents. However it is really only a handful of temperate races of the cavity dwelling honey bees, *Apis mellifera* (e.g. Italian and Carniolan bees) and *Apis cerana* (e.g. Chinese and Japanese bees) that restrict their swarming behaviour to reproduction.

The giant honey bees as well as some, notably African, races of the Western Honey, most species and races of Asian honey bees and the dwarf honey bees all form absconding swarms. They swarm not only reproductively but also in response to dearth and in response to disturbance and predation. Absconding behaviours are canvassed more fully in the chapter on giant honey bees.

Global distribution of honey bees[318]

Honey bees of the *Apis* genus are eusocial animals, that is they are socially highly-organised raising brood cooperatively and intergenerationally and are specialised in their division of labour. They range naturally across Africa, Europe and Asia, are absent at high latitudes, but are not native to the Americas and the Pacific nations of the New World: the original distribution of *Apis mellifera* and *Apis cerana* is depicted in Figure 2.5[319].

Figure 2.5
Native range of the Western Honey Bee (*Apis mellifera*) [left Africa-Europe] and of the Asian Honey Bee (*Apis cerana*) [right Asia-China-USSR].

The oldest records of honey bee species (*Apis*) appear at the Eocene–Oligocene boundary 34 million years ago[320]. Hepburn and Radloff[321] argue that more recent adaptations of the honey bees must be considered in terms of speciation and radiation of modern *Apis* species, that is in a Pleistocene context. That epoch spans the period from 2.5 million years ago up until the present unofficially recognised Anthropocene.

Charles Michener's *The Bees of the World* recognises three large groups of extant taxa (see *The giant honey bees* for a simple listing of existing known honey bee species):
- small species with single exposed combs; dances on expanded horizontal base of comb: *Apis florea* Fabricius, *Apis andreniformis* Smith;
- large species with single exposed combs; dance on vertical curtains of bees or on comb: *Apis dorsata* Fabricius, *Apis laboriosa* Smith, *Apis dorsata binghami* Cockerell, *Apis breviligula* Maa. The last two appear to be allopatric segregates of *Apis dorsata*; and

- medium-sized species with multiple combs in cavities; dance on vertical surfaces of combs in the dark: *Apis mellifera* Linnaeus, *Apis cerana* Fabricius, *Apis koschevnikovi* Buttel-Reepen, *Apis nigrocincta* Smith, *Apis nuluensis* Tinget, Koeniger and Koeniger, *Apis indica* Fabricius.

As noted elsewhere, the last two (*Apis nuluensis* and *Apis indica*) have only recently been recognised as specifically distinct from *Apis cerana*.

Biogeography of *Apis* species

The most wide-spread honey bee species, *Apis mellifera*, *Apis cerana* and *Apis dorsata*, have a number of geographically delineated races and halotypes, their intraspecific differences being supported by both morphometric and genetic studies. The most geographically restricted species belong to the cavity dwelling honey bees, *Apis koschevnikovi*, *Apis nigrocinta* and *Apis nuluensis* (or *Apis cerana nuluensis*) and *Apis indica* have diverged from *Apis cerana* as separate species in relatively recent times while a few giant honey bees have similarly separated from *Apis dorsata*. They, the giant honey bees, include the Giant Mountain Honey bee *Apis laboriosa* inhabiting high mountainous terrain on the southern Asian subcontinent, the Giant Phillipines Honey Bee *Apis breviligula* and a giant honey bee from the *Sulawesi* nominated *Apis dorsata binghami*. The primitive giant honey bee subgenus is reviewed separately.

Overall the greatest biodiversity of honey bee species is found in the South China Sea region, that is in and around the Phillipines, Kalimantan and the Sulawesi and is where the ancestral genus is likely to have originated and radiated from.

With the rise of agrarian societies during the Holocene (the geological epoch spanning the last 11,500 years), honey bees have been transported far and wide: Temperate races of *Apis mellifera* and *Apis cerana* have been employed extensively for pollination and for honey production.

The contemporary range of the *Apis* genus is shown in Figure 2.6. While many social bee species, including *Apis* species (other than *A. cerana* and *A. mellifera*), have also been exploited for both honey production and pollination services, their geographic range is typically contracting due to habitat destruction and over exploitation. Most *Apis* species, as well as *Apis* subspecies of *Apis mellifera* and *Apis cerana*, have a very high swarming and absconding propensity. Together, these factors have limited their usefulness and, in some cases, their very survival is threatened.

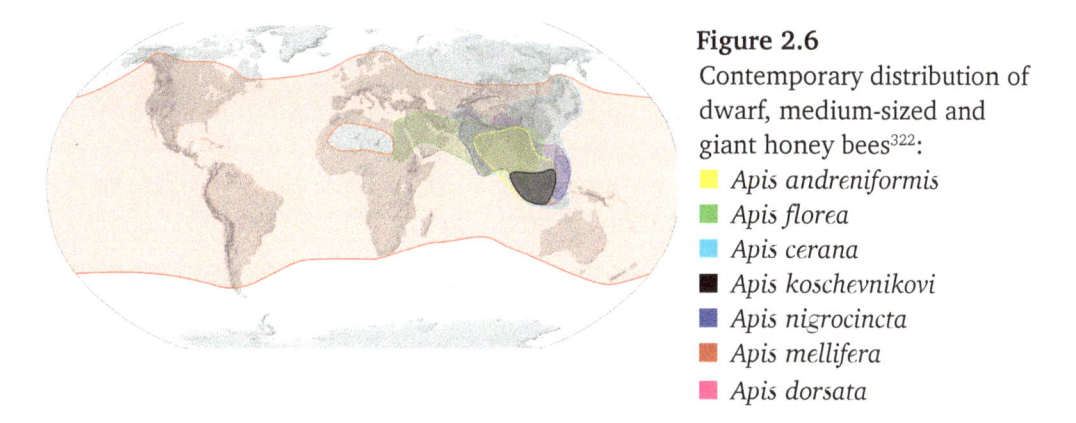

Figure 2.6
Contemporary distribution of dwarf, medium-sized and giant honey bees[322]:
- *Apis andreniformis*
- *Apis florea*
- *Apis cerana*
- *Apis koschevnikovi*
- *Apis nigrocincta*
- *Apis mellifera*
- *Apis dorsata*

Natural range of species and subspecies of honey bees

Honey bees of the modern era fall into three subgenera, the less socially complex giant honey bees (Megapis) that nest on a single exposed comb, the advanced cavity dwelling honey bees (Apis) that include Asian and Western honey bees on multiple combs and the least socially evolved dwarf honey bees (Micapis) also restricted to a single comb but confined to much of the Old World tropics. Both *Apis cerana* and *Apis mellifera* have extended their original range into temperate and arid climes by taking advantage of their sheltered nesting habit, by hoarding and storing honey (carbohydrate) and pollen (protein) in brief times of abundance and by complex absconding and reproductive swarming behaviours.

The contemporary delineation of honey bee species and subspecies is based on morphometric and genetic studies that take into account their geographic origins. Cervancia[323] describes eight species of honey bees: *Apis mellifera* Linnaeus, *Apis cerana* Linnaeus, *Apis nuluensis* Tingek, Koeniger, Koeniger, *Apis nigrocincta* Smith, *Apis andreniformis* Smith, *Apis florea* Fabricius, *Apis dorsata* Fabricius and *Apis laboriosa* Smith.

So how many distinct honey bee species are there? Arias and Sheppard[324] suggest there is sufficient phylogenetic evidence to indicate that there are ten distinct living species, namely the giant bees (*Apis dorsata*, *Apis binghami* and *Apis laboriosa*), dwarf bees (*Apis andreniformis* and *Apis florea*), and cavity-nesting bees (*Apis mellifera*, *Apis cerana*, *Apis koschevnikovi*, *Apis nuluensis* and *Apis nigrocincta*).

With the more recent addition of the Giant Philippine Honey Bee (*Apis breviligula*) and the Indian Plains Honey Bee (*Apis indica*)[325], we can conclude that there are twelve separate honey bee species. As neither a taxonomist, nor a molecular geneticist nor an evolutionary biologist, I've simply come up with a candidate listing (see Table 2.2 'Extant honey bees' in the section on *The giant honey bees*).

Honey bees harbour a number of parasitic arachnids, so-called mites reviewed under *Phoretic honey bee mites*. Bacterial, fungal, microsporidian, protozoan and viral diseases affecting *Apis* species are often of obscure origin and their associations with hymenoptera (bees, wasps and ants) are many. They are simply listed among honey bees pests and diseases under *Bee pests and diseases*.

The dwarf honey bees (subgenus Micrapis)

There are two dwarf honey bees, the more western and the more widely distributed *Apis florea* and the more eastern *Apis andreniformis*. Micrapis are considered the most primitive of honey bees, constructing a single horizontal comb.

Apis florea ranges from continental Asia and Africa, southeastern Asia: Thailand, Iran, Oman, India, Myanmar, and some parts of China, Cambodia, and Vietnam and forested regions of the Middle East (Figure 2.7).

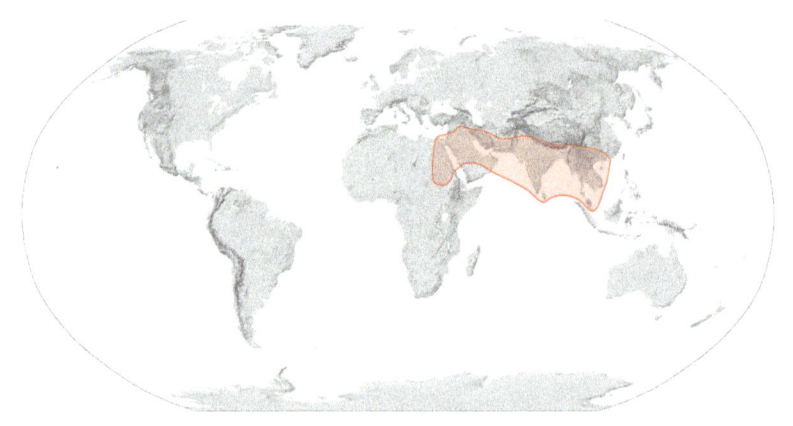

Figure 2.7　　*Apis florea* distribution[326].

Apis andreniformis is found more easterly though it is sympatric with *Apis florea* in eastern India through SE Asia down to the end of the Malaysian Peninsula. *Apis andreniformis* tends to live at higher altitudes, its habitat being principally in Southeast Asia: southern China, India, Burma, Laos, Vietnam, Thailand, Malaysia, Indonesia, and the Philippines (Figure 2.8).

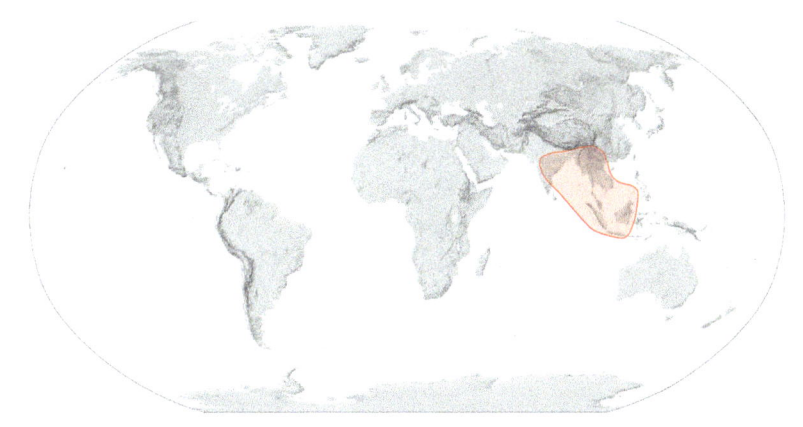

Figure 2.8 *Apis andreniformis* distribution[327].

Medium-sized, cavity-dwelling honey bees (subgenus Apis)

The medium-sized, cavity-dwelling species comprise a cluster of four or five species. The best known and most widely distributed species, *Apis cerana* and *Apis mellifera*, are of common, if disputed evolutionary origin. Many of their nest structural, pheromonal and behavioural characteristics are shared, although there are also many subtle differences[328]. There are three, perhaps four, other geographically distinct species, *Apis indica*, *Apis koschevnikovi* and *Apis nigrocincta*, (also *Apis nuluensis*, if it is regarded as a separate taxon), all closely allied to *Apis cerana*. This species complex has been well researched and the breakdown of subspecies follows essentially that of Engel[329].

Apis cerana, the Asian Honey Bee is widely distributed with eight, geographic subspecies (Figure 2.9):

Their parasites include the microsporidian *Nosema ceranae*; and the arachnids *Varroa destructor* (two genotypes have crossed over to *Apis mellifera*), *Varroa jacobsoni*[330] and *Acarapis woodi* that has crossed over from the Western Honey Bee.

The distribution of *Apis cerana* subspecies, like that of the Western Honey Bee, *Apis mellifera*, is remarkable:
Apis cerana cerana (the Asian Honey Bee) – Afghanistan, Pakistan, north India, and south along the central deserts and mountain ranges, across most of central and southern China, along the eastern edge of Asia up to Korea and Ussuria and south to northern Vietnam.
Apis cerana heimifeng (the Black Chinese Honey Bee) – Central China at relatively high elevation: northern Sichuan Province, southwestern Gansu.

Province, and Eastern Qinghai Province.

Apis cerana japonica – (the Japanese Honey Bee) – Japan, Thailand.

Apis cerana javana (the Javanese Honey Bee) – Java to Timor, now also PNG, Irian Jaya and NE Australia.

Apis cerana johni (the Sumatran Honey Bee) – Sumatra.

Apis cerana skorikovi (the Himalayan Honey Bee) – Central and east Himalayan mountains (Himalayan Uplift).

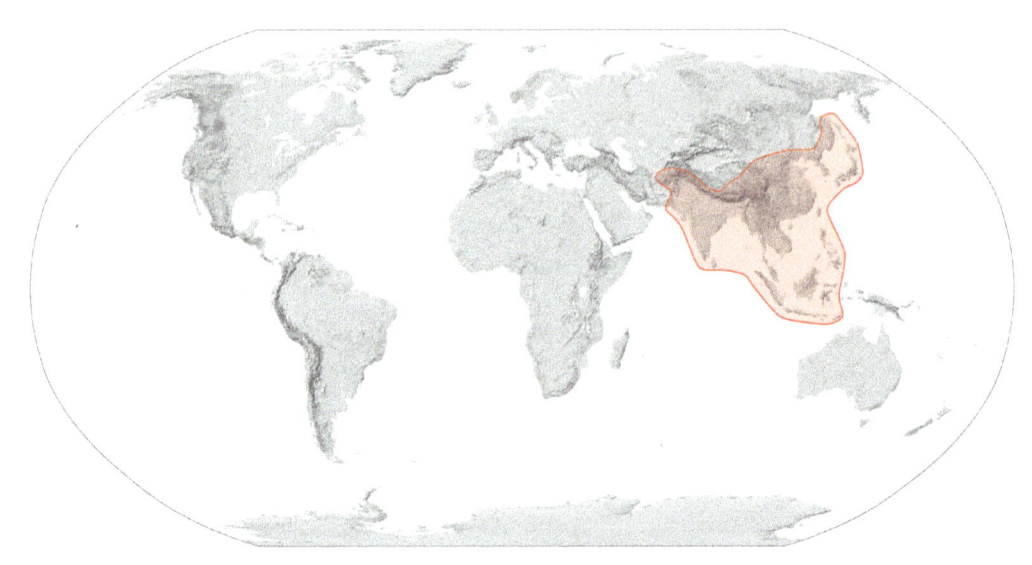

Figure 2.9 Morphometric clusters of the Asian Honey Bee, *Apis cerana*[331]. This map does not delineate the range of the more recently assigned species *Apis indica* (Indian Plains Honey bee) or *Apis nuluensis* (Malaysian Mountain Honey bee of Sabah).

Apis indica (Plains Honey Bee of India)

Originally designated *Apis cerana indica*, its range extended from South India, Sri Lanka, Bangladesh, Burma, Malaysia, Indonesia (presumably excluding Java and the archipelago east) and the Philippines. While *Apis indica* is clearly delineated from *Apis cerana*, the range of the Plains Honey Bee has not yet been formally defined.

Apis nuluensis (Malaysian Mountain Honey Bee)

The Malaysian Mountain Honey Bee is restricted to a mountainous Mt Kinabalu region of Sabah (1400-4100 m) in the state of Sabah in East Malaysia (Figure 2.10). Its status as a separate taxon remains uncertain. This particular bee, together with *Apis nigrocincta,* is host to *Varroa underwoodi*.

Figure 2.10   *Apis cerana nuluensis* distribution[332].

Apis nigrocinta (the Philippine Honey Bee) – Sulawesi and the nearby islands of Sangihe, and the Phillipines (Mindanao) (Figure 2.11). *Apis nigrocinta* is parasitised by *Varroa underwoodi*.

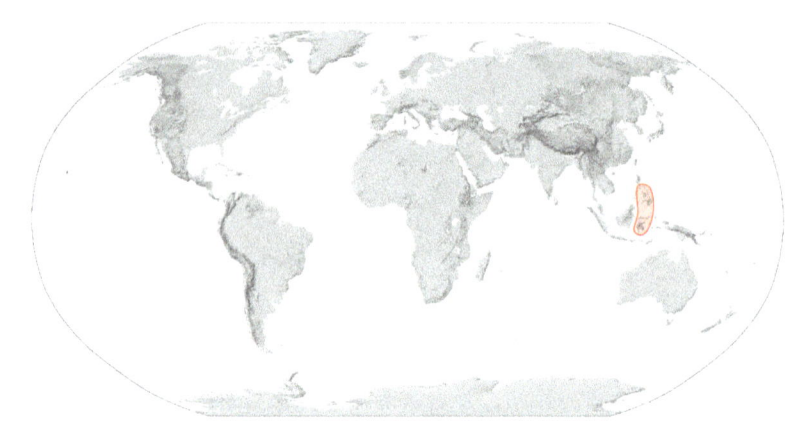

Figure 2.11   *Apis nigrocincta* distribution[333].

Apis koschevnikovi (Koschevnikovi's Honey Bee) (Figure 2.12). This species is the natural host of *Varroa rindereri* but has spread to *Apis cerana* and is now in Irian Jaya and Papua New Guinea[334]). Its natural range includes the Malaysian Peninsular, Sumatra and also Kalimantan where it is sympatric with *Apis cerana nuluensis* (*Apis nuluensis*). It is not domesticated and is probably more under threat from habitat destruction than to have extended its range.

Figure 2.12   *Apis koschevnikovi* distribution[335].

***Apis mellifera* (Western Honey Bee)** The Western Honey Bee is widespread in the Old World being found naturally in Africa, in the Middle East and in Europe.

Twenty eight, originally geographically isolated, subspecies have been described:

Apis mellifera adami
Apis mellifera adansonii
Apis mellifera anatoliaca
Apis mellifera artemsia
Apis mellifera capensis
Apis mellifera carnica
Apis mellifera caucasia
Apis mellifera cecropia
Apis mellifera cypria
Apis mellifera iberiensis
Apis mellifera intermissa
Apis mellifera jemenitica
Apis mellifera lamarckii
Apis mellifera ligustica
Apis mellifera litorea

Apis mellifera macedonica
Apis mellifera meda
Apis mellifera mellifera
Apis mellifera monticola
Apis mellifera pomonella
Apis mellifera remipes
Apis mellifera ruttneri
Apis mellifera sahariensis
Apis mellifera scutellata
Apis mellifera siciliana
Apis mellifera simensis
Apis mellifera sosimii
Apis mellifera syriaca
Apis mellifera unicolor

Whitfield and co-workers have described the origins of the Western Honey Bee (*Apis mellifera*) and its European, Asian and African races classifying them into four branches[336] based on the pioneering morphological studies of Ruttner and co-workers (Figure 2.13). A follow up study by Cridland, Tsutsui and Ramíre[337] further elucidates the origins of the Western Honey Bee and builds on this hypothesis. Brother Adam elaborates on the characteristics of many characteristics of various subspecies from his extensive travels in Europe and North Africa[338].

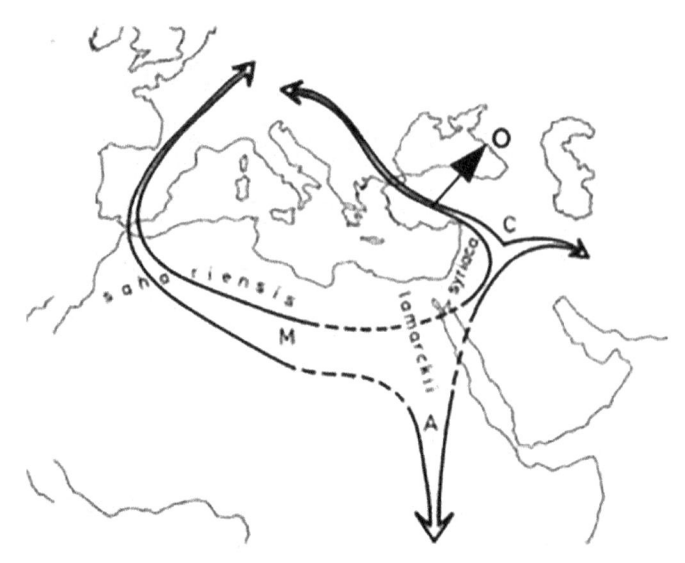

Figure 2.13 Morphometric clusters of the Western Honey Bee subspecies, *Apis mellifera*: African subspecies (branch A), northwestern European subspecies (branch M), southwestern European subspecies (branch C) and Middle-Eastern subspecies (branch O)[339].

Northern European subspecies

Apis mellifera mellifera (the German Black Bee or the European Honey Bee) ranges from central Europe north of the Alps to western southern France, southern Sweden, British Isles.

Southern European subspecies

Apis mellifera artemsia (the Russian Steppe Honey Bee) – Central Russian Steppes
Apis mellifera carnica (the Carniolan Honey Bee) – Carniola region of Slovenia, the Eastern Alps and northern Balkans: Yugoslavia, Romania
Apis mellifera cecropia (the Greek Honey Bee) –
Greece and surrounding Aegean Islands
Apis mellifera iberiensis (the Iberian Honey Bee) – Iberian Peninsula (Spain and Portugal)
Apis mellifera ligustica (the Italian Honey Bee) – Italian Peninsular
Apis mellifera macedonica (the Macedonian Honey Bee) – Northern Greece (Macedonia and Thrace), Republic of Macedonia
Apis mellifera ruttneri (the Maltese Honey Bee) – Maltese Islands
Apis mellifera siciliana (the Sicilian Honey Bee) – Sicily

Apis mellifera sosimii (the Ukrainian Honey Bee) – Ukraine and northern Caucasus Mountains

Middle Eastern and Asian subspecies

Apis mellifera adami (the Cretian Honey Bee) – Crete
Apis mellifera anatoliaca (the Anatolian Honey Bee) – central region of Anatolia in Turkey and Iraq extending as far east as Armenia
Apis mellifera caucasia (the grey Caucasian Honey Bee) – the Caucasus Mountains
Apis mellifera cypria (the Cyprian Honey Bee) – Cyprus
Apis mellifera meda (the Median Honey Bee) – Iran, Iraq, southern Turkey and Northern Syria
Apis mellifera pomonella (the Tien Shan Honey Bee) – Tien Shan Mountains in Central Asia
Apis mellifera remipes (the Yellow Armenian Honey Bee) – Armenia
(This subspecies is similar to and possibly the same as *Apis mellifera intermissa*.)
Apis mellifera syriaca (the Syrian Honey Bee) – Near East and Israel

African subspecies

Apis mellifera adansonii (the West African Honey Bee) – Niger, Senegal, Zaire
Apis mellifera capensis – (the Cape Honey Bee) South Africa
Apis mellifera intermissa (the Tellian Honey Bee) – northern coast of Africa as far west as Morocco, Libya, Tunisia bordered by the Atlas Range
Apis mellifera jemenitica (the Arabian or Nubian Honey Bee) – Somalia, Uganda, Sudan, Yemen
Apis mellifera lamarckii – (Lamarck's Honey Bee) – Nile valley: Egypt and Sudan
Apis mellifera litorea – (East African Coastal Honey Bee) – e.g. Kenya
Apis mellifera monticola (the East African Mountain Honey Bee) – High altitude mountains of East Africa:Kenya, Tanzania: Mt Elgon, Mt Kilimanjaro, Mt Kenya and Mt Meru
Apis mellifera sahariensis (the Saharan Honey Bee) – Moroccan desert oases of Northwest Africa, southern side of Atlas Range
Apis mellifera scutellata (the African Honey Bee) – widespread from South Africa to Somalia in eastern Africa. Its introduction to the Americas is reviewed by Winston[340]
Apis mellifera simensis (the Ethiopian Honey Bee)
Apis mellifera unicolor (the Madagascan Honey Bee) – Madagascar

The giant honey bees (subgenus Megapis)

There are three, perhaps four, species of giant honey bee. *Apis dorsata* is widespread across the Indian subcontinent to Southeast Asia: China, Indonesia, India, Pakistan, and Sri Lanka and some parts of the Philippines.

The *Apis dorsata* group has evolved with geographic isolation. It is reviewed in *The giant honey bees* where their distribution and their biology is explored:

Apis dorsata (the Giant Honey Bee) – widespread across most of Indian subcontinent and much of southern Asia
Apis dorsata binghami (the Sulawesi Giant Honey Bee) – Sulawesi
Apis breviligula (the Philippine Giant Honey Bee) – Philippines excluding western Palawan Island group
Apis laboriosa (the Giant Mountain Honey Bee or Himalayan Cliff Honey Bee) – Vietnam, Myanmar, Laos, southern China, eastern India (Nagaland, near Saramati Mountain)

Ancestral honey bees

There are a number of extinct honey bee species known from the fossil record, one from the New World (*Apis neartica*)[341]. These species fall into three subgenera, two unrelated to extant species groupings.

Fossil *Apis* species (subgenus Cascapis)

Apis armbrusteri - (Miocene) Germany, Europe
Apis neartica – (Miocene) Nevada, North America

Fossil *Apis* species (subgenus Megapis)

Apis lithothermaea – (Miocene) Japan

Fossil *Apis* species (subgenus Synapsis)

There are six well-documented extinct honey bees:
Apis cuenoti – (Oligocene) Europe
Apis henshawi (Henshaw's Honey Bee) – (Oligocene) Creské Stredhori Mountains of Czech Republic, Europe

Apis longtibia (the Long-legged Honey Bee) – (Miocene) Shandong Province, China

Apis miocenca (the Chinese Miocene Honey Bee) – (Miocene) Shandong Province, China

Apis petrefacter (the Petrified Honey Bee) – (Miocene) Creské Stredhori Mountains of Czech Republic, Europe

Apis vetustus (the Aged Honey Bee) – (Oligocene) Germany, Europe

Putative fossil *Apis* species

Other now extinct species have been cited in the literature, though their actual past existence is less certain:
Apis catanensis
Apis enigmatica
Apis trigona

The giant honey bees[342]

Part I The sting is in the tail

In an extension to the review of the *Global distribution of honey bees*, my attention was diverted to a world of bees in microcosm, the tropically limited range of giant honey bees.

These giant bees have been around for a long time having a lineage that dates back to the mid Miocene (Figures 2.14; 2.15). That was some 14-16 million years ago and that was way before primitive forms of *Homo sapiens* first appeared just 300,000 ybp (years before present).

Figure 2.14   Site of fossil *Apis lithohermaea* Iki Island, Japan.

The *Apis dorsata* group are the whoppers of the honey bee kingdom: they have the largest body mass, the longest and largest stings, by far the largest venom glands and one extra sting barb. They are also exceptionally aggressive and have distinctive behaviours that ward off the most unwelcome intruder.

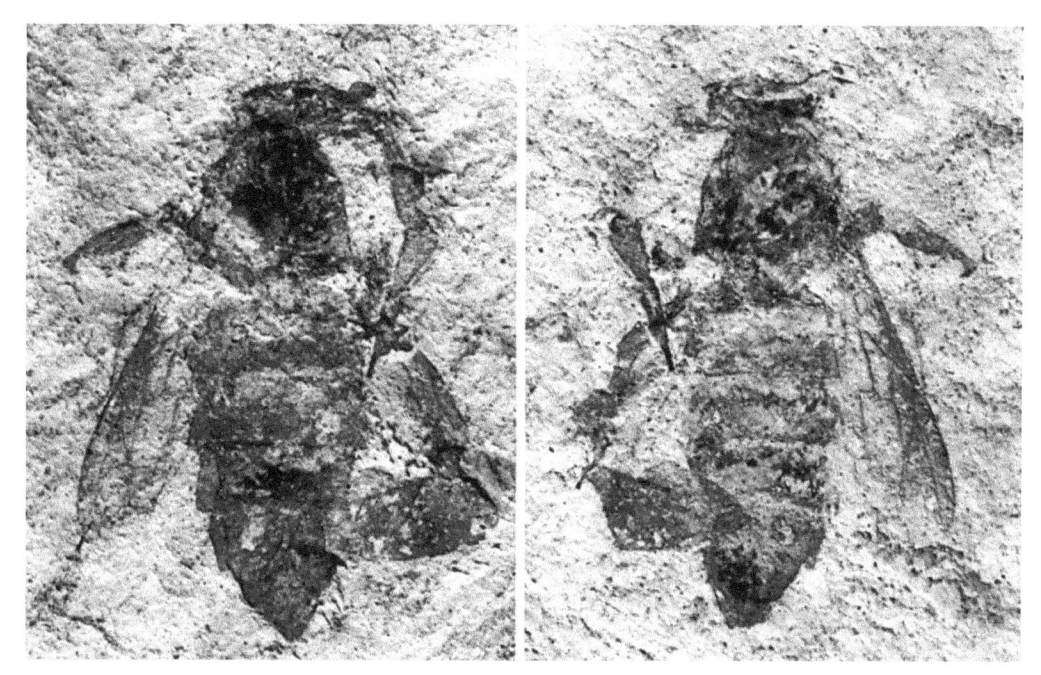

Figure 2.15 Photomicrographs of *Apis lithohermaea*[343].

Engle also traces the evolutionary pathway and affiliation with other honey bee species signalling that they evolved alongside the cavity-dwelling honey bees but after the more primitive dwarf honey bees (Figure 2.16).

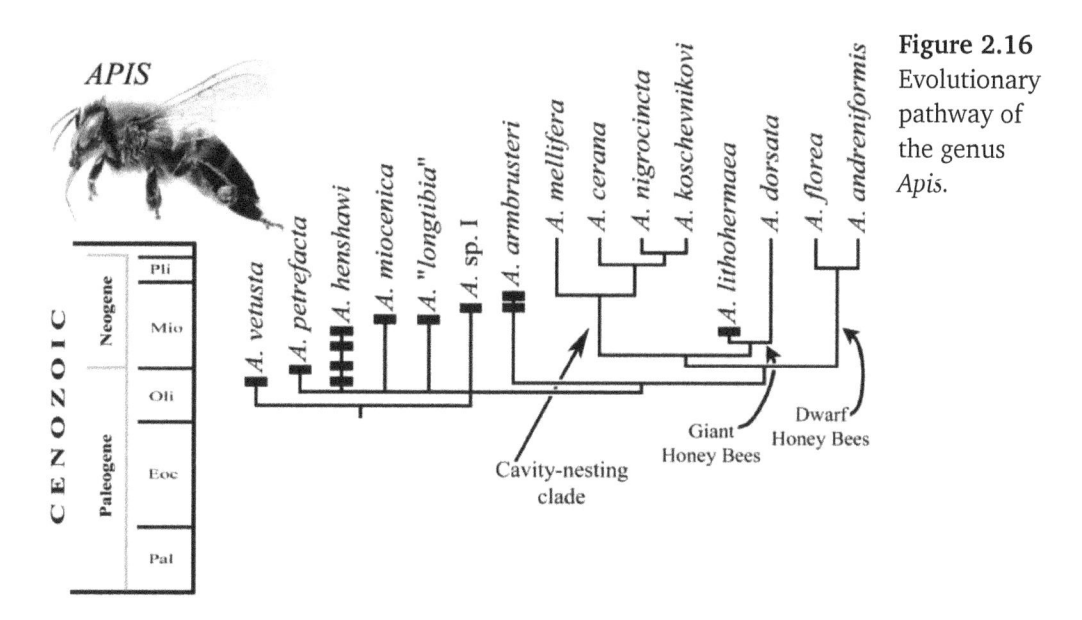

Figure 2.16 Evolutionary pathway of the genus *Apis*.

The enigmatic giant honey bees

Charles Michener's *The Bees of the World*[344] lists one hundred and thirty eight scientific names for honey bees. While some are well known to palaeontologists many are extinct. However the majority are simply old names for the ten or maybe twelve different honey bees currently recognised (Table 2.2).

Apini tribe (*Apis*)	Species	Common names	Distribution
Megapis [Open nested giant honey bees]	*Apis dorsata* [*Apis dorsata dorsata*]	Giant Honey Bee	Widespread from India and to most of southern Asia below ~1200 m
	Apis dorsata binghami	Giant Sulawesi Honey Bee	Sulawesi Island, Indonesia
	Apis laboriosa	Himalayan Cliff Honey Bee/ Giant Mountain Honey Bee	Himalaya and mountainous Vietnam and Laos nest above ~1200 m and foraging up to 4100 m
	Apis breviligula	Giant Philippine Honey Bee	Philippines excluding the Palawan Islands
Apis (Synapis) [Cavity dwelling medium sized honey bees][345]	*Apis cerana* (~5 races)	Asian Honey Bee	Indian subcontinent to most of southern Asia
	Apis koschevnikovi	Koschevnikovi's Honey Bee	Malaysian Peninsula, Sumatra and Kalimantan
	Apis nigrocincta	Philippine Honey Bee	Sulawesi and the Philippines
	Apis cerana nuluensis (*Apis nuluensis*)	Malaysian Mountain Honey Bee*	Sabah (1500-3400 m) in Kalimantan
	Apis indica	Plains Honey Bee	Southern India (below ~450 m)
	Apis mellifera (~55 races)	Western Honey Bee	Southern Europe, Middle East and Africa

Micrapis [Open nested dwarf honey bees]	Apis florea	Red Dwarf Honey Bee	Continental Asia and Africa, southeastern Asia: Thailand, Iran, Oman, India, Myanmar, and some parts of China, Cambodia, and Vietnam and forested regions of the Middle East
	Apis andreniformis	Black Dwarf Honey Bee	Southeast Asia: southern China, India, Burma, Laos, Vietnam, Thailand, Malaysia, Indonesia, and the Philippines

Table 2.2 Extant honey bees.
* On the basis of detailed morphometric measurements *Apis nuluensis*, found at high elevation (1500-3400 m) in Sabah, has sometimes been accorded separate species status.

The Giant Honey Bee is widespread in India and Asia but does not extend eastward to Timor, Irian Jaya, Papua New Guinea or to the Pacific (Figure 2.17).

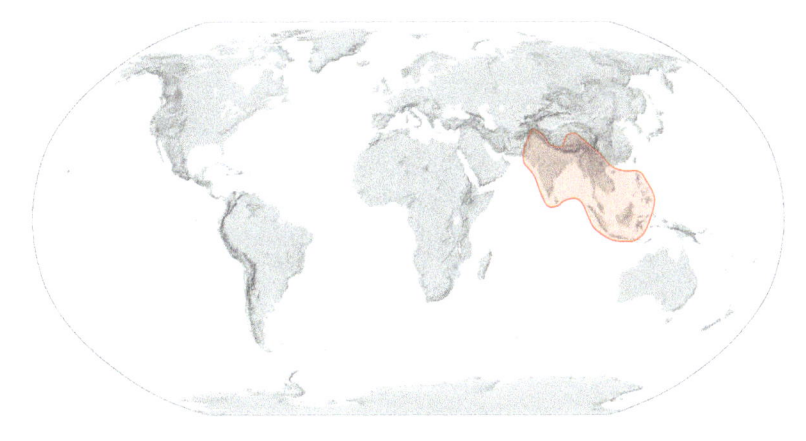

Figure 2.17 Geographic range of the giant honey bee *Apis dorsata*[346].

Until relatively recently there was thought to be a single extant giant honey bee, *Apis dorsata*. It is spread right across the tropical Indian and southern Asian subcontinents. There are now known to be a total of three, perhaps four, different species of giant honey bee. First to be recognised as a new species was the Himalayan Cliff Honey Bee or Giant Mountain Honey Bee, *Apis laboriosa*. It is well recognised for its adaptation to seasonal nesting and foraging of summer flora to altitudes up to 4100 m and for its being the largest of all honey bees (Table 2.3). Then came a recent genetic study according the long isolated Giant Philippine

Honey Bee, *Apis breviligula*, separate species status[347]. On the back burner is the also isolated Giant Sulawesi Honey Bee, *Apis dorsata binghami*. It, like the other two species and Asian mainland *Apis dorsata,* has different abdomen and stripe patterning but its genetic makeup is not yet sufficiently clear to determine whether it, too, has been isolated long enough to give it separate status.

Species	Subgenus	Body length (mm)	Body weight (mg)	Forewing length (mm)	Tongue length (mm)
Apis dorsata	Megapis	17-20[348]	120	12.7-14.6	6.6
Apis laboriosa	Megapis	~30[349]	165	~13.6	7.1
*Apis mellifera**	Apis	13-16[350]	85	8.5-9.4[351]	6.1
Apis cerana	Apis	~10[352]	65	7.4-9.0	5.4
Apis florea	Micrapis	7-10[353]	23	6.0-6.9	3.3

Table 2.3   Defining morphometric features[354] of representative worker honey bee subgenera (approximate values)[355].
*Honey bee sizes vary between season, subspecies type and other factors such as nutrition and that European races of *Apis mellifera* tend to be larger that those found in Africa.

Other defining features of the giant honey bees are evident at the colony (superorganism) level (Table 2.4) though there are few published studies on the nest and colony characteristics of the giant Sulawesi and Philippine honey bees.

Species	Optimal nest cavity size (L)	Population size	Comb size (lxw mm²)	Comb area (cm²)	No. cells per nest[356]	No. combs	Colony density
*Apis mellifera**	~45	20,000-60,000	–	2500	100,000	~8	2-15/km²
*Apis dorsata***	–	36,600 ± 29,100	1500 x 700	3200 ± 2550	23,300 ± 18,700	1	Up to 100 per tree
Apis laboriosa	–	Definitive estimates not known	1500 x 1000	–	–	1	Up to 100 per tree

Apis breviligula***	–	–	–	–	–	1	Do not aggregate
Apis cerana****	~6.7$^\Psi$	6900 ± 3400	–	2800 ± 2000	34,600 ± 24,600	~6	Do not aggregate
Apis florea	-	6300 ± 5000	170x 110	200 ± 150	4700 ± 3400	1	Clumped up to ~360/km²

Table 2.4 Giant honey bee colony population metrics (numbers rounded) compared with other sub genera[357].

> ^a   *Apis mellifera* colonies are larger and less numerous per square kilometre in Europe than those found in Africa.
> **   In giant honey bees workers and drones are raised in the same sized cell.
> ***   Studies of the biology of the Giant Philippine Honey Bee, *Apis breviligula* and the Giant honey bee of the Sulawesi *Apis dorsata binghami* are few though their biology appears similar to that of widespread *Apis dorsata dorsata*. Both commonly nest in tall trees, but *Apis dorsata binghami* is less likely to aggregate in large numbers (several to ten for *Apis dorsata binghami* and up to around one hundred for *Apis dorsata dorsata*.
> **** Akratanakul[358] proposes a 20 L super for keeping *Apis cerana* as opposed to a 40 L box for keeping *Apis mellifera* bees.

Overall the giant honey bees are becoming more widespread in the tropics but are absent from Central America, the Pacific Islands, Australia, New Zealand, Irian Jaya and from Papua New Guinea. The giant honey bees are the natural hosts of parasitic mites belonging to the genus *Tropilaelaps*[359].

The most defining feature of the different members of the *Apis dorsata* group is that they are geographically separated where allopatric speciation, genetic drift due to long isolation, has occurred. The overall picture is one of the large geographic range of *Apis dorsata* (*Apis dorsata dorsata*) and island occupation by the variants (Figure 2.18).

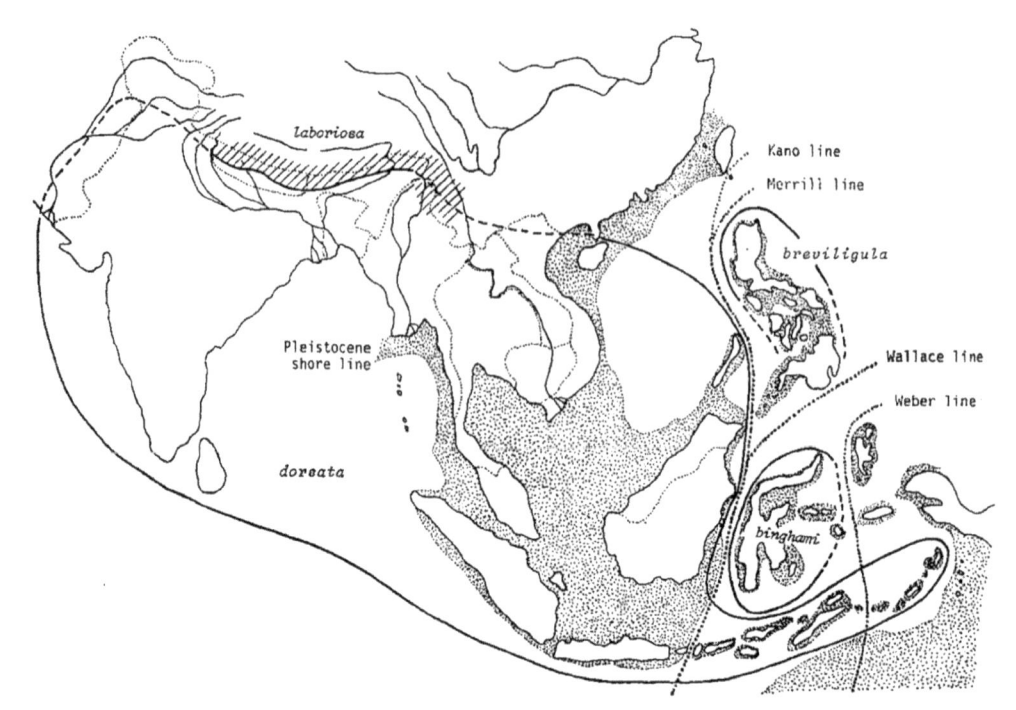

Figure 2.18 Geographic distribution of four forms of the *Apis dorsata*-group (Megapis)[360].

County distribution of giant honey bees and association with co-evolved *Tropilaelaps* parasites is shown below (Table 2.5).

Giant honey bee	Geographic range	Parasite host association
Apis dorsata [Apis dorsata dorsata]	Indian subcontinent to Southeast Asia#: Afghanistan, China, India, Indonesia (except the Sulawesi), Philippines (Palawan Islands only), Kenya, Laos, Malaysia, Myanmar, Bangladesh, Nepal, Pakistan, South Korea, Sri Lanka, Thailand and Vietnam	*Tropilaelaps mercedesae* Now also spread more widely, e.g. to *Apis mellifera* in Papua New Guinea. *Apis dorsata* also hosts the external parasites *Acarapis externus* and *Acarapis dorsalis* but their origin is uncertain. *Acarapis externus* is found in Australia but has not been accorded pest status.
Apis dorsata binghami	Indonesia: Sulawesi	*Tropilaelaps clareae*
Apis breviligula	Philippines: Luzon, Mindanao, Bohol, Cebu, Laguna and Mindanao excluding western Palawan Islands	*Tropilaelaps koenigerum* Philippines excluding Palawan Islands Also hosts *Tropilaelaps clareae*[361]
Apis laboriosa	Bhutan, China (Yunnan and Tibet), India, West Bengal, Laos, Myanmar, Nepal, Sikkim and Vietnam	*Tropilaelaps thaii* Lower Himalaya

Table 2.5   Geographic distribution of giant honey bees and associated *Tropilaelaps* parasites.

There is a limited literature on giant honey bee pheromones (see *Let us spray: Honey bee pheromone chemistry*). We know that Nasonov glands of giant honey bees produce alarm pheromones. In the cavity dwelling Apis sub genus, they function primarily to facilitate bee aggregation, for example forming part of the glue that keeps swarming bees together and that assists bees in orienting themselves at colony entrances.

In a study conducted by Shearer and coworkers, we learn that principle component of queen substance, 9-oxodec-trans-2-enoic acid – we might call it hive glue – is produced in comparable quantities by *Apis dorsata, Apis cerana* and *Apis mellifera* queens[362]. As also noted elsewhere this and associated queen mandibular gland pheromones are inimical to the cohesion and functioning of the brood nest and functions in other ways in maintaining colony homeostasis: the roles of queen substance are as disparate as control of drone comb and queen cell construction, control over the production of both supersedure and swarm queens and for queen retinue and swarm cohesion.

The self-assembly processes in honey bees is easily recognised in phenomena such as when bees cluster together when they are dislodged from frames outside the hive, when honey bees swarm and in their capacity to coordinate defence activities in response to hive invasion by predators. This topic is reviewed under *Let us spray: Honey bee pheromone chemistry*. Yet the range of behaviours differs distinctly between dwarf, cavity nesting and giant honey bees are well illustrated in the disparate strategies for colony self defence adopted by the different honey bee groups.

Colony defence

Apiaries comprising domesticated western and Asian honey bees – European and a few South African races of *Apis mellifera* and Japanese and Chinese races of *Apis cerana* – are afforded a large measure of protection by the hive we provide them. All wild colonies of cavity dwelling honey bees gain further protection from well insulated hollows, from their proclivity to employ small defendable colony entrances and from some other behavioural traits, notably Asian honey bees heat balling of insect predators.

To varying degree all honey bees employ stinging to fend off vertebrate predators (bears, mice, people, skunks, gophers...), their venom having much the same potency across all *Apis* species. However the amount of venom they can deliver and their proclivity to sting varies widely.

The dwarf honey bees employ strategies such as camouflage (hiding nests in leafy undergrowth), applying sticky exudates to branch attachment to repel ants and retreat and readily abscond in response to wholesale predation. Giant honey bee nests, on the other hand, are in the open and quite conspicuous, though typically inaccessible to non flying predators. They build their nests high in large trees (Figure 2.19), under building eves or attach their single comb to rock overhangs and have a lethal sting apparatus (Table 2.6).

This all said the giant honey bee defence lies as much in their large sting and venom apparatus as in their aggression behaviour. This makes the giant honey bees such dangerous animals.

Jerzy Woyke and coworkers[363] in assessing the best way to handle giant honey bees noted that:

> The extraordinary defense behavior of Apis dorsata is well known. Roepke (1930)[364] described an A. dorsata attack, during which the worker bees followed him for 500 m. When Lindauer (1956)[365] was attacked, the bees followed him for 2 km. Near Anuradhapura, Sri Lanka, three water buffalos were killed by A. dorsata. According to Morse and Laigo (1969) there is no question that this is the most ferocious stinging insect on earth.

Open nesting bees – both the dwarf and giant bees – form a protective barrier at their point of attachment to the substate effectively removing the risk of predation by ants. Other behaviours such as hissing and extending the size of the curtain and bee shimmering, the latter two particularly pronounced in giant honey bees, are effective in preventing attack by predatory insects such as wasps[366].

One detail of a behavioural trait in both the Giant Honey Bee and in the Giant Philippine Honey Bee is upward abdomen flipping (the aforementioned shimmering) captured on camera by Kastberger, Weihmann, Zierler and Hötzl[367]. This behavioural trait helps regulate brood nest temperature of the bee curtain protecting the open nest and would appear effective in warding off wasp predators. Short video clips[368] demonstrate dorso-ventral abdomen flipping response of *Apis dorsata* colony to wasps.

Species	Subgenus	Sting		Number of acute barbs	Venom (Dufour) gland
		Length (mm)	Width (mm)		Volume per worker bee (µL)
Apis dorsata	Megapis	2.70	3.23	11	218
Apis mellifera	Apis	2.55	2.92	10	138
Apis cerana	Apis	2.00	2.34	10	43
Apis florea	Micrapis	1.34	1.65	10	27

Table 2.6   Defence characteristics of representative honey bee species[369].

To conclude our introduction to these giant bees there is of course a long history of their association with human settlement in southern Asia and the problems they present. A favourite childhood recollection of the strife these creatures can inflict is told in the *Red Dog* yard spun by Rudyard Kipling in his second jungle book[370].

Figure 2.19 Giant Kajoolaboo (*Tetrameles nudiflora*) or Silk-cotton tree (*Ceiba pentandra*) hosting *Apis dorsata* colony in high canopy at Preah Khan Temple near Siem Reap in Cambodia mid November 2014.
Photo: Alan Wade

A cautionary tale

When marauding hordes of the Red Hunting Dog or Dhole[371], *Cuon alpinus*, of the Deccan Plateau descended from the Seeonee Hills they did not reckon on the fearsome giant honey bee (*Apis dorsata*) festooning the cliffs of the Northern Indian Waingunga River Gorge:

> *The split and weatherworn rocks of the gorge of the Waingunga had been used since the beginning of the Jungle by the Little People of the Rocks—the busy, furious, black wild bees of India; and, as Mowgli knew well, all trails turned off half a mile before they reached the gorge. For centuries the Little People had hived and swarmed from cleft to cleft, and swarmed again, staining the white marble with stale honey, and made their combs tall and deep in the dark of the inner caves, where neither man nor beast nor fire nor water had ever touched them. The length of the gorge on both sides was hung as it were with black shimmery velvet curtains, and Mowgli sank as he looked, for those were the clotted millions of the sleeping bees...*

As he listened he heard more than once the rustle and slide of a honey-loaded comb turning over or falling away somewhere in the dark galleries; then a booming of angry wings, and the sullen drip, drip, drip, of the wasted honey, guttering along till it lipped over some ledge in the open air and sluggishly trickled down on the twigs.

The pursuing Red Dog pack, led by Mowgli, plunged into the gorge:

Overhead they could hear furious short yells that were drowned in a roar like breakers—the roar of the wings of the Little People of the Rocks. Some of the dholes, too, had fallen into the gullies that communicated with the underground caves, and there choked and fought and snapped among the tumbled honeycombs, and at last, borne up, even when they were dead, on the heaving waves of bees beneath them, shot out of some hole in the river-face, to roll over on the black rubbish-heaps.

We can now turn to the biology of each of the giant honey bees those thrown up by the complex biogeography of the South China Sea and the South Asian mainland. This region is the cauldron of evolution of not only the giant honey bees but also of all honey bees.

Part II The tale is in the sting

As we have seen – given enough time and aided by geographic isolation – the giant honey bee has evolved to become rather different looking creatures (Figure 2.20).

Giant Honey Bee	Giant Cliff Honey Bee	Giant Sualwesi Honey Bee	Giant Philippine Honey Bee
Apis d. dorsata	*Apis laboriosa*	*Apis d. binghami*	*Apis breviligula*

Figure 2.20   The Giant Honey Bee kaleidoscope.

Their biology, shaped by their disparate environments over a long timeframe, is also distinctly different giving us yet more insights the remarkable plasticity of honey bees.

The Giant Honey Bee *Apis dorsata* **(*Apis dorsata dorsata*)**

The general biology of The Giant Honey Bee (*Apis dorsata*) is broadly canvassed by Randall Hepburn and Sarah Radloff[372] and, in a more contemporary fashion, by Oldroyd and Wongsiri[373]. Paar and coworkers[374] track the genetic variability and mating behaviour of north Indian populations of this bee examining the propensity of many colonies to aggregate at one location, a feature not shared by cavity dwelling bees. These bees track seasonal sources of food, governed by the phenology (seasonal pattern of blooming) of flowering plants. They might be said to have invented migratory beekeeping.

The giant honey bees – all species – are notable in their propensity to migrate rather than simply to abscond in response to dearth or disturbance. The dwarf honey bees, the Asian honey bees and many African races of *Apis mellifera* do likewise. As already noted, from the earliest studies of the biogeography of The Giant Honey Bee[375], it was apparent that there were a second species related to *Apis dorsata,* the high elevation and more northerly Giant Mountain Honey Bee, *Apis laboriosa*[376].

The type species, *Apis dorsata* [*Apis dorsata dorsata*], migrates seasonally over vast distances, very notably some 50 km across the Melaka Straight between Sumatra and the West Malaysian peninsula and up to 200 km in Sri Lanka. On these long journeys colonies bivouac at favourable locations returning to the principle nesting site or to the shelter they had reproductively swarmed from. The parent species *Apis dorsata* ranges from western India, eastwards below the Himalayas to the whole of continental southern Asia including across much of the Indonesian archipelago. As noted under *Global distribution of honey bees* the giant honey bees decamp to more seasonally favourable sites on an annual cycle.[377, 378, 379]

All the giant honey bees, despite their propensity to migrate, can store large quantities of honey (Morse reports more than 50 kg) under favourable seasonal conditions. For example it has had a seemingly long history of being farmed for honey production on the lower Mekong River delta (Figure 2.21)[380].

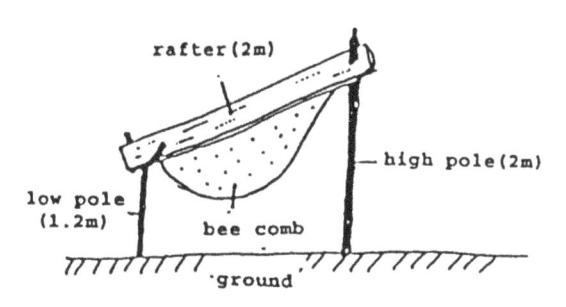

Figure 2.21 *Apis dorsata* rafter farming in the *Melaleuca cajuputi* forests of South Vietnam.

The Giant Mountain Honey Bee (Giant Himalayan Cliff Honey Bee) *Apis laboriosa*

Mountain dwelling *Apis laboriosa* is the largest of all the giant honey bees. Its self imposed geographic isolation, namely its adaptation to high altitude living, is both remarkable and extraordinary.

Initially considered to be confined to the lower slopes of the Himalaya, its natural range is now known to extend east to the mountainous region of North Vietnam[381] (Figure 2.22). Its migratory pattern is, however, strikingly different to that of *Apis dorsata*. While its colonies aggregate in similarly large numbers it does so in inaccessible cliff overhangs (Figure 2.23) while the common and widespread giant honey bees nest under eaves of dwellings and in large trees.

The ups and downs of Apis laboriosa[382]

As already noted the altitudinal migration of *Apis laboriosa* is a defining characteristic of this cliff-dwelling honey bee giant. In his doctoral studies Benjamin Underwood examined the behaviour and energy balance of high-altitude survival signalling that it nests in the open, perched high on cliffs:

> *One of the keys to colony survival at such high altitudes is seasonal migration. Nest sites within the subalpine zone (above approximately 2800 m) are occupied for a maximum of only four months in summer (June through September), while those within the warm-temperate zone (1200 to 2000 m) may be used for as long as ten months of the year (February through November).*
>
> *...a combination of poor foraging conditions and low colony stores also leads laboriosa colonies to migrate. In the steep, narrow valleys inhabited by these bees, the seasonal abundance and distribution of floral resources vary widely within short distances, and colonies need not move more than 10 to 20 km to migrate between cliff sites separated by as much as 2000 m in altitude. There they survive the cold winter months of December and January by huddling near the ground.*

Underwood's observations are corroborated by a study undertaken by Fred Dyer and Tom Seeley that compared differences in thoracic flight temperatures in *Apis florea*, *Apis cerana*, *Apis mellifera* and *Apis dorsata* with ambient air temperatures[383].

In this atypical – for honey bees – hibernating state they rely on camouflage and an extremely low metabolic rate to survive returning to the same lofty overhangs over summer.

Underwood further reports on the Nepalese flight performances of *Apis laboriosa*, *Apis cerana* and *Apis mellifera* and their ability to regulate their flight efforts in response to expected gains and associated costs as demonstrated by their being presented with different sugar level rewards.

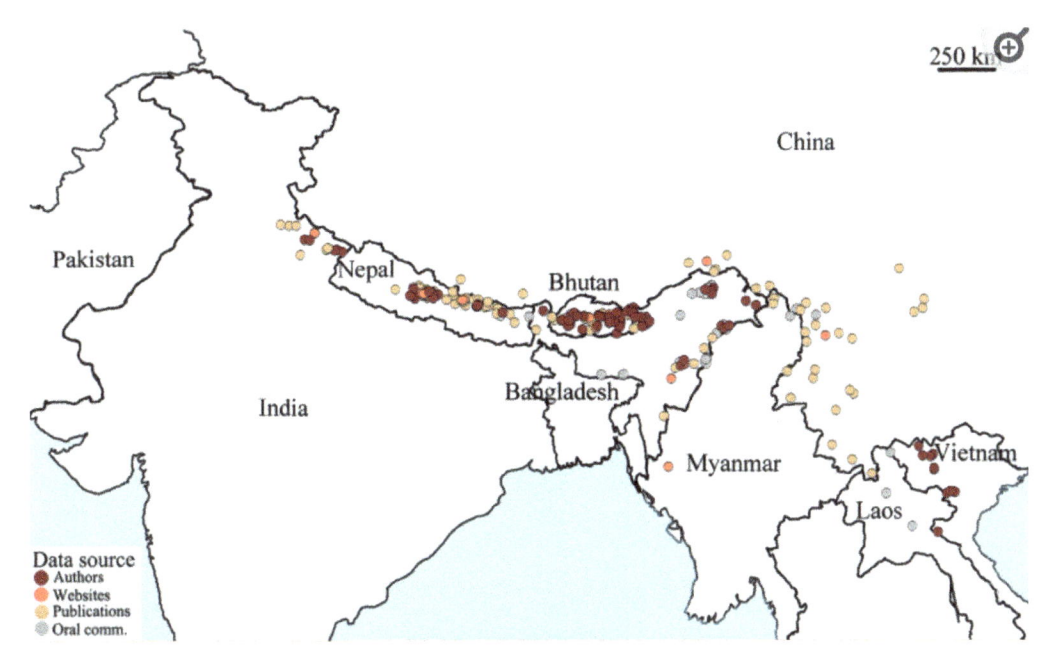

Figure 2.22   Giant Mountain Honey Bee (*Apis laboriosa*) distribution.

All this means is that bees – within the limits of their physiological makeup and their seasonal propensity to migrate – adjust their behaviour so they only collect nectar if it is energetically profitable to do so. Underwood suggests that *Apis laboriosa* may be grouped with *Apis dorsata* and *Apis florea* [the common dwarf honey bee] as a relatively low-powered, open-nesting bee, in contrast to the more highly powered cavity nesters, *Apis cerana* and *Apis mellifera*.

Two other follow up studies underscore Underwood's findings. Roubic and coworkers[384] report a nesting range in Nepal of 2500-3000 m where bees forage up as high as 4100 m. They further note that *Apis laboriosa* build nests under rock ledges in deep, vertical river valleys, likely to avoid predation by bears noting nests constructed at lower elevations may well be occupied year round.

Figure 2.23   *Apis laboriosa* nest congregation overhang in Bhutan[385].

Woyke, Wilde and Wilde[386] made a separate followup study in the Himalayan low altitude warm zone in the winter of 1999. Referencing Underwood they noted that:

> ...dropping temperatures make even the lower altitude cliff sites inhospitable and colonies must leave the cliffs. They survive December and January in the forest in energy-efficient, comb-less clusters near the ground.

The Woyke group study was confined to the altitudinal zone between 1250 and 1500 m identifying sites with multiple (6-53) nests. They observed that some of the colonies were active, noting that bees collecting pollen and water and found that comb collected by honey hunters had brood in all stages. Perhaps this confirms Roubic's findings of year round occupation of some lower elevation nests. From their limited time observations, the Woyke team concluded that:

> ...colonies were not preparing themselves for abandoning the combs and for migration. Because colonies migrate when all or almost all brood emerges, those colonies would not migrate at least within three weeks [sic that] is

till 10th of January. However, the amount of brood present and the young development stages, suggests that the colonies would occupy their combs for [a] longer period.

Clearly the migratory pattern of this honey bee is complex and contingent upon local microclimatic conditions and availability of forage. As Woyke surmised:

...colonies from the cool areas above 2000 m migrate for winter into the lower warm areas, where they pass some time in comb-less clusters. However, established colonies in the warm areas, below 1500 m do not abandon their nest in winter, but remain on their combs rearing brood during that season.

Since they found abandoned dark comb at sites in their warm observation area they postulated that, in summer, some or all colonies may abandon their combs in the warm zone and migrate to higher altitudes.

The studies may also be confounded by the extensive traditional honey hunting by Himalayan communities, now exacerbated by ecotourism, where there appears to be a decline in colony occupation of overhang congregation shelters[387].

A fallout of this trend would also appear to be a declining pollination service to orchards. Batra[388] further notes that colonies of the Giant Mountain Honey Bee in northern India may migrate in response to predation by the hornet, *Vespa mandarinia*.

The Giant Sulawesi Giant Honey Bee *Apis dorsata binghami*

The Giant Sulawesi Honey Bee (*Apis dorsata binghami*) differs from its Asian mainland cousin, *Apis dorsata* (*Apis dorsata dorsata*), in having a longer tongue and longer wings[389] but, as already noted, whether it can be classified as a separate species is yet to be resolved[390].

All the forms resemble,
Yet none is the same as another;
Thus the whole of the throng
Points at a deep hidden law.

Johann Wolfgang *von Goethe* (1749–1832)[391]

Starr, Schmidt and Schmidt[392] review the status of the Giant Sulawesi Honey Bee (*Apis dorsata binghami*) in the regional context of both the Giant Philippine Honey Bee (*Apis breviligula*) and the common Giant Honey Bee (*Apis dorsata dorsata*). The debate on its position harks back to 1953 when Maa delineated the giant honey bees as three separate species[393]. Here it is worth noting that the Giant Mountain Honey Bee, *Apis laboriosa* was not recognised until 1980[394]. Contemporary taxonomists require more genetic information to ascertain whether *Apis dorsata binghami* should be accorded separate species status.

The Giant Sulawesi Honey Bee is restricted naturally to the Sulawesi and surrounding islands. Nagir, Atmowidi and Kahono[395] describe the characteristics of its nest, its nesting trees and its nesting behaviour in the Moros forests of the southern Sulawesi. Of the of one hundred and fifty two colonies they located (17 active nests, 85 abandoned combs) they found eleven nests at 0-11 meters, forty nests at 11-20 meters and fifty one nests at more than 21 meters above ground level and found them in thirty four different types of trees. Nests were all firmly attached to sound woody branches, all sloping from the main trunk by less than 60°. All nests were camouflaged by lianas and foliage.

Apart from a few images of nests and bees there appear to be no other reports of the natural history of this enigmatic honey bee.

The Giant Philippine Honey Bee *Apis breviligula*

This species, with a black rather than the normal yellow and black *Apis dorsata* abdomen[396], is confined to the main islands in the Philippines (Figure 2.24) long separated from the Asian mainland so evolved to become a new species. It does not occur on Palawan Island to the west that was once connected to Kalimantan and the Asian mainland. The giant honey bee in the Palawan Islands is the Asian mainland species *Apis dorsata dorsata*.

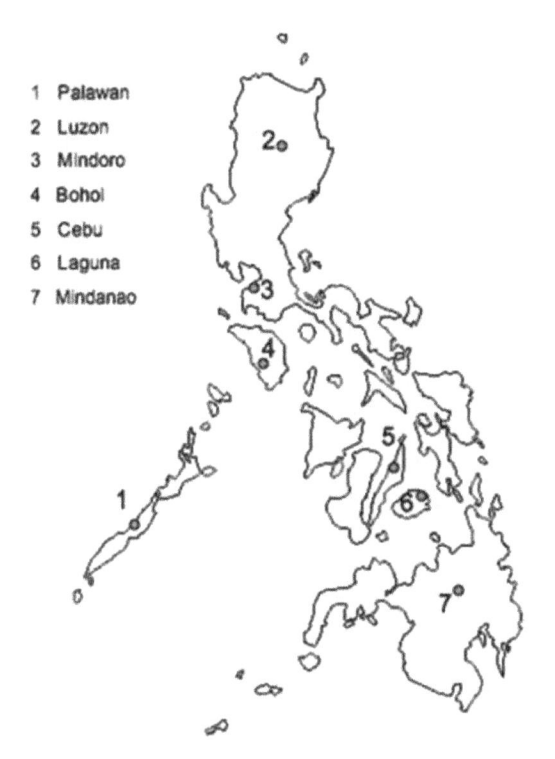

1 Palawan
2 Luzon
3 Mindoro
4 Bohol
5 Cebu
6 Laguna
7 Mindanao

Figure 2.24 Giant Philippine Honey Bee (*Apis breviligula*) distribution excluding the Palawan Islands.

In an early description of the nest of the Giant Philippine Honey Bee, *Apis dorsata*, now known as a distinct striped giant honey bee *Apis breviligula*, Roger (Doc) Morse gives a rare insight into its biology[397]. Of its temperament he notes:

> *Apis dorsata (sic breviligula) is the most ferocious of stinging insects. It is not uncommon for five to ten per cent of a nest population to attack an intruder within a few seconds of being disturbed. The workers will pursue a disturbing man or animal for long distances; the bees will pursue enemies into shaded areas to a greater extent than other Apis.*
>
> *Nest populations varied from a low of 1000 to a high estimated at 70,000 in the thirty nests examined. Drone populations approximated those in Apis mellifera colonies.*
>
> *…Between 80 and 95 per cent of the population may be used, depending upon weather conditions, to build a curtain of bees, several bees thick, around the comb. There is a bee space, or working space, between the curtain of bees and*

> *the comb. The bees in the curtain are inactive, hanging head upwards. When a colony swarms the curtain may be only one bee thick over the brood, but queen cells which are left behind are well protected with a heavy curtain of bees.*

Further notes signal that the giant honey bee performs its dance on the bee curtain though in a restricted area. From his study Morse concluded that aspects of honey bee biology, such as nest temperature control and the concomitant well-defined brood rearing cycle, would have evolved early in the development of the genus. Nevertheless there are many subtle differences in the likes of swarming behaviour that set the giant honey bees apart from those of other honey bees[398]. Giant honey bees not only swarm to form new colonies nearby, they also abscond once the colony becomes weakened and they also migrate, typically twice per year – once to a distant locale when floral resources decline, then returning to either the swarmed location or to the parent hive vicinity – when the season returns to a honey gathering mecca.

Future of giant honey bees

Survival and spread

Before we leave the giant honey bees, letting them live their own lives, we might reflect and be concerned about their ultimate survival. They are extremely prone to habitat destruction (e.g. deforestation), honey hunting and shifts in climate. They also perform an essential function in pollinating tropical flora, crops and orchards. Their decline would see both the bees and the communities dependent of their ecosystem service as major losers.

It would however be naïve to think giant honey bees might be welcome elsewhere. Their spread – for example to the Near East, to the central Americas or to our region – presents a very real existential threat to tropical ecosystems and may cause panic – they are much more a risk to people that common or garden bees – in the general community. Further their potential to host *Tropilaelaps* (tropi) mites that would spread to and threaten our honey bee and regular sideline industry should not be underestimated. *Tropilaelaps mercedesae* – hosted by *Apis dorsata* – has gradually spread from the tropics and wreaked havoc in honey bee (*Apis mellifera* and *Apis cerana*) apiaries in colder climes. Worryingly, *Tropilaelaps* have been found to both outcompete and displace *Varroa* in hives co-infected with both parasites.

Role in spread of tropi mites

The grooming behaviour of the giant honey bees, as well as their propensity to migrate – resulting in breaks in the breeding cycle – greatly reduces parasite impact. The most coherent accounts of *Tropilaelaps* mites, and their associations with honey bees and their impacts, is provided by Anderson and Morgan[399] with a more detailed analysis of the ecology and life history of parasites provided by de Guzman et al[400].

There are four *Tropilaelaps* species (*Tropilaelaps mercedesae* and *Tropilaelaps thaii*) distinctly separated from *Tropilaelaps clareae* and *Tropilaelaps koenigerum*[401]. How, where and when they evolved is only now being unravelled. However some recognition of the geographical variability of their distribution and their association with different giant honey bees will be key to understanding their expanding impact not only on the cultivated Asian and Western honey bees but also to all other honey bees.

de Guzman et al[402] note that until fairly recently the two most common species were not clearly delineated suggesting that previous studies on *Tropilaelaps clareae,* conducted in areas east of the Wallace line, including New Guinea, be attributed to *Tropilaelaps mercedesae,* whereas those west of the Wallace line be referred to *Tropilaelaps clareae*. Together they present the main threat to the global apiary industry.

The *Tropilaelaps mercedesae* mite with its indigenous host, *Apis dorsata,* is widely distributed on the Asian mainland excluding the Palawan Islands in the Philippines. The propensity of *Apis dorsata* to migrate long distances and for colonies to aggregate in large numbers, up to 100 per tree, may account for this mite being so widely distributed. These mites now infect other species, *Apis laboriosa, Apis cerana* and introduced *Apis mellifera* across Asia.

The native host of the related *Tropilaelaps thaii* is the Giant Mountain Honey Bee *Apis laboriosa* from the Himalaya region.

Tropilaelaps clareae presents a similar threat. For example it infests *Apis mellifera* colonies across the Philippines[403] (except on the Palawan Islands). Its sparse distribution has been attributed to the isolation of its adapted hosts, the Giant Sulawesi Honey Bee (*Apis dorsata binghami*) and the Giant Philippine Honey Bee

(*Apis breviligula*) and to the fact that unlike *Apis dorsata*, *Apis breviligula* colonies do not aggregate[404].

Tropilaelaps koenigerum, related to *Tropilaelaps clareae,* is the smallest mite of this genus and is also a parasite of *Apis dorsata* in mainland Asia and Indonesia. Both *Tropilaelaps koenigerum* and *Tropilaelaps thaii* appear not to affect the Western Honey Bee[405].

That *Tropilaelaps* may devastate the existing beekeeping industry more radically than has *Varroa* should give us all considerable room to reflect.

Phoretic honey bee mites[406]

> *Phoresis or phoresy is a non-permanent, commensalistic interaction in which one organism (a phoront or phoretic) attaches itself to another (the host) solely for the purpose of travel... Phoresis is rooted in the Greek words phoras (bearing) and phor (thief).*

There are of the order of a million species of mites, the majority of which are un-described. While the vast majority of these mites are free living (macrophages and detrivores) many others are less benign. A number are serious pests of vertebrates, plants and insects including, of course, bees.

Phoretic mites are hitchhikers. They attach themselves to their host, in our very special case to honey bees. This aids in both their dispersal and survival. Relatively few bee-associated mites are parasitic, many being commensal (mutually beneficial) beehive occupants.

Honey bee mites, like all arachnids, are wingless so must crawl into, or hitch a ride to, beehives to consume comb, bee debris and fungal material (Table 2.7). As noted by Hepburn and Radloff[407]:

> *Parasitic mites represent only a minor fraction of the diversity of Acari associations with honeybees. Most Acari found in the nests of honeybees usually have a saprophagous lifestyle and feed on fungus-infected debris in the hives, dead bees and sometimes pollen (kleptophages).*

Mite order	Mite species	Bee-mite association
Astigmata	*Forcellinia faini*	Scavenger of bee and insect debris and fungi
Prostigmata	*Pseudacarapis indoapis*	Pollen feeder in *Apis cerana* colonies
Mesostigmata	*Melichares dentriticus*	Predator of scavenger mites
	Melittiphis alvearius	Globally distributed pollen feeding mite in *Apis mellifera* colonies
	Afrocypholaelaps africana and other species	Flower mites phoretic on honey bees
	Neocypholaelaps spp	Flower mites phoretic on honey bees

Table 2.7 Commensal mites associated with honey bees. There are many others, see for example Refaei and coworkers[408]. Mites of Australian native bees are also known[409].

Mites as arachnids

We can start by examining the larger group to which the mites belong. Mites are all arachnids and all are jointed, eight-legged invertebrates. Arachnids are easily distinguished from insects: insects have six legs, wings and antennae. Next time you pick up a spider or a scorpions – or remove a tick – check it out to make sure it does not have wings or a fewer or greater numbers of legs.

Fleas have their own way and so do mites:

> *The vermin only tease and pinch*
> *Their foes superior by an inch.*
> *So, naturalists observe, a flea*
> *Has smaller fleas that on him prey;*
> *And these have smaller still to bite 'em.*
> *And so proceed ad infinitum.*
>
> **On poetry – A rhapsody**
> Jonathon Swift[410]

The arachnids – aráchni is spider in Greek – have a close affinity with the Chelicerae (Figure 2.25) having descended from their marine ancestors. They are related to present day marine horseshoe crabs and sea spiders. The Chelicerae are in turn a subphylum of the vast invertebrate group Arthropoda – jointed leg invertebrates – that encompasses the insects, crustacea, centipedes and millipedes.

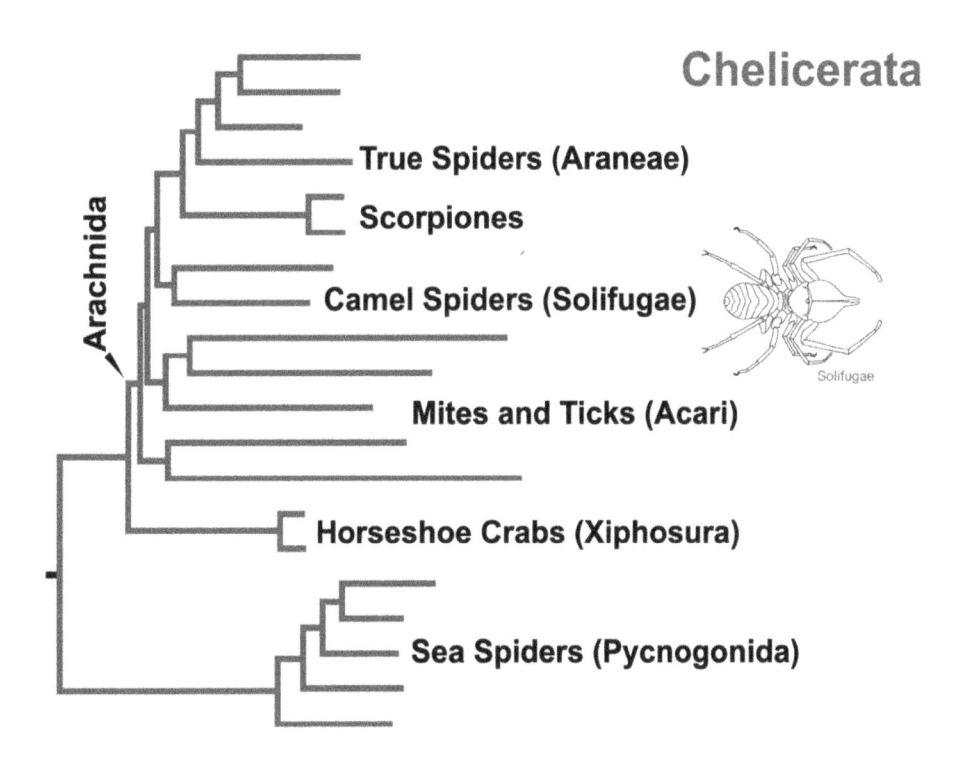

Figure 2.25 Relationship of arachnids such as mites to jawed or fanged arthropods (Chelicidae).

Drilling further down we discover that the mite group belong to a major subclass of arachnids, the Acari (Figure 2.26). The Acari comprise two distinctive lineages, the superorders Acariformes and the Parasitiformes. Together, the Acari are clearly delineated from spiders, scorpions, daddy longlegs and solifuges shown to the right of the chart.

There are a multitude of mites found in habitats as diverse as the deep oceans, deserts, poles and the upper atmosphere. So when you hear beekeepers talk about 'the mite' they are understating their diversity.

```
                    Arachnida
                [8-legged arthropods]
    ┌──────────┬─────────┬────────┬──────────┬──────────┐
   Acari    Araneae  Scorpionidae Opiliones   Solifugae
Ticks and    Spiders   Scorpions  Daddy longlegs  Solifuges: camel
  mites                           & harvestmen   spiders,
                                                 sun spiders &
                                  Other orders   wind scorpions
                                  Pseudocorpionies
                                   Pseudoscorpions
                                  Thelyphonida
                                   Whip scorpions
                                  Ricinulei
                                   Hooded tickspiders
                                  Palpigradi
                                   Microwhip scorpions
```

Acariformes   *Parasitiformes*

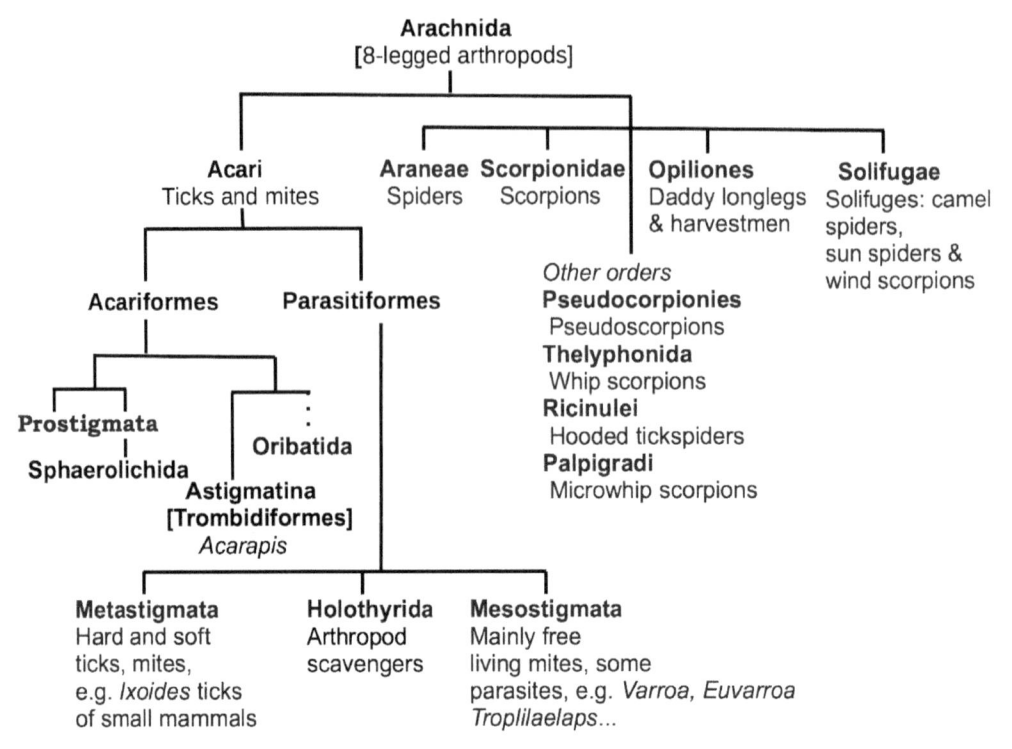

Prostigmata
 | Oribatida
Sphaerolichida
            Astigmatina
           [Trombidiformes]
              Acarapis

Metastigmata   Holothyrida   Mesostigmata
Hard and soft   Arthropod   Mainly free
ticks, mites,   scavengers   living mites, some
e.g. *Ixoides* ticks   parasites, e.g. *Varroa, Euvaroa*
of small mammals   *Troplilaelaps...*

Figure 2.26 Simplified arrangement of the major orders of 8-legged arachnids (Class: Arachnida).

In examining this assemblage I recall a visit I made to the Jenolan Caves back in the late 1960s. An enthusiastic guide, eager for us to recall the difference between stalactites and stalagmites, provided the ultimate mnemonic: 'When the mites come up, the tights come down'.

Phoretic mites

Mites are a plague on vertebrates and arthropods alike and honey, bumble, stingless and solitary bees are no exception. Many arrive as uninvited hitchhikers, the so-called phoretic mites.

Unlike the myriad honey bee pests that can literally fly in the front door – hive beetles, *Braula* fly, wax moths, and vespid wasps and hornets – the honey bee phoretic mites rely on normal behaviours of bees, robbing, foraging and swarming to gain hive entry. There is little beyond culling infested colonies and drastic treatment you can do to keep these mites at bay. A good rule of thumb is to locate your bees well away from other, often poorly managed, apiaries that your healthy bees will be inclined to raid.

Predatory honey bee mites, all phoretic, fall into the Astigmatina and Mesostigmata arachnid orders[411]. Their natural hosts are listed in Table 2.8. The mites have evolved to feast on haemolymph (bee blood and fat bodies) but, with the exception of the tracheal mite, their native hosts have developed fairly effective defence strategies such as grooming and biting behaviour. A notable example of successful coexistence is that of *Varroa* mite has on its well-adapted original host *Apis cerana*.

Mite order	Mite family	Parasite species	Original host species
Astigmatina	Tarsonemidae	*Acarapis woodi* [Tracheal mite]	*Apis mellifera*
		Acarapis externus *Acarapis dorsalis* [External mites][412]	Not known but found on *Apis dorsata, Apis cerana* and *Apis mellifera* in Asia
Mesostigmata	Varroidae[413]	*Varroa destructor* [*Varroa* mite]	*Apis cerana*
		Varroa jacobsoni *Varroa rindereri* *Varroa underwoodi*	Likely *Apis cerana* *Apis koschevnikovi* *Apis cerana*
		Euvarroa wongsirii *Euvarroa sinhai*	*Apis andreniformis* *Apis florea*
	Laelapidae[414]	*Tropilaelaps clareae* *Tropilaelaps mercedesae* *Tropilaelaps koenigerum* *Tropilaelaps thaii*	*Apis breviligula* and *Apis dorsata binghami* *Apis dorsata* *Apis dorsata* *Apis laboriosa*
Prostigmata	Erythraeidae	*Lepus ariel*	*Apis mellifera scuttelata*

Table 2.8   Phylogenic relationships of predatory mites with honey bees.

Broadly speaking:

▸ Acarapis mites that appear to have evolved with *Apis mellifera*;
▸ *Varroa* mites have evolved within the Asian honey bee group that includes *Apis cerana*;

- *Euvarroa* mites have evolved with the dwarf honey bees;
- *Tropilaelaps* mites have evolved with the giant honey bees; and
- *Leptus* mites in Africanised bees in Brazil that parasitise a range on invertebrates.

The mighty mites

Mite-bee host relationships are extremely complex, not least in terms of their cross over to other honey bee species. For example, *Apis koschevnikovi* is the natural host of *Varroa rindereri* that has spread to *Apis cerana* and to *Apis nuluensis* but not as yet to *Apis nigrocincta* or to *Apis mellifera*. Excellent overviews of the interactions between these mites and the honey bees in the Asian region are provided by Chantawannakul et al[415] and by Warrit and Lekprayoon[416].

Varroa mite jumping species to unadapted *Apis mellifera* is an exemplar of animal-parasite imbalance. Ebola, Hendra and Corona viruses moving from bats to vertebrate hosts, not just to human beings, may ring a bell.

That brings us to the serious honey bee pest status of phoretic mites (Table 2.9). Each group of mites is considered in turn.

Acarapis (Acarine) mites

The *Acarapis* honey bee parasite group belongs to the Acariformes arm of the Acari. The origin of the tracheal mite *Acarapis woodi*, the putative agent of Acarine (tracheal mite) disease, is unknown but it appeared on European honey bees, where it was first thought to be problematic, on the Isle of Wight in 1904. It was not formally identified as a tracheal mite until 1922. Sammataro, Gerson, and Needham[417] argue that its subsequent finding that it is more widespread dispels the notion that it suddenly evolved in Europe. More recently Acarine disease has also have been reported to have had a devastating impact on the Asian Honey Bee reportedly causing destruction of *Apis cerana japonica* colonies in Japan. A more recent review of the literature indicates that the mite, is a natural host of *Apis mellifera* and is relatively benign.

Mite name	Mite species	Mite pest status
Acarine Disease, Tracheal Mite, Isle of White Disease	*Acarapis woodi*	Natural pest of *Apis mellfera*
External Acarapis Mite	*Acarapis dorsalis* *Acarapis externus*	Minor pests of *Apis dorsata, Apis cerana* and *Apis mellifera*
Varroa Mite	*Varroa destructor*	Serious global pest of *Apis mellifera* and *Apis cerana*
Varroa Mite	*Varroa jacobsoni* *Varroa rindereri* *Varroa underwoodi*	Pests of *Apis cerana, Apis nigrocincta, Apis koschevnikovi* and *Apis nuluensis*
Euvarroa Mite	*Euvarroa wongsirii* *Euvarroa sinhai*	Pests of native *Apis florea* and *Apis andreniformis*
Tropilaelaps Mite, Asian Mite	*Tropilaelaps clareae* *Tropilaelaps koenigerum*	Serious pests of *Apis mellifera* and *Apis cerana* but spread is currently limited to southern Asia
Tropilaelaps Mite, Asian Mite	*Tropilaelaps mercedesae* and *Tropilaelaps thaii*	Appear to be mainly pests of native *Apis breviligula* and *Apis laboriosa* but have spread mainly to other *Apis dorsata* subspecies. Importantly *Tropilaelaps mercedesae* has crossed over to *Apis mellifera*

Table 2.9   Parasitic mites of honey bees.

The evolutionary origins of the related ectoparasites *Acarapis dorsalis* and *Acarapis externus* found in southern Asia are similarly obscure. They, too, appear to have a wide natural distribution being found on *Apis dorsata, Apis cerana* and introduced *Apis mellifera*. Their observed impact on honey bees has been minimal.

Varroa mites

Varroa destructor

Varroa mites have evolved with the Asian Honey bee, *Apis cerana,* and its close relatives, *Apis nululensis, Apis nigrocincta* and *Apis koschevnikovi*.

Varroa destructor, whose Asian mainland natural host is *Apis cerana*, appears to have taken 50-100 years to host shift to *Apis mellifera* and done so at least twice.

Of the eighteen known *Varroa destructor* halotypes only the common Korean halotype and the less common Japan-Thailand halotype found in this region and in the Americas have had a critical impact on *Apis mellifera*.

This said, the global impact of *Varroa destructor* (*Varroa* mite) has been devastating amplified in large measure by bee viruses. Of these Deformed Wing Virus has been most widely cited as the prime cause of colony collapse.

Varroa mite appears to have been much less destructive to some African races of the Western Honey Bee. Interestingly Tom Seeley[418] has shown that while the mite initially decimated wild honey bee colonies in New York State, (i.e. after its arrival), colony numbers quickly returned to pre-*Varroa* incursion levels. Seeley attributes their resurgence to wide separation of wild bee nests – where robbing is less problematic – frequent swarming and to the relatively small size of wild colonies where the potential to build up large mite populations is minimal.

Efforts to breed *Varroa* resistant strains of honey bees has met with limited success while attempts to relocate pockets of *Varroa*-tolerant bees has so far been similarly unsuccessful. One awaits the outcome of ABC news reports of David Briggs[419] working to artificially inseminate *Varroa*-resistant material into Australian queen bee stock to make beekeeping here *Varroa* ready.

Varroa jacobsoni

Varroa jacobsoni, also a parasite of *Apis cerana*, is widespread but appears to have crossed over to *Apis mellifera* in Papua New Guinea although this assessment has now been revised. *Varroa jacobsoni* now represents more than one species and may include an *Apis mellifera* pathogenic halotype of *Varroa destructor*.

Varroa jacobsoni is native to Sumatra, Java, Kalimantan and surrounding islands. It is now widespread across central and southern Thailand, Malaysia, Indonesia, Palawan Island and Papua New Guinea as well as the Central Americas. *Varroa jacobsoni*'s natural hosts are *Apis cerana* and the Philippine Honey bee *Apis nigrocincta*. It is now well established as a parasite if *Apis mellifera*, for example of *Apis mellifera scuttelata* hybrids in Brazil.

Varroa rindereri and *Varroa underwoodi*

Varroa rindereri, whose natural host is *Apis koschevnikovi*, is found in Malaysia and in western Indonesia including Kalimantan. It is not known to have crossed over to other species and was, for a long time, thought to be identical to *Varroa jacobsoni*.

Varroa underwoodi, whose natural host is *Apis cerana*, is known from bees found in the Sulawesi and Kalimantan. It has bee found on *Apis nuluensis* and *Apis nigrocincta*. It has also been found on, but has not reproduced on, *Apis mellifera* in Papua New Guinea.

Euvarroa mites

Euvarroa species are the natural hosts of the dwarf honey bees, *Apis florea* and *Apis andreniformis*.

Euvarroa sinhai's natural host is *Apis florea*. It has been reared on *Apis mellifera* and on *Apis cerana*.

Euvarroa wongserii's natural host is the southeastern Asian dwarf honey bee *Apis andreniformis* but it is now found on the more westerly and the more widely distributed *Apis florea*.

Tropilaelaps mites

There are four *Tropilaelaps* species. All are naturally hosted by the giant honey bee *Apis dorsata* of which there are four recognised geographical variants. *Apis laboriosa* is one of two very distinctive forms of giant honey bee. How and when the different *Tropilaelaps* species and the regional variants of *Apis dorsata* evolved or diverged is unknown, but some recognition of the geographical variability of their distribution may be key to understanding their impact on both the cultivated Asian and Western Honey Bee.

Apis laboriosa, the Giant Mountain Honey Bee, is the biggest extant (living) honey bee. Its isolation and its adaptation to high altitude has accorded it separate species status. This bee never harvests abandoned nest wax and hive materials and avoids the same rock attachment on return to the same cliff overhang after overwintering in forests at lower altitude.

The other really distinctive species is *Apis breviligula* found in most parts of the Phillipines. It has also been accorded separate species status[420].

The grooming behaviour of the giant honey bees, as well as their propensity to migrate – resulting in breaks in the breeding cycle – greatly reduces parasite impact.

The most coherent account of *Tropilaelaps* mites, and their associations with honey bees and their impacts, is provided by de Guzman et al[421]. Quoting from elsewhere they note ominously:

> *Varroa and Tropilaelaps mites have coexisted in Apis mellifera colonies in Asia for more than fifty years and that Tropilaelaps mites are considered to be the more dominating and reproductively successful parasites of Apis mellifera than Varroa mites.*

Lilia de Guzman and coworkers go on to note that *Tropilaelaps mercedesae* and *Tropilaelaps thaii* are closely related being genetically separate from similarly closely related *Tropilaelaps clareae* and *Tropilaelaps koenigerum*.

The widely distributed *Tropilaelaps mercedesae* is found across mainland Asia [Afghanistan, China, India, Indonesia (except the Sulawesi), Kenya, Laos, Malaysia, Myanmar, Nepal, Pakistan, Papua New Guinea, South Korea, Sri Lanka, Thailand and Vietnam] and, more ominously for Australia, in the Palawan Islands in the Phillipines in the South China Sea and in islands neighbouring Australia [Indonesia excluding the Sulawesi in Indonesia and in Papua New Guinea (PNG)]. In PNG it is found on *Apis mellifera* despite the region being free of giant honey bees.

According to former CSIRO scientist Denis Anderson and Matthew Morgan *Tropilaelaps clareae* appears to be much more restricted being found only in the Sulawesi in Indonesia and the Phillipines outside the Palawan Islands[422]. *Tropilaelaps mercedesae,* with which *Tropilaelaps clarae* has been confused, is considered to be the mite most threatening to the global beekeeping industry.

With this the United States Department of Agriculture's Samuel Ramsey agrees. In extensive study of this mite in Thailand[423], he concluded this 'tropi' mite is potentially a much greater threat than *Varroa* to the global beekeeping industry. Where they are found to simultaneously infect *Apis mellifera* colonies *Tropilaelaps* out-compete *Varroa*. The well known and only effective control means of controlling *Tropilaelaps* is to break the brood cycle, a measure hardly compatible with building colonies for the honey flow.

Tropilaelaps koenigerum is a parasite of *Apis breviligula* (Asia including Indonesia except the Sulawesi). It is now also a parasite of *Apis laboriosa* and *Apis mellifera*.

Tropilaelaps thaii is a parasite of *Apis laboriosa* and originated in Vietnam. It is now widely distributed in South and Southeast Asia and is found mainly in forested areas of Nepal, Malaysia and Singapore.

The jury is out

If you are a little confused as to which mite infects which honey bee and the future of honey bees, you are in good company. I am of the view that some of the lesser-known honey bee species are themselves threatened, mainly from loss of habitat and human predation. I have trouble concurring with the widespread view that *Apis mellifera* (the Western Honey Bee) and perhaps also *Apis cerana* (the Asian Honey Bee) are under imminent threat, despite widespread managed colony losses.

The apiculture industry is certainly seriously impacted, but the continued survival and recovery of wild populations of these keystone species suggests that they are adapting to the parasites. The greater risk to bees would appear to revolve around well intended but perhaps mistaken use and reliance on miticides for their control.

Randy Oliver from *Scientific Beekeeping* is of the view that chemical control of mites is exacerbating their potential impact. In much the same way antibiotic control of *Paenobacillus larvae,* the causitive agent of American Foulbrood, only masks its survival and spread. So use of miticides to control *Varroa* mite is breeding mite resistance and giving free reign to development of more virulent mite-associated viruses.

A few parting thoughts:

- Most recent interceptions of exotic honey bees and their mites to Australian shores has been associated with commercial shipping from fairly distant ports, one now realised in the establishment of the mite detected around

the port of Newcastle in June 2022[424].
- However, Papua New Guinea and Indonesia to our near north bristle with exotic honey bees and pest mite species, so the risk of their island hopping through the Torres Strait and by boat from Nusa Tengarra, the Indonesian southeastern islands, remains existential.

Beetle mania[425]

In May 1967 Sergeant Pepper's lonely hearts club band hit the number one spot on the Top Forty charts: Beatlemania was rampant[426]. The 2002 arrival of Small Hive Beetle [SHB] (*Aethina tumida*) from sub Saharan Africa saw the beginnings of an Australian beetle mania. By then SHB was already very well established in the southern United States arriving there as early as late 1996[427]. And as for Beatles, *With a little help from my friends* rings sour when a bee hive takes in unfriendly beetles.

Now twenty years on southeastern Australia has experienced consecutive uncharacteristically wet and humid summers induced by warm El Niño sea temperature conditions in the north west Pacific. Already the bane of much of the eastern coastal zone, anecdotal reports indicate SHB is now creating significant problems for beekeepers on the Southern Tablelands of New South Wales. Former Canberra Region Beekeepers club president Dermot Asis Sha'Non, a noted collector of nuisance hives around Canberra, remarked:

> 'I've never seen the numbers of SHB that I've been seeing recently before this. I've lost a few recovering cut-out colonies to them. It hasn't happened in previous seasons.'

To date most of the world remains free of the sub-Saharan African Large Hive Beetle [ALHB], rather two beetles, *Oplostomus fuligineus* and *Oplostomus haroldi*, and many colder climes remain unaffected by SHB though it has just established itself in Australia's southern island of Tasmania. So what do we know about SHB and how can we best manage it and what is the impending threat of ALHB.

A couple of years ago the former editor of *The Australasian Beekeeper* Des Cannon was at a bee forum in New Zealand. He was there to give a talk on the hazards presented by small hive beetle, when a South African speaker at the forum, Mike Allsopp, got up to say:

> 'You think you have a problem with SHB.'

> 'You haven't encountered the large African hive beetle (ALHB). It is like a tractor hauling a harvester mowing combs in your hive.'

Mike has been joined by Sydney University's Ben Oldroyd in surveying the existential threat of this mega beetle to Australian beekeeping and to crops reliant on bee pollination of which more anon. So add beetle surveillance to your disease check list and be aware that moths can take a similar toll.

Small Hive Beetle SHB

Almost 6000 SHB larvae can be raised on a single frame of brood. Larvae mature in moist soil generally close to – but anywhere up to 100 m away from – the hive. Emerging adults can fly ten kilometres, again generally less, guided largely by honey bee alarm pheromone. Here they infest new hives or augment the populations of beetles already present.

Nicholas Annand from the New South Wales Department of Primary Industries describes the key options for managing SHB reproducing a diagram clearly outlining beetle life cycle (Figure 2.27)[428]:

> *Unlike wax moth larvae which produce webbing (the damage left is dry), SHB larvae chew through combs, causing the honey to ferment, resulting in 'sliming' of the combs...*

> *SHB are capable of prolific multiplication. Under laboratory conditions 80 SHB can become more than 36,000 adult SHB by day 63.*

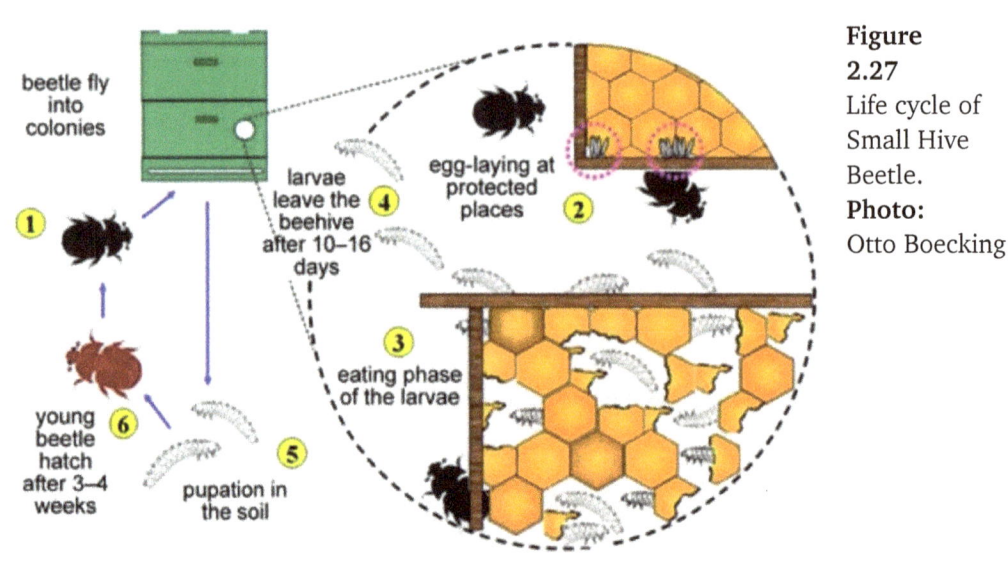

Figure 2.27 Life cycle of Small Hive Beetle. **Photo:** Otto Boecking

© Dr Otto Boecking LAVES Institut für Bienenkunde Celle, Germany – 2005

Entrapment

While weak hives are certainly more susceptible to SHB infestation in the normal course of events, and with Canberra's usual hot dry summers, the beetles have rarely gained a foothold.

A good case for intercepting flying beetles by any means possible is to employ fruit-fly style lure traps[429] located in the general vicinity of beehives. A more targeted approach is to lay a chux wipe on top of frames. However beetles trapped in the fiber also ensnare some bees. A superior strategy – one currently used by Des Cannon – is to place a carpet square under the mesh screen bottom board where it serves also to monitor beetle numbers.

Many beekeepers we know ignore the occasional, say up to half a dozen or so, beetles they find squishing the few that appear on top bars before they scurry away into the comb matrix. Others opt for fairly passive control relying on look-alike mini stormwater-gully style traps (Figure 2.28) that are inexpensive and that you can purchase in bulk on-line. You spoon in diatomaceous earth (a swimming pool filter medium available from pool shops and some stock and station agents) or simply pour in a tablespoon of cooking oil. Workers chase beetles into the traps where their spiracles clog up and they die.

A more effective cassette-style 'Apithor' trap (containing the insecticide fipronil) works well and is safe to use provided it is handled well[430]. Place the trap smooth side down – to prevent harbourage of beetles and wax moths under the trap and mark date of placement on the trap using a marker pen. While the manufacturers claim an effective three month period, one probably suited to warm moist coastal apiary locales, Doug Somerville from the NSW Department of Primary Industries provided informal advice that they be replaced each spring at least for the NSW Southern Highland climate.

Figure 2.28
SHB gully trap inserted between frame top bars.

More innovative approaches such as that recently reported by Neil Richie[431], involving under-floor traps, avoid the use of chemicals but require regular maintenance.

In our experience annual trap replacement works well. Beetles rarely ever appear and by late autumn the weather is too cold for beetles to migrate and to become established.

The gooey mess produced by beetles is toxic to humans so handing slimed combs and gear (Figure 2.29) should be conducted with gloves and a protective mask. As Rod Bourke[432], the local New South Wales biosecurity officer, points out, cleaning up hives overtaken by the beetle is not much fun. Take Rod's advice by intervening early by euthanising bees, then opting for a total clean out whenever beetle larvae present themselves en masse. I can attest to this after finding a hive taken over by the beetle co-infected with American Foulbrood. The seething mass of larvae and their spilling out onto un-diggable rocky soil and having to saturate the surrounds with a registered insecticide was a gargantuan struggle.

> *Ooey gooey was a worm*
> *And a mighty worm was he*
> *He sat upon a railroad track*
> *And a train he did not see*
> *Ooey Gooey!*
>
> Traditional Scout ditty

Since the life cycle, eggs to adult, can be as little as 4-6 weeks, there may be as many as six generations per year under favourable conditions. Beetles can also breed on stored hive combs and hive debris around hives and in bee sheds so, as with wax moths, it is best to extract honey quickly and to avoid leaving hive scrapings around the apiary. Little wonder that weak hives can easily fall prey to SHB and that whole apiaries can be at risk if good hygiene is not maintained. Requeening hives with strains that produce hygienic behaviour workers, keeping hives strong and removing excessive amounts of gear are all strategies that limit the risk of severe colony damage.

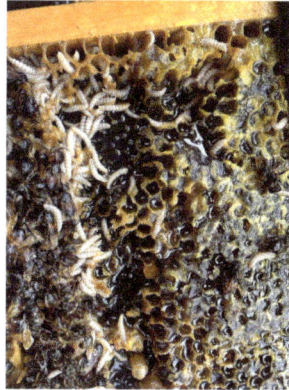

Figure 2.29
Small Hive Beetle hive infestation:
(a) tell-tale slime oozing from infected hive; and
(b) beetle larvae spoilage of stored honey.
Photos:
Peter and Jenny Robinson

But what measures can you take if you have failed to take preventive action and find patches of brood overrun by a seething mass of beetle larvae? Peter and Jenny found a single hive that had become queenless – probably weakened in the process – that already had several slimed-out frames. What to do?

First we considered trying to requeen the unit after replacing the damaged frames proposing to achieve this by inserting several frames of sealed brood along with a spare queen at hand. This may well have worked but we had six other healthy hives and there was a small risk that disease may have been a factor in the colony's dwindling. In any case there was the more tangible risk that small hive beetle eggs were present in large numbers on sound frames and anyway the queenless bees were exceptionally cranky. This is something that visitor Alan discovered when, to his great discomfort, he turned up suit-less to help appraise the situation.

With discretion being the better part of valour, Jenny and Peter euthanised their likely old bees by shaking them into a tub of water laced with a squirt of dishwashing detergent. Any few bees that escaped simply joined other healthy hives. The remaining 'good frames' were given the freezer treatment.

There is no registered hive treatment for SHB in Australia so put that can of fly spray away!

Beetle behaviour

Ellis and Ellis at the University of Florida have produced a contemporary account of small hive beetle behaviour[433]:

> *Investigators have shown that small hive beetles fly before or just after dusk and odors from adult bees and various hive products (honey, pollen) are attractive to flying small hive beetles. Some investigators have suggested that small hive beetles also may find host colonies by detecting the honey bee alarm pheromone. Additionally, small hive beetles carry a yeast (Kodamaea ohmeri) on their bodies that produces a compound very similar to honey bee alarm pheromone when deposited on pollen reserves in the hive.'*

These workers outline their spread and their then 2010 distribution (Figure 2.30) and refer back to studies by Neumann and Elzen[434] describing the biology of SHB and their much earlier spread in Africa where they appear to be a minor pest.

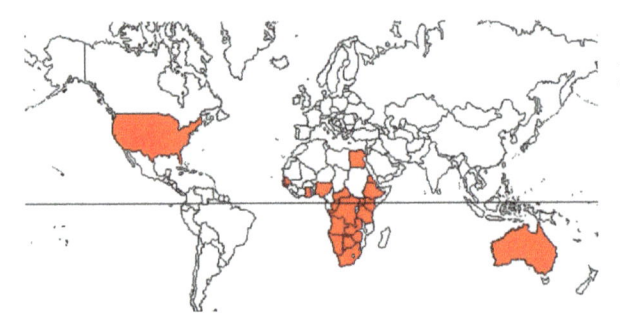

Figure 2.30
Distribution of Small Hive Beetle (2010).

African Large Hive Beetle ALHB

Two species of ALHB are native to sub-Saharan Africa. Oldroyd and Allsopp have made some keen observations on the beetles in their African habitat. Based on a Kenyan study and their observations in South Africa and Botswana, it appears likely that of the two species *Oplostomus haroldi* has a preference for wetter coastal areas and that *Oplostomus fuligineus* is more common in dryer grazing land where dung is abundant.

The measures required to exclude this beetle monstrosity include providing a beetle proof hive entrance (Figure 2.31), and a perforated hive bottom board to allow mature larvae to migrate and pupate in a dung-filled subterranean trap. The authors also survey food preferences for ALHB signalling a low preference for sand, cow dung, local soil and straw and a strong preference for horse dung along with a strong dietary preference for brood comb, a slightly lower preference for pollen, a much reduced preference for comb honey and a low preference for fruit. Destructive large hive beetles attack the colony engine, brood areas that colonies depend on to replace and maintain worker honey bee populations.

Adult ALHB prey on bee brood, whereas larvae and pupae live in dung[435]. Giant hive beetle poses an existential threat to all beekeepers worldwide: these behemoths – no they are not moths – pose a danger also drawn attention to by BeeAware[436].

Figure 2.31
African Large Hive Beetle at a hive entrance.
Photo: Ben Oldroyd

Beetles in other environments

Do hive beetles just affect honey bees? SHB has been reported to successfully cycle on fruit but reports of such damage are sparse and it appears to be principally only a problem for European honey bee colonies kept in warm moist zones. There SHB wipes out feral colonies. And stingless bees, while not subject to wax moth predation, appear also to have good defences against SHB[437].

> *Small hive beetle, Aethina tumida Murray, is a parasite of social bee colonies and has become an invasive species, raising concern of the potential threat to native pollinators in its new ranges. Here, we report the defensive behavior strategies used by workers of the Australian stingless bee, Austroplebeia australis Friese, against the small hive beetle. A non-destructive method was used to observe in-hive behavior and interactions between bees and different life stages of small hive beetle (egg, larva, and adult). A number of different individual and group defensive behaviors were recorded. Up to 97% of small hive beetle eggs were destroyed within 90 min of introduction, with a significant increase in temporal rate of destruction between the first and subsequent introductions. A similar result was recorded for 3-day-old small hive beetle larvae, with an increased removal rate from 62.5 to 92.5% between the first and second introductions. Of 32 adult beetles introduced directly into the 4 colonies, 59% were ejected, with the remainder being entombed alive in hives within 6 h. Efficiency of ejection also significantly increased between the first and third introductions. Our observations suggest that A. australis colonies, despite no previous exposure to this exotic parasite, have well developed hive defences that are likely to minimize entry and survival of small hive beetles.*

While keeping African Large Hive Beetle out of Australia remains a priority, I can report that I, and many other beekeepers, are working hard to ensure that small hive beetle control is well practised in our apiaries, noting for example that:

> *Of six healthy colonies in Peter and Jenny's hives, one other appeared to have become queenless and soon succumbed to SHB. Peter and Jenny have upgraded their integrated pest management system, using Apithor hive beetle traps installed on the floor of each hive and adding disposable beetle traps with diatomaceous earth on the bars of the top supers.*

Where rust and moth doth corrupt[438]

Old comb is an easy meal for wax moths. To the beginner or unwary, the wax moth is the ultimate destroyer. Open your stored combs to find a mass of tangled web and pupated moths buried deep into much weakened gear and you will get a clear message. Any beekeeper who knows the ropes will tell you that wax moth is the dustbin of all abandoned wild and forgotten beehives, the great cleanser and clearer of diseased bee gear.

The truth lies somewhere half way between these maxims. There is some suspicion that wax moths may vector bee pests such as chalkbrood. It is also true that the seasoned beekeeper will be occasionally caught out and find his treasured store of combs in a parlous webbed state.

Unguarded bee comb is easy prey to wax moths:

> *Lay not up for yourselves treasures upon earth,*
> *where moth and rust doth corrupt,*
> *and where thieves break through and steal.*

Matthew 6:19

Waxing lyrical on moths

But there isn't just one wax moth, there are several:

The Lesser Wax Moth, *Achroia grisella,* is often found on combs in competition with the Greater Wax Moth. In practice it is outcompeted though it is well known for burrowing beneath the cappings of sealed brood where it leaves a characteristic snail trail. Gently lift the webbing with a hive tool – the white or off-white caterpillar will be found on the comb surface at the end of the trail. The Lesser Wax moth fares rather better in colder climates, but in competition with the Greater Wax Moth, it is often relegated to scavenging insect frass and wax on the hive floor. Screened bottom boards better ventilate hives and are effective in sifting hive debris attractive to moths.

The Greater Wax Moth, *Galleria mellonella,* is particularly destructive (Figure 2.32)[439]. It has gained a favourable press in its reputation as an eater of plastic, welcome news to the waste industry. However rest easy: they eat polyethylene products but not the other polymers that go into making plastic frames, plastic foundation and plastic beehives.

Other moths (Table 2.10) have proved problematic for stored comb materials elsewhere in the world and a few, such as the Bumble Bee Wax Moth and the Fruit Pollen Moth have also caused considerable damage in honey bee hives. The Bee Moth and the Indian Meal Moth are both found in Australia. Both Mediterranean flour and Indian meal moths can destroy stored honey bee products while the Fruit Pollen Moth can attack honey bee combs.

Keeping the moth at bay

In the apiary: While wax moths may weaken honey bee colonies, they are more a pest than any real threat to bees[440]. Good apiary hygiene – collecting wax and propolis when working hives hives and collecting material from cleaning up hive gear in the shed – will go a long way to preventing establishment of a wax moth reservoir.

Honey bee associated moths	Species
Greater Wax Moth	*Galleria mellonella*
Lesser Wax Moth	*Achroia grisella*
Bee Moth/ Bumble Bee Wax Moth	*Aphomia sociella*[441]
Fruit Pollen Moth	*Vitula edmansii*[442]
Mediterranean Flour Moth/ Mill Moth*	*Ephestia kuehniella*[443]
Indian Meal Moth	*Plodia interpunctella*[444]

Table 2.10   Honey bee moth pests.
*Flour moths are a pest of stored honey bee products.

In the hive: Strong bee hives headed by young productive queens, largely free of diseases such as chalkbrood and sac brood virus and with good nutrition, will never fall prey to the moth[445]. Bees recognise moths and their larvae and remove them from the hive. In healthy hives wax moth is only ever found in areas not subject to routine patrol, for example between hive mats and frame top bars, in bottom board debris or where too much gear is left on the hive. I have ever only encountered several handfuls of wax moths over the many years I have kept bees. Keeping tractable bees that are easy to inspect and maintain and judicious supering practice has made wax moth control a no brainer.

Figure 2.32   After dinner. Canberra Region Beekeepers bee shed 5 May 2022.

In storage: Never assume old comb is safe in storage. Always treat your gear especially when putting good combs away for winter.

The simplest option is to freeze combs overnight: this will kill all moth life stages. Don't leave as under warm conditions old comb or even comb with stored honey can be destroyed in a few weeks. Then securely seal them and repeat treatment if a few moths appear. Combs stored in airy spaces with some daylight (wax moths do best in dark confined spaces) and out of reach of rodents can be stored without treatment. Alternatively combs can safely stored below 5 ^{0}C but you may not have the luxury of cool rooms.

Fumigation with carbon dioxide, sulfur dioxide, glacial acetic acid and phosphine (generated from aluminium phosphide tablets purchasable from some stock and station agents) are widely employed, leave no residue but are an occupational hazard if employed in a confined space. Glacial acetic acid, by way of example, is not only an excellent fumigant, killing all life stages of wax moth and small hive beetle, but is also effective in destroying European Foulbrood (*Mellisococcus pluton*) spores. But though effective, it must be handled with gloves and it will corrode metal fittings, nails and screws if used for longer than needed. For all fumigants, read the material data safety sheet and always work in the open.

Some old treatments, notably using p-dichlorobenzene (PCB, familiar in public urinals) and ethylene dibromide are no longer used or indeed legal because of concern about their toxicity and their leaving residues in wax.

Biological treatments notably spraying dried combs with *Bacillus thuringiensis* extracts, widely employed to control horticultural pests, have proved ineffectual in killing the Lesser Wax Moth while some formulations are toxic to bees.

Perhaps the only other sound advice is to avoid placing too much gear on a hive and to ensure that your colony is as strong as you might reasonably anticipate. Weak colonies headed by queens beyond their use-by date ask for moth infestation.

Pests and diseases of social honey bees

Pests and diseases in divided into two parts:

- Part I Pests and diseases of honey bees; and
- Part II Australian native bee pests.

The listings are simply those of biological agents that I have come across either in my experience of keeping of bees or have read about. They comprise a web of microbial and larger invertebrate consumers and vertebrate predators that have exploited the rich resources that social bees amass. As such they a simple reference listing and not intended to be a treatise on the problems they may or may not present to keeping of healthy bees.

Part I Honey bee pests and diseases[a]

Common Name	Species names	Status
Fungi		
Chalkbrood[446]	*Ascosphaera apis*	✓
Stone Brood	*Aspergillus fumigatus, A. flavus, A. niger*	✓ (rare)
Microsporidia		
Nosema Disease, Nosemosis[b]	*Nosema apis*	✓
Nosema Disease, Nosemosis	*Nosema ceranae*	✓ (recent)
Protozoa		
Amoeba Disease	*Malpighamoeba mellificae*	✓
Gregarine Disease	Gregarinidae (*Monoica apis, Apigregarina stammeri, Acuta rousseaui, Leidyana apis*)	✓
Lotmaria Disease[447]	*Lotmaria passim*	✗
Neogregarine Disease[448]	*Apicystis bombi*	
Trypanosomatid parasite, Crithidia[d]	*Crithidia mellificae*	
Bacteria		
American Foulbrood (AFB)	*Paenibacillus larvae*	✓
European Foulbrood (EFB)	*Mellisococcus plutonius*	✓ (not WA)
Half-Moon Disorder	*Bacillus coagulans* (aetiology uncertain)	✗ (NZ only)
Powdery Scale Disease	*Paenibacillus larvae* ssp. *pulvifaciens* [*Bacillus pulvifaciens*]	✗ (USA and Mexico only)
Septicaemia	*Pseudomonas aeruginosa*	✓
Spiroplasma; May Disease	*Spiroplasma apis, Spiroplasma melliferum*	✗

Serratia[449]	*Serratia marcesens* strain *sicara*	✗
Insects		
Ants (many species)	*Iridomyrmex purpureus* (meat ant)	✓
Australian Sap Beetle[450]	*Brachypeplus basali*	✓
Braula Fly[e]	*Braula coeca*	✓ (Tas only)
European Wasps (German Wasp, Common Wasp; English Wasp)	*Vespula germanica, V. vulgaris*	✓
European beewolf	*Philanthus triangulum*	
Giant Willow Aphid[f]	*Tuberolochnus salignus*	✓
Honey Bees (exotic) and associated parasites[g]	*Apis cerana, A. indica, A. koschevnikovi, A. nigrocincta, A. nuluensis, A. m. capensis, A. m. scutellata* [Cavity-dwelling Honey Bees]; *A. dorsata, Apis d. binghami, A. laboriosa. A. breviligula* [Giant Honey Bees]; *A. florea* and *A. andreniformis* [Dwarf Honey Bees]	*Apis cerana* now established in Cairns area of northern Australia *Varroa destructor* incursion in Newcastle area in June 2022
Hive Beetles[h] African Large Hive Beetle Small Hive Beetle[451]	*Oplostomus fuligineus, Oplostomus haroldi* *Aethina tumida*	✗ ✓ (not WA)
Hornets (European Hornet, Asian Hornet, Oriental Hornet, Lesser Banded Hornet, Yellow-legged Hornet)	*Vespa crabro, V. orientalis, V. affinis, V. mandarinia, V. velutina* ssp. *nigrithorax*	✗ (Asian Hornet, Yellow-legged Hornet are of major concern)

Moths[452]
  Greater Wax Moth[453]
  Lesser Wax Moth
  Bumble Bee Wax Moth
  Mediterranean Flour Moth
  Fruit (pollen) Moth
	Galleria mellonella
	Achroia grisella
	Aphomia sociella
	Esphestia kuehniella
	
	Vitula edmansae
		✓
		✓

Phorid Fly	*Apocephalus borealis*	✗
Vertebrate predators		✓
Cane Toad	*Rhinella marina*	✓ (northern Australia)
Mouse	*Mus musica*	✓
Pied Currawong	*Strepera versicolor*	✓
Rainbow Bee Eater	*Merops ornatus*	✓ (mainly northern Australia)
Arachnids (8-legged mites)		
Acarine Disease, Tracheal Mite	*Acarapis woodi*	✗
External Acarapis Mite	*Acarapis dorsalis,* *A. externus* (syn *A. vagans*)	✗ ✓
Euvarroa Mite[i]	*Euvarroa wongsirii* *Euvarroa sinhai*	
Spiders (e.g. Red Back Spider)	*Latrodectus hasselti*	✓
Tropilaelaps Mite, Asian Mite[j]	*Tropilaelaps clareae;* *T. koenigerum, T. mercedesae;* *T. thaii*	✗
Varroa[j] Mite[454]	*Varroa destructor; V. jacobsoni*	✗
Leptus Mite	*Leptus ariel*	✗
Viruses[455]		
ABPV	*Acute Bee Paralysis Virus* (ABPV)	✗ [concern for Australia]
IAPV (related to ABPV)	*Israeli Acute Paralysis Virus* (IAPV)	✓

KBV (related to ABPV)	*Kashmir Bee Virus (KBV)*	✓
BQCV	*Black Queen Cell Virus (BQCV)*	✓
CBPV[k]	*Chronic Bee Paralysis Virus (CBPV)*	✓
CWV	*Cloudy Wing Virus (CWV)*	✓
DWV[456]	*Deformed Wing Virus (DWV)*	✗ [concern for Australia]
FV (AmFV)[l]	*Filamentous Virus (FV) (Apis mellifera Filamentous Virus)*	✓
Sac Brood, SBV[m]	*Sac Brood Virus (SBV)*	✓
SBPV (also designated SPV)	*Slow Bee Paralysis Virus (SBPV)*[457]	✗
ALPV[n]	*Aphid Lethal Paralysis Virus (ALPV)*	
BSRV	*Big Sioux River Virus (BSRV)*	
BBV[o]	*Berkeley Bee Virus (BBV)*	
BV-X and BV-Y[p]	*Bee Virus X and Bee Virus Y (BV-X; BV-Y)*	
LSV-1[q]	*Lake Sinai Virus-1 (LSV-1)*	✓
LSV-2	*Lake Sinai Virus-2 (LSV-2)*	✓
ABPV	*Arkansas Bee Picona-like Virus (ABPV)*	
AIV[r]	*Apis Iridescent Virus (AIV)*	
ARV-1 and ARV-2[458]	*Rhabdovirus 1 and 2 (ARV-1; ARV-2)*[459]	
ABV-1 and ABV-2	*Bunya-like Virus (ABV-1 and ABV-2)*	
AFV	*Apis Flavivirus (AFV)*	
ADV	*Dicistro-like Virus (ADV)*	
ANV	*Apis Noravirus (ANV)*	
KV	*Kakugo Virus (KV)*	

Notes
a The listing is oriented to Australian honey bee maladies but is otherwise focussed on known pests, diseases and parasites that affect *Apis* species.
b *Nosema bombi* is a parasite of bumblebees.
c Affects some species of *Bombus* and has been occasionally detected in *Apis mellifera*.
d *Crithidia bombi* is a common bumblebee parasite.
e Other known species of *Braula* not found in Australia and are not considered to be bee parasites are *B. kohli, B. orientalis, B. pretoriensis* and *B. shmitzi*.
f The aphids produce melezitose trisaccharide honeydew sugars from willows and poplars that form a hard candy and that bees cannot metabolise. This honeydew is concern only as it uses hive storage space and may be a contributing factor to bees starving.
g *Apis andreniformis* sometimes sympatric with *Apis florea* but probably developed allopatrically where their ranges then converged. Together they may harbour a range of bee diseases but do not harbour arachnid mites of concern to *A. mellifera*. The giant honey bees are the original sources of *Tropilaelaps* spp. The biogeographic origins of each of the giant honey bees may be important for understanding the importance of *Tropilaelaps* as a honey bee disease. Relatives of *Apis cerana, A. koschevnikovi* (Koschevnikovi's Honey Bee of Kalimantan-Sumateran-Peninsular Malaysia) is sympatric with, and closely related to *A. cerana* while the Philippine honey bees, *Apis nuluensis* and *Apis nigrocincta* are all of interest as the *Apis cerana* group are the original vectors of *Varroa*. *A. koschevnikovi* is the vector of *Varroa rindereri*.
h SHB is now widespread in Australia but not in Western Australia where it is presently confined to the Kimberley.
i *Euvarroa wongsirii* and *E. sinhai* are both parasites of *Apis andreniformis*. *Euvarroa sinhai*, but not *E. wongsirii*, is also a parasitise *Apis florea*. *Euvarroa sinhai* is known to have reproduced on *Apis mellifera*.
j The original *Tropilaelaps* hosts are the Asian bees *Apis dorsata* (*Apis dorsata dorsata* and *Apis dorsata binghami*), *Apis breviligula* and *Apis laboriosa*. *Tropilaelaps clareae* (Philippines and Sulawesi (except Palawan Island), Luzon Island) is a parasite of *Apis cerana, Apis dorsata, Apis florea, Apis dorsata laboriosa* and *Apis mellifera*); *Tropilaelaps koenigerum* (Asia including Indonesia except Sulawesi) is a parasite of *Apis breviligula*. It is now also a parasite of *Apis dorsata, Apis laboriosa* and *Apis mellifera*; *Tropilaelaps mercedesae*; (Asia except Sulawesi including Papua New Guinea) is a parasite of *Apis cerana, Apis dorsata*, possibly *Apis dorsata laboriosa* and *Apis mellifera*); *Tropilaelaps thaii* (Vietnam) is a parasite of *Apis laboriosa*.
k Sometimes associated with Chronic Paralysis Satellite Virus (CBPSV) of unknown symptomology.
l Turns haemolymph milky white but is otherwise symptomless.
m *Sac Brood Virus* is closely related to *Thai Sacbrood Virus* that adversely affects *Apis cerana*.
n ALPV is a virus of pest aphid species that appears to be transmitted to bees through

honeydew. ALPV has some affinities with BSRV and BBV and, while numbers in bees build up with honeydew, their impact on bees has not been ascertained though they impact on aphid populations.

o BBV and ABPV occur together and are symptomless.

p BV-X and BV-Y are serologically related viruses that are symptomless in adult bees and do not multiply in honey bee larvae and pupae: BV-X is associated with the winter diarrhorea-inducing protozoan *Malpighamoeba mellificae* [see protozoan diseases] while BV-Y is associated with *Nosema apis*.

q Closely related species that are in turn related to CBPV. They are seemingly symptomless and likely related to BV-X and BV-Y. Two other strains of *Lake Sinhai Virus* have also been isolated.

r Symptoms of AIV are similar to those of CBPV affecting flight in adults.

Part II Australian native bee pests and diseases[a]

Common Name	Species names	Status
Social bees [*Tetragonula* and *Austroplebeia*]	**Disease Organism [Host Species]**	
Fungal diseases		
Chalkbrood	*Ascosphaera* sp. (probably *Ascosphaera apis*) (*Tetragonula carbonaria*) [uncommon]	
Leafcutter Bee Chalkbrood	*Ascosphaera aggregata*	✗
Bacteria		
Shanks Brood Disease[b, 460]	*Lysinibacillus sphaericus* (different 'strains' for *Tetragonula carbonaria* and *Austroplebeia australis*)	✓
Insects		
Assassin Bugs Orange Assassin Bug	*Gminatus australis*	✓
Bembix Wasps/Sand Wasps	*Bembix flavipes* (*Austroplebeia* males) *Bembix moma* (stingless bees and other insects) *Bembix musca* (*Tetragonula* mainly males) *Bembix tuberculiventris* (*Austroplebeia*, *Tetragonula* and solitary bees at flowers)	✓ ✓ ✓ ✓

Braconid Wasp
  Stingless Bee
  Braconid Wasp
	Syntretus trigonaphagus	✓

Green Ant	*Oecophylla smaragdina*	✓
Hive Syrphid Fly[c]	*Ceriana ornata australis*	✓

Hive Beetles
  Native Hive Beetle
  Small Hive Beetle (SHB)
	Brachypeplus spp.
	Aethina tumida
		✓
		✓

Phorid flies
    Hive Phorid Fly
    Phorid flies (unspecified)[d]
	Dohrniphora trigonae (*Tetragonula* spp.)
	Apocephalus including *Apocephalus borealis, Melaloncha, Pseudohypocera*
		✓
		✗

Wax Moths
    Greater Wax Moth
    Lesser Wax Moth
	Galleria mellonella
	Achroia grisella
		✓
		✓

Vertebrate predators		✓
Asian House Gecko	*Hemidactylus frenatus*	✓
Cane Toad	*Rhinella marina*	✓
Arachnids (8-legged mites)		
Pollen Mite	*Cerophagopsis trigona*	✓
Silken Web Spiders		✗

Common Name	Species names	Status
Solitary bees	Disease organism [Host species]	
Fungal diseases		
Blue Banded Bee Chalkbrood	Unknown (*Amegilla chlorocyanea, Amegilla* spp)	
Chalkbrood (Carpenter bees[e])	*Ascosphaera apis* (*Xylocopa californica* ssp.), Arizonensis sp. (*Xylocopa* spp?)	
Leafcutter Bee Chalkbrood	*Ascosphaera aggregata* (*Megachile quinquelineata, Megachile nigrovittata*)	✗
Neogregarine Disease[f]	*Apicystis bombi*	
Insects		
Bembix Wasps	*Bembix moma* (stingless bees and other insects including solitary bees)	✓
Cuckoo Bee[g]	*Thyreus spp* (Blue banded bee, *Amegilla* spp.)	✓

Notes

a This listing is oriented to pests and diseases of Australian native stingless bees restricted to two genera, *Tetragonula* and *Austroplebeia* and to a few known pests of native solitary bees.

b Note the species is a common soil and gut bacterium, the toxins of which are similar to those produced not only by this genus of bacteria, but also the *Paenibacillus* genus that includes the causitive agent of AFB.

c A wasp mimicking hoverfly nest parasite of *Tetragonula*.

d Includes species such as North American *Apocephalus borealis*, a number of *Melaloncha* widely distributed in the USA and South America, parasitoid wasps that parasitises bumble bees, honey bees and paper wasps. Of these *Pseudohypocera kerteszi* is a renowned stingless bee predator.

e May affect Australian native carpenter bees, *Xylocopa* spp.

f Affects some species of *Bombus* and has been occasionally detected in *Apis mellifera*.

g *Amegilla cingulata* nests are parasitized by the neon cuckoo bee *Thyreus nitidulus*.

II
Beekeeper Sketches

Beekeeper Sketches is a sampler of enigmatic beekeepers dating from the 1890s, all well known in their time but that have now all but fallen off the radar. We all know of Lorenzo Langstroth, François Huber, the Root fraternity, Clayton Farrar, John Holzberlein, Mykola Haydak, Thomas Seeley, Colin Butler, Mark Winston and many more. However the also exceptional contributions and progressive style of these beekeepers very much need to be brought to the fore.

E.W. Alexander –
The North American inventor of the two-queen hive[461]

In tracking the career of this enigmatic beekeeper, one might be intrigued to find that he only ever signed himself off as E.W. Alexander. His feisty correspondents had, without fail, referred to him in like formal manner. For it is in mentioning a name that one gets closer to that individual and can sketch that person as a real life human being. Alexander almost fails that humanity test.

But to say that E.W. Alexander [1845–1908] was anything other than an extra ordinary beekeeper would have been a gross understatement. So inventive was he that H.H. Root, a celebrated editor of *Gleanings in Bee Culture*, engaged Alexander to chronicle his findings made over a long forty years[462]. In his preface to his collection of *Alexander's writings* Root noted that:

> *While at times he may have seemed unorthodox, yet it must be remembered that he occupied a locality where conditions were peculiar, not to say remarkable. He was the only bee-keeper in the United States who was ever able to manage from 700 to 800 colonies all in one yard.*

But Alexander's writings convey a deep sense and sensibility when it came to all aspects of keeping bees. His many inventions aside, he provided this practical counsel under a June 1906 heading entitled *Leaving the well-beaten path, and the consequences*:

> *In regard to the wisdom of cautioning beginners about leaving the beaten paths too far, and following what may in their localities turn out after all to be a phantom, I wish to say that, when I was a boy, a very small minority of bee-keepers left the well-beaten path of setting their best colonies over a brimstone-pit in order to get a little honey, and adopted the more humane way of cutting a little out of the sides of their hives in order that they might save their bees for another year; and I could never see any phantom about that.*
>
> *I can well remember many years ago a small minority that left the well-beaten path of:*
> *...box hives, and in their place adopted movable-comb hives;*
> *...keeping black bees, and in their place keep only good strains of Italians;*
> *...squeezing their honey through a bag, and in its place adopted the improved honey-extractors of to-day; and*

...producing their surplus comb honey in coarse hemlock boxes holding 15 or 20 lbs. apiece, and adopted the nice attractive section of the present day...

History shows us, in thousands of instances, where minorities have been in the right, and were a target for the arrows of critics who only followed in their wake and drifted with the masses.

I have avoided the temptation to enumerate his many practical inventions or to repeat his heroic pioneering exploits in establishing the first two-queen hives in North America. Instead I suggest you might wish to peruse a few of his letters to *Gleanings in Bee Culture*[463] that established him as the first ever beekeeper to successfully maintain a few-to-many queens in the one honey bee colony. Alternatively you might like to heed the advice of the editors of the *British Bee Journal* and never contemplate operating a hive with more than a single queen[464].

In one apiary sporting 237 colonies (Figure 2.33) Alexander made the claim that all of his colonies were so strong that all bees returning to his apiary could get back safely to hives – maybe not all to their hive – even at night. Alexander only ran powerful hives. Any below par were automatically relegated to nucleus apiaries and treated that way.

Figure 2.33   Colonies boiling over with bees in Alexander's apiary, May 1907.

Isaac Hopkins reports on one of Alexander's more outlandish claims to invention (Figure 2.34)[465]:

> It gave me much pleasure some seven months after the publication of the first edition of this bulletin, wherein I had suggested the adoption of this process, to find that the well-known E.W. Alexander, one of the most extensive and experienced beekeepers in the world, was working on the same method. His articles on the subject in Gleanings, early in 1906, created quite a sensation among beekeepers in America, some of whom rather fiercely criticized him and his method, and in reply he wrote, 'But I do say that the man who has had experience, and has the necessary storage-tanks, can ripen his honey after the bees commence to cap it so that it will be just as good as if left with the bees all summer'.

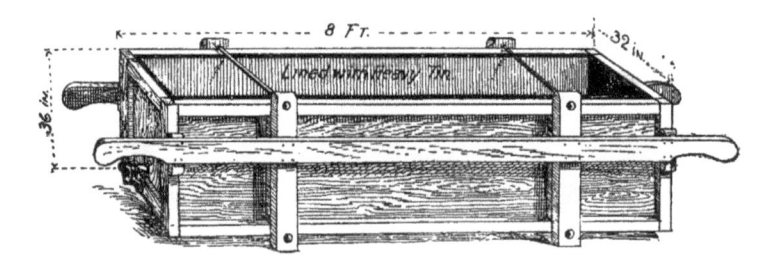

Figure 2.34   Alexander's storage and evaporation tank.

Roger Morse in his short history of the Empire State Honey Producers' Association[466] records Alexander as having produced 2300 boxes of honey from 90 hives in 1873 and that his son Frank hosted the Association's annual picnic at his father's Delanson Apiary August 1921, that is some thirteen years after Alexander's death.

And as a surprise I did eventually discover E.W. Alexander's first name[467]. Enigmatic as Alexander was, he was probably the most revered and successful beekeeper of his day.

George Wells – The British inventor of the doubled hive[468]

In compiling a book on running hives with an extra queen[469] I received an unsolicited letter from a Sarah Austin from Yorkshire in the United Kingdom. Sarah is the great great granddaughter of George Wells [1835–1908], the inventor of the doubled hive.

Double hives – pairs of honey bee colonies sharing common supers – were the early forerunner of two-queen hives. The two-queen hive would not appear until fifteen years later when, even more remarkably, it was found that an extra queen that could be accommodated in the same hive. It was not till almost a century later that it was found that a second queen could be reliably established in the same brood nest[470]. The official announcement of Wells invention was made to the *Quarterly Conversazione* of the British Bee-Keepers' Association in late March 1892[471]. To follow came independent commentary on that famous meeting from the Editors of the *The Bee-Keepers' Record and Adviser*. Their reports[472] reflect meticulous detail of the discovery:

> *In carrying out his method, Mr Wells, in the summer of 1890, divided the combs, young bees, and brood of three supered stocks which had swarmed into nine nuclei – three from each. Thus nine nuclei were formed, and each of these did fairly well: the young queen in all cases hatched out [sic emerged] from the single queen-cell left, and was duly fertilised. In the autumn of that year one of these nucleus stocks was united with each of the nine strong stocks divided as described, and, with one exception, they wintered safely and came out very strong the following spring, with large populations quite early in the season. When supering-time came round a sheet of excluder zinc was place over the double set of frames, and the bees of both lots allowed free access to the supers.*

In a follow up article in May 1892 they went on to detail how to operate bees on the Wells principle.

In an early response to the 1892 discovery of the Wells system dozens of enquiries were made by readers of *The Bee-Keepers' Record and Adviser* a monthly journal devoted to practical bee-keeping. *The Record* championed practical elements of the Wells System, Wells himself responding splendidly to queries tended by readers. In one foray he explained how he established his double colonies from swarms containing more than one queen[473]:

> *So great an interest has been evinced in what is known as the Wells System that we print below some replies made by Mr Wells to the queries of several correspondents of the BJ (British Bee Journal), who ask for information on certain points not quite clear to them.*

The editors of *The Record* also provided a salient summary of the Wells System, notes of which have enhanced my doubled hive management. It would not be an understatement to conclude that 1890s beekeeping was a blazing Wells fever, evidenced not only by the tidal wave of correspondence but also an optimistic and lengthy op-ed penned shortly after Wells announced his method to startled beekeepers[474]:

> *…How much of the activity displayed is due to the publicity given to the double-queen method in the pages of the Bee Journal and the Record it is not for us to say; but on all sides we hear of Wells hives being either already at work, or of preparations being made for using such so soon as the best form of hive for the purpose has been definitely decided on. All this cannot fail to be gratifying to Mr Wells, whose only object in bringing the subject forward has been to benefit his fellow bee-keepers…*

> *Our idea of finality will have been realised when all bee-keepers, suitably located, can tot up an average of over 150 pounds [~70 kg] per colony in a season, as Mr Wells has done.*

Even by today's standards, any colony making this amount of honey in a backyard would probably be said to have done exceptionally well.

One early report on the success of the doubled hive came from a J.N.R. in Lancashire[475]:

> *This year I have taken 108 lb extracted honey from my one Wells hive, while my four single stocks have yielded only 135 lb among them. In 1894 I took more than double the quantity of honey from my Wells hive than I did from any two single hives.*

Nevertheless it is also hard to adjudge the mood and style of beekeeping that pervaded the cottager beekeeping fraternity at the close of the 19th Century. A measure of the ethos is reflected in a number of enquirers seeking advice on how to move bees from a rudimentary containers to Wells hives, an era when beekeepers relied on swarms to provide new queens and bees to populate their hives and where skep beekeeping was still practised.

In response to a Chiddingley beekeeper who had tried his hand at harvesting honey from a tree hollow and who was now considering starting with a Wells hive[476], the Editor of *The Bee-Keepers' Record* diplomatically counselled:

> *Since our correspondent has only just started bee-keeping, we first strongly advise his acquiring a little more experience in bees before starting with a double queen, or Wells hive. So many beginners have failed in managing bees on this system that we cannot in justice encourage more failures.*

In the same edition of the *The Bee-Keepers' Record* we hear that a Paul Pugh from Pennal, Machynlleth sought advice on how to transfer a stock of bees from an old box into a Wells hive:

> *Nothing is said in the above as to our correspondent's knowledge of bees or their management, and, in case of his being only a beginner, we venture to doubt the wisdom of adapting the Wells system until some experience has been gained. Apart from this word of caution, however, we advise that the bees be allowed to swarm naturally, and the swarm hived in one compartment of the Wells hive...*

Late Victorian beekeeping

The late Victorian era witnessed a game-changing transition from skep to moveable frame hives. Wells was an early exponent of frame hives, keeping his bees in standard frame hives albeit in oversized brood chambers. He ventured the notion the honey bees needed ample space for a productive queen to lay into. Wells is recorded as having kept up to around fifty of such long hives.

George and Esther Wells had twelve children. Great great granddaughter Sarah Austin recounts the milieu of their late 19th Century family home:

> *George Wells was a Calvinist and applied their strict rules to his family. Playing cards were regarded as the Devil's playthings and were barred. The girls however rebelled in secret and would play in their bedroom'. Sundays were clearly observed religiously for G.W. as he states in [the] BBJ [British Bee Journal] that visitors are welcome to come and see his apiary by arrangement*

after 5pm during the week, but at no time on a Sunday. My impressions are that G.W. was an honest man, firm but fair, whilst also being a warm and loving husband, father and grandfather, and possessing a clever wit and humour...

Sarah Austin's grandfather Len Austin, writing his autobiography, conjures up the harsh at the same time idyllic existence of the Wells family[477]:

With so many of the Wells children leaving home, the financial situation of the parents was greatly improved and George started to make provision for his retirement by purchasing six cottages as an investment. In addition to this he had a fairly good addition to his income from bees.

At Bluebell Hill we left the main road and walked down a steep chalky path by the side of an enormous chalk pit used... in the cement trade at Millbay. Eccles was finally reached where we were to stay for one week with our Grandparents Wells. The bungalow smelt lovely, a combination of honey, apples and flowers, the honey room adjacent to the kitchen where separating was carried out being the source of the pervading aroma. Flowers abounded in the garden, front and back, and draped over all the front was a huge grapevine bearing luscious bunches in profusion... There were rosy apples to pick up, late strawberries, raspberries, and best of all many trees of large Morello cherries trained fanwise on the walls and fences. About fifty double hives in two rows were on one side of the garden and the air was full of the sound of bees.

In a followup October 1894 meeting of the British Bee-Keepers' Association Quarterly Converzatione[478] Wells elaborated on details of his doubled hive scheme. To follow this meeting were an enormous volume of notes and correspondence recorded not only in the *British Bee Journal* but also *The Record*.

Members of the British Bee-Keepers' Association even caught the train down to Aylesford to inspect his apiary[479] noting that his bees were kept in immaculate condition. Wells' hives were later featured in the *British Bee Journal* (Figure 2.35)[480].

Figure 2.35   Mr Geo Wells's apiary, Eccles, near Aylesford, Kent 1899.

There was much peripheral banter about the revolutionary Wells hive. As William Woodley, seemingly the other most successful beekeeper of his time, recorded in March 1893[481]:

> *The Wells hive and system is still to the fore... That the system is engrossing the minds of beekeepers, the pages of both bee papers, BBJ and Record, are witness, and I have no doubt appliance makers who advertise the Wells hives get many queries on their working. I, myself, have received several lately on the subject, but my only source of information on the practical working of the Wells system has been the published correspondence in the Journal.*

Woodley was well connected to the beekeeping establishment not to mention royalty and had Britain's largest bee-farm (Figure 2.36)[482].

By no means all other beekeepers of the era were enamoured by Wells' method of operating bees, not least those expressed by William Herrod-Hempsall[483]:

> *Mr. Wells could ... work this system, but he was the only man who ever succeeded in doing so ... those who purchased Wells' hives very quickly adapted them for chicken coops or dog kennels.*

Wells' findings[484], particularly the details of operation of his hives, are amply explained in the pages of both journals from 1892 almost up until his untimely death in 1908.

Figure 2.36   Mr Wm Woodley's home-apiary, Beedon near Newbury circa 1892.

We also now have a pamphlet written by Wells[485] – rescued in mysterious ways by *Northern Bee Books* – that sets out in great detail how he had made his discovery occasioned by his keen observation, diligent attention to detail and inventive spirit.

The pamphlet explains in the simplest terms just how the doubled hive system – fully functional hives with two queens – had evolved. Wells began by describing an ingenious pre-1890 system for overwintering two queens together. He divided his fourteen-frame full depth Langstroth hive bodies equally with a tight fitting central fine wire screen. Each section housed a separate single-queen colony with its own entrance but the colonies benefitted by sharing each other's warmth and came away quickly in late winter.

He then experimented further replacing the screen division with a punched reinforced thin pine board sheet, his so-called Wells Dummy, finding that overwintering bees clustered tightly around this central division board. Wells went still further overlaying the double-brood chambers with oversized queen excluders in early spring topping those with shallow coffin supers fitted out with small comb sections.

It is also hard to realise that reliable queen excluders were really only readily available from around this time and upon which schemes such as that developed by Wells now depended. Prominent amongst the purveyors of the queen excluder were those Gilbert Doolittle[486], C.W. Dayton[487] and Dr G.L. Tinker[488] who recognised their facility in compartmentalisation of hives into brood nest and storage zones -- providing sufficient room for a young queen to lay – and providing facility to operate a second queen.

The American beekeeper Dr C.C. Miller, famed for his *Fifty years among the bees*[489] was asked to venture an opinion as to the new method of management inaugurated by Mr West [sic Wells] in England. In his review he distills the Wells scheme focussed very much on the novelty of queen excluding gizmos[490]:

> *The plan, in brief, is to have a perforated division-board in the centre of a hive, the perforations being queen excluding, a queen in each half of the hive, and a queen-excluder placed over the brood-chamber and under the supers. Thus the workers are allowed to comingle freely, while each queen is kept in her own side of the house.*

Wells maintained powerful hives, aided and abetted by maintaining extra queens year round. His hives produced a useful surplus on early blossom flow while other beekeepers in his district were still building their bees for the main honey flow. His doubled hives, bees on steroids, out-performed the side-by-side overwintered colonies he had previously united to make an early head start. The Wells hives were a cause célèbre: Wells had become an instant, if humble as well as dignified, folk hero.

Highways and Byways of Beekeeping - Alan Wade

The Bee-keepers' Record
AND ADVISER.

A MONTHLY JOURNAL DEVOTED TO PRACTICAL BEE-KEEPING.

No. 28. [New Series.] MAY, 1892. [Vol. III.

WORKING TWO QUEENS IN EACH HIVE.

In our last issue, a promise was made to give fuller details of the system of working an apiary with two queens in each hive, as adopted and carried out by Mr. Wells, of Aylesbury, Kent, and explained by him at a recent *conversazione* of the British Bee-keepers' Association. We therefore propose to give details of the plan sufficiently explicit for those of our readers who may choose to make a trial of it.

Since the publication of the proceedings at the meeting referred to in our weekly, the *B. J.*, several communications have reached us on the subject from bee-keepers more or less interested in it, some expressing high approval of the plan, and a few others seeming inclined to pooh-pooh it, as something quite old, which has already been tried and found wanting. That it is not altogether new—in principle, at least—was freely admitted at the meeting; besides, it is well known to bee-keepers of any extended experience, that the bees of two colonies may readily be got to work amicably in one surplus chamber, so arranged as to extend over both brood nests. But these twin colonies did not prove a success, for reasons which we need not here enter into; it is sufficient to say the practice of working bees on that plan has dropped out of use.

No doubt this failure has given rise to the misgiving expressed by those who make light of, or who fear for the permanent success of, Mr. Wells' plan; but, in considering the matter, it must be borne in mind exactly what that gentleman stated, instead of wandering off into more elaborate theories as to the possibility of 'having young queens mated in hives with a laying queen,' with no interruption of the maternal duties of the latter. Neither must we take it that the plan consists of so joining two stocks that the bees of both may help to fill the same set of supers.

What was set forth and explained by Mr. Wells was his method of carrying out a simple expedient by means of which he gained the important advantage of having two laying queens wintered in each hive, in order to secure doubly strong stocks of bees in spring, thereby reaping the full benefit of the early harvest; and when it is stated that from the whole eleven hives experimented on an average return of 125 pounds of honey was last year taken, few will question the fact that the experiment was a success. Now, when we remember that at certain seasons surplus young queens may be had at the cost of little trouble and no expense beyond the provision for keeping such hives as may on occasions be adaptable for nucleus colonies, it will be seen that the advantages of the plan are open to all whose hives are so constructed as to be suitable for carrying it out.

To do this strictly on the lines laid down by Mr. Wells—and we strongly advise no departure therefrom—we must follow the narrator's own proceedings as given by himself. He told the meeting of his difficulty in preventing swarming, but added that when this occurred he found little or no stoppage of work in the supers or diminution in the quantity of honey stored, by hiving the swarm on frames of foundation, on the old stand, and replacing all the supers. The combs of the original brood nest were then divided into lots of three or four combs, and these lots, each having a good queen-cell on one or other of the combs, were set on separate stands some distance from the original location; and, the young queens having safely hatched and mated, eventually became the second string from which, so helped, to secure the ultimate success of the plan, as follows:—
In the autumn one of these nuclei was

added to each of the old stocks, the two lots being kept apart by the means explained on p. 40 of our last issue. There being, however, already two queens in each hive, it will be obvious that at the time of the second 'joining up' the oldest or least valuable of the two must be removed, and the bees of the hitherto divided brood chamber allowed to intermingle, which, of course, they will do amicably enough. The dummy is then placed between the combs of the old colony and those of the nuclei, and so each year a succession of young queens is maintained in each hive in the natural course of procedure.

The possibility of having two laying queens in the same hive has been several times referred to in the *Record*, and in view of making readers acquainted with the subject in all its bearings we—in addition to the article by Dr. Tinker on p. 45, last month—insert in this issue a paper written by Mr. C. W. Dayton, for our contemporary, *Gleanings*. But while admitting the analogy between the views expressed in these articles, and the case with which we are now dealing, it is very desirable that the two should, for the present at least, be kept apart and gauged each on its own merits. Both Dr. Tinker and Mr. Dayton, as well as Mr. Doolittle, have in view mainly the rearing of young queens for sale, and the novelty of their plans consists in accomplishing this, as well as securing the fertilisation of the queens so raised, in full colonies with laying queens without interruption of the labour, maternal or otherwise, of such stocks as are used for the purpose.

Mr. Wells, on the other hand, simply asked that his hearers should rear a few surplus queens at a time when such are easily available, instead of allowing them to be killed or cast out—as they so often are from swarmed hives—to keep these in nucleus hives till wanted, and in the autumn add them to strong stocks in the manner described. The importance to British bee-keepers of having large populations of working bees at a given time is too well known to need comment; but if anything were needed to give emphasis to this fact, it is surely furnished in Mr. Wells' account of what took place in his own apiary last year. He said that, while his own bees were storing so rapidly, those of a neighbour, whose bees were situate only about forty yards away from his own, gathered food enough for wintering on, but did not get an ounce of surplus honey. 'Another neighbour, about a quarter of a mile off, had three frame hives and six skeps; but he obtained no honey at all from them, and had to feed considerably. In both the above cases one queen only was kept in each hive.' One remarkable circumstance was, that the crop of sainfoin from which his (Mr. Wells') bees gathered the greater portion of the honey, was situated on the other side of his neighbour's grounds, and furthest away from his apiary, so that the bees had to fly over the ground where the hives that yielded nothing were placed, in order to get at the forage. And, while the bees of the latter were doing almost nothing, there was a continuous stream of his (Mr. Wells') bees going to the honey and back again.

In conclusion, it may be said that the best argument in favour of Mr. Wells' plan is the success which, in his hands, has attended the carrying of it out; and, while we cannot pretend to say that equally good results will follow its adoption in every case, it will be useful as well as instructive if such readers as can conveniently make trial of it next year will, during the coming swarming season, preserve a few surplus queens in nuclei in order to utilise them as directed, and report results.

Stan Hughston – Stringy Hughston's coffin hives[491]

In another unlikely tale of rivers of honey, I recently stumbled across two extraordinary beekeepers. One of them is Dave Flanagan. These days he runs forty or so hives to pollinate almonds as a sideline business and says full time beekeeping as a way of making a living – and not living at home – is for the birds. The other beekeeper is Stan Hughston of whom we know little except from some encounters by Des and from what Dave has told Alan about him and the fact that he entered folklore when he became better known as Stringy.

Stringy Hughston ran an amazing enterprise, some 2400 tripled hives – ten frame hives with three queens each – on the Paroo River in the north western corner of New South Wales. Dave worked with him in the seasons of 1990 and 1991 so is well placed to relate the goings on in Stringy's enterprise.

Doubled and tripled hives

Doubled and tripled hives are simply a row of single-queen hives supered over a common queen excluder or queen excluders. Gouget[492] is the only other beekeeper that I've found that has operated tripled hives though doubled hive setups are well known and stretch back to the early 1890s[493].

We illustrate how a doubled hive was set up in Alan's back yard. To start a strong hive was split in autumn. The splits were juxtaposed and each split was requeened with a caged queen (Figure 2.37).

Setting up paired colonies this way allows each split to come away quickly by late winter jump starting the traditional practice of splitting and requeening hives in mid-to-late spring.

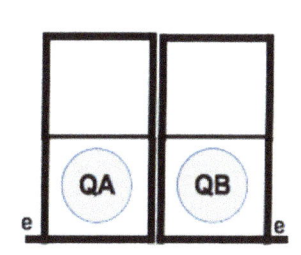

Figure 2.37 Overwintering paired hive setup:
e = entrance; x = queen excluder; QA,
QB autumn introduced queens 8 March 2021.

Then in early spring the overwintered single hives were united by placing a sheet of newspaper and a queen excluder over the two brood nests and adding supers that formed a common nectar ripening and honey storage zone (Figure 2.38).

a   b

Figure 2.38 Spring doubled hive set up 2 September 2021:
e = entrance; x = queen excluder;
QA, QB = autumn introduced queens;
(a) overwintering setup; and
(b) colony reorganised in early spring.

In practice we find that bees in two deck single-queen hives move up into the upper super over winter. Since we employ eight frame gear we repurpose the lower (now empty) super as a honey super and expand the single brood box by adding a half-depth super (as illustrated here) to increase the available space for the young overwintered queen to lay. Note that we have also provided a third hive entrance to the upper storage chambers to avoid the need for house bees receiving nectar from foragers to traverse the brood nest to facilitate more ready honey storage.

Sid Murdoch's super coffin hives at Manjimup and Ken Gray's six-brood chamber hives working Wandoo eucalypt woodland[494] are other examples of apiaries worked on the multiple queened hive principle. Bee historian Eva Crane records her encounter with these operations[495]:

> *I saw multiple-queen hives in commercial use in Western Australia in 1967, that extended horizontally, instead of vertically as skyscrapers do; they were called coffin hives. The unit consisted of a row of six 8-frame Langstroth brood boxes, separated by dividers, their flight entrances facing alternately to either side of the row. A single super (honey-storage chamber), holding 50 frames, extended across all six brood boxes, each of which had its own queen excluder.*
>
> *Alternatively a three-decker outfit was used, with a row of six separate honey supers below the 50-frame super…*

The Paroo triples

In researching doubled and multiple queened hives, Alan got chatting to Dave Flanagan who had worked with Stringy in the early 1990s using tripled queen hives.

Stringy's operation comprised around 7200 (2400 tripled) ten frame working brood chambers. This required a centralised backup apiary, needed to provide a steady and ready supply of building and replacement brood chambers. These hives were generated by splitting strong ten frame colonies and introducing a caged queen to each new colony (Figure 2.39).

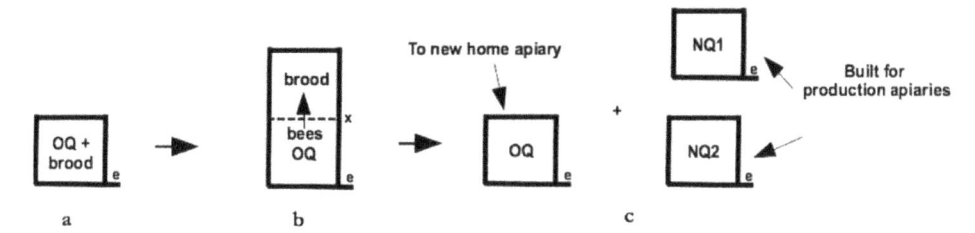

Figure 2.39 Single hive establishment:
e = entrance, x = excluder, OQ = old queen; NQ = new queen:
(a) strong colony;
(b) bees were shaken into empty super, brood and stores moved up; and
(c) brood, stores and bees were distributed and a caged queen introduced to each split.

The scale of this nursery operation is reflected in Des's first recollection of meeting Stringy:

> I watched him once at a Tocal beekeeping field day demonstrating how he used a post hole auger on a tractor to mix sugar syrup and pollen patties. That was the first time I met him.

The main apiary operation was spread some ninety kilometres upstream and a further ninety kilometres downstream of the tiny town of Wanaaring (population 49) in New South Wales Channel Country. Each apiary pod comprised 120 hive pallets, two tripled hives to the pallet. All up we've calculated there must have been ten apiary sites each running 240 tripled hives powered by 7200 queens. Each colony was dynamically managed both for honey production and for regular brood box replacement.

The essential set up was that of three single-queen single ten frame colonies strapped together (Figure 2.40). These were constituted as tripled hives by addition of thirty five frame full depth coffin supers over excluders united using the normal newspaper method. The coffin supers were dynamically managed using standard under-supering and blower techniques to remove bees from honey supers.

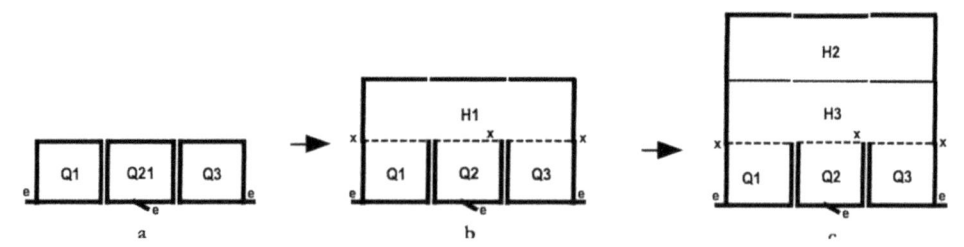

Figure 2.40 Coffin hive setup and operation: e = entrance, x = excluder, Q = queen; H = honey excluder:
(a) three single-queen ten-frame colonies strapped together;
(b) triple queen hive established by papering on a 35 frame coffin hive; and
(c) each tripled hive operated dynamically by under-supering and progressively removing fully capped honey supers.

The harvest

Dave did not really have anything to do with the honey extraction plant in Wanaaring. His job was to cart away extracted stickies, work the floodplain apiaries on regular five-day round trips and bring back as many coffin hives filled with honey as quickly and efficiently as possible (Figure 2.41). Dave tells us that Stringy employed as many in the town who were willing to work in the honey extraction plant. Perhaps one clue as to the scale of the operation is reflected in another of Des's recollections:

> I didn't know Dave Flanagan but did know Stringy Hughston quite well. The last time I saw Stringy was at the Capilano factory at St Marys: he was 86 and had just driven a semi load of honey down from Wanaaring. He told me at one stage the reason he went to the big 35 frame supers was so that he could use a machine loader to lift them off, figuring it was better OHS [occupational health and safety] wise or words to that effect. I don't think OHS existed as an acronym at the time.

Figure 2.41 Dave Flanagan transporting coffin hives on the Paroo River New South Wales in 1990.
Photo: Dave Flanagan

Des also relates a story about Stringy that tells us that his life was not entirely devoted to bees:

> *I must also relate a very funny story about the NSW Conference in 1998 when we shaved his new beard (the first time he ever grew one) off with electric horse clippers to raise money for charity – the humour was in what he did when we squirted tomato sauce over his front to make out that we had cut his throat. He knew we were going to do it in advance, and jumped off the stage yelling we had cut his throat, and it was worse than The Man from Ironbark. We had to chase him around the conference dinner and drag him back screaming. His poor wife thought we had cut his throat and almost fainted.*

After this, maybe you are wondering how on earth anyone except a tall tale teller might have got a name like Stringy. In Des's words:

> *I also found out indirectly how he got his nickname when my truck broke down near Bemboka, and the local garage organised a lift home with a tyre distributor. This guy asked me if I knew Stringy, and said that he (the tyre guy) had carted semi loads of honey tins out of a Stringybark flow in Tumut in 1952, and that Stan Hughston had made so much honey on that flow that he was known as 'Stringy' ever after.*

On hearing this tale, Dave Flanagan responded by telling Alan it was wet that year – most of southeastern Australia experienced once in a century floods in the early 1950s – and that most beekeepers left the area. But of course Stringy stayed and got ten 60 lb tins of honey per hive. That was an amazing amount of honey and anyway, despite the fact that Stringy was running single-queen hives at that time, the name stuck for a lifetime.

As parting thought you might ask why anyone would bother running a hive with more than one queen. John Holzberlein[496] signals that for not too dissimilar two-queen hives, and for about 50% more work with a similar amount of gear, one can obtain the same amount of honey as a beekeeper running double the number of hives:

> *The two-queen system develops a higher degree of efficiency, since the larger the population the more honey is stored per bee. It permits the operation of large colonies without the danger from swarming that results when single-queen colonies are permitted to attain large populations before the honey flow begins.*

The reality is of course that you need some beekeeping acumen to run hives with an extra queen and that their operation is really only suited to exceptionally productive or permanent apiary sites. That doubled or tripled hives are more or less swarm proof is signalled by the fact that failure of a queen does not result in the queen in that chamber being superseded. That unit simply fails entirely as the queen pheromone titre from any remaining queens mitigates against queen replacement.

And a disclaimer: *Northern Bee Books* in West Yorkshire have recently published a book by Alan on doubled and two-queen hives[497]. It provides more details of Stringy's Paroo River venture and about ways beekeepers can get a whole lot more honey out of their bees.

Don Peer – The wisdom of lost beekeeping practice[498]

First and foremost Don Peer knew how to produce honey. As Ron Miksha recalls[499]:

One of the legendary beekeepers of western Canada, Don Peer, a Nipawin beekeeper with an entomology PhD, once told us at a bee meeting, 'If I were king of the world, I'd make a law that every beekeeper had to own one more super for each hive of bees'.

Bees need comb space to hold wet nectar. Dr Peer was astonishingly successful. At first, he ran two-queen colonies from packages. According to Dr Eva Crane (from her book *Making a Beeline*[500]), Don Peer's hives (Figure 2.42) made up to 40 pounds [~18 kg] a day:

I saw his outfit and stood on the back of a truck to reach the top supers. Such tall hives made him switch back to single-queen hives, but even then he stacked supers as high as he could reach. 'Bees need space', he said...

Figure 2.42 One of Don Peer's apiaries in bee heaven. Image ECT-0373, Copyright courtesy of the Eva Crane Trust.

In a subsequent note Miksha notes a record of Eva Crane meeting Don[501]:

> *Doug took me further east to Nipawin to visit Dr Don Peer whom I had met in 1953 when he was a graduate student in Madison, Wisconsin. He had now developed large-scale beekeeping on scientific lines, and had 1000 or more hives. He bought packages of bees each spring and made two-queen colonies from pairs of them. Each of these had 90 to 100,000 bees by July, and could store 20 kg of honey a day from the main flow – mostly from legumes, alfalfa and fireweed.*

These small testaments given, it seemed worthwhile to set out to unearth the contribution that this renowned late 20th Century apiarist brought to beekeeping practice. Discovering much about him has proved a challenge as he, as a commercial apiarist, published little. But his reputation and acumen matched the likes of more famous researcher-beekeepers: François Huber, Johann Dzierżoń, Clayton Farrar, Eva Crane, Colin Butler, E.W. (Egbert) Alexander, George Wells, George Demaree, Eugene Killion, Robert Laidlaw, Robert Banker, Roger Morse, Tom Seeley, Winston Dunham, John Holzberlein, Mykola Haydak, Gilbert Doolittle, Moses Quinby, Lorenzo Langstroth, Floyd Moeller, John Hogg and dozens more.

Ron Miksha, an associate of Don, hints at the success of the Peer outfit[502]:

> *We stood on cold concrete in near darkness, facing row after row of neatly painted honey drums, Don Peer and I. Don pointed to a label, 'Water White. 16.6% Moisture. 664 Pounds' Three hundred forty two barrels of water white honey. Almost a quarter of a million pounds [133 tonnes].*
>
> *'What are you going to do with all this honey, Doctor Peer?' I asked. This was twenty years ago, I was an earnest young man. Don Peer, the only commercial beekeeper I knew with a PhD in entomology, shrugged his shoulders. I looked to him for marketing advice. I took out a notebook. Began to write.*
>
> *'I guess we'll sell whatever we can't eat,' he said.*
>
> *Of course, Don finally delivered the information I needed. He gave me the names of packers and brokers and he told me what the honey market had been doing over the past few months. But all of this was long ago, at his shop in Nipawin, Saskatchewan. In those days, I was producing a hundred thousand pounds [45 tonnes] of honey a year and it was never hard to sell the stuff.*

Following Don's advice, I would call a few packers in the east, send a small sample, consider the offers. A semi-truck would soon arrive at my shop and the honey disappeared within the truck's steel walls. I never had a problem selling to Paul Doyon or Jack Grossman or Elise Gagnon. These people sent cheques when they said they would, paid the money they promised. Every time.

Don Peer published several articles arising from his doctoral dissertation[503] all relating to estimating the mating range of honey bee queens and to the number of times a queen mates. Using genetically recessive Cordovan – leather coloured Italian – drones and queens, Peer was able to demonstrate that queens mate – sometimes in multiple mating flights – on an average of seven times and that they readily cross-mate at distances 9.6 km (6 miles) apart. More contemporary studies[504] signal that the average is more like seventeen matings but varies widely (1-59 matings). In a subsequent study Peer[505], using an isolated mating yard, was able to show that mating success dropped off as the range increases, mainly signalled by delay in queens commencing to lay. This may be interpreted as queens mating to volume of semen never achieved in a few matings:

Genetically-marked virgin queen honey bees were located at various distances up to 22.5 km (14.0 miles) from an apiary stocked with genetically-marked drones in an area containing only these experimental bees.

Some matings occurred across distances up to 19.3 km (10.1 miles). With increasing distances from the drone source a decreasing percentage of queens mated successfully. Queens located at the drone source, 6.1 and 9.8 km (3.8 and 6.1 miles) distant, began laying at approximately the same time. Those located 12.9 km (8.0 miles) distant began laying later and those at 19.3 km (10.1 miles) later still.

Like E.W. Alexander[506], Peer was an inveterate inventor. Here he is (Figure 2.43) with two honey gates no doubt fabricated to manage his giant honey harvests.

Figure 2.43   Peer with two honey gate designs.
Image ECT-0375, Copyright courtesy of the Eva Crane Trust.

Writing about the exigencies of controlling bee shed fires, Miksha[507] also reports:

> A beekeeping colleague in Nipawin, Saskatchewan, Dr Don Peer, years ago built a very long narrow honey shop: 'Just in case of a fire', he told me.

And writing about bee losses, Miksha writes of Peer[508]:

> I am aware that winter losses for most North American beekeepers are higher than keepers can afford. It used to be expected that 15% of colonies would die each winter, according to statements made in the 1970s by commercial beekeeper and research scientist Dr Don Peer. Losses of 30% are frightening. Especially when that is an average. It means some beekeepers lost a whole lot more. For some operators, there is no recovery. Since those 30% averages have been sustained in most parts of the USA, research is conducted to try to find the cause. Everything from varroa mites to neonicotinoids to genetic weaknesses have been implicated.

Peer describes[509] how he learnt to overwinter bees in Saskatchewan winters, seasons that can plummet to -40 °C. Peer's was an era where beekeepers killed their bees using hydrogen cyanide gas starting afresh with package bees each April.

In summarising Peer's method Walker reports:

> *In September the colony is fed, a bottom entrance reducer is put in position with a hole over the top entrance of the hive. In October, pre-cut fibreglass insulation is packed around a group of four hives, and the whole is covered with black building paper. There is a gap in the insulation over each top entrance hole, and this is finally covered with a piece of plywood in which there is a 10x1.9 cm slot allowing moisture to escape from the hive. Another piece of plywood tied to the top of the hive serves as a roof. Of colonies overwintered this way, about 90% have survived, and in the following year their honey production has been 50-125% higher than in control packages.*

Clearly overwintering bees successfully was becoming a strategy superior of restarting hives each spring with package bees.

On the character of Peer, Allen Dick writes in a late 2003 report[510]:

> *I recall, some thirty three years ago, when I was first about to set up beekeeping in Alberta, writing Dr Peer asking advice on my (somewhat simplistic and idealistic) plans. He wrote back a long and very thoughtful letter explaining where I would go wrong, and how I could make my plan work.*
>
> *I've never forgotten that, and if there is one experience that has caused me to want to share what I have learned in beekeeping – once I got to the point where I had something worth sharing – it is that example of neighborliness to a (stupid?) unknown kid in Alberta by Dr Peer, whom I had never met at that time.*
>
> *I did meet him later on a number of occasions, and know he was a man of strong opinions, and some of his opinions I could not support. I guess I learned something from that too. Nonetheless, I always did respect him.*

Earlier Eva Crane recalls her encounter[511] with Peer (Figure 2.44) in rather more detail:

I found Nipawin a stimulating place mentally: Mr Hamilton was very knowledgeable, and almost next door lived Dr DF Peer, whom I had last met in 1953 in Madison, doing research work under Dr CL Farrar—whose methods of management he both practises and preaches. Don Peer left his career in bee research several years ago, because he could get a much higher income by producing honey. More than almost anyone I met, he seemed to have thought out the biological and economic implications of colony development and honey production in the near-optimum conditions in parts of the Prairie Provinces.

…Don Peer uses two-queen colonies on Dr Farrar's system[512]. He gets 2-lb. packages about 23rd April, and uses two per hive, the division board between them being later replaced by a queen excluder. By 23rd June or so the 16 000 bees in the two packages have produced 70 000 or 80 000; by July there are 90 000 or 100 000 in each hive. A thousand or more hives, in apiaries of 25-35, are spread over an area perhaps 50 x 30 miles [80 x 50 km], parts of which are also used by other beekeepers. During my visit Dr Peer gave a party for seed producers and local officials, to discuss pollination problems, and to show them round his honey house; this was a model plant, which made a great impression on all the visitors. I was distressed to learn that, although it is only just completed, it will shortly be destroyed when a dam is built across the river.

Steve Taber, a renowned north American queen breeder and researcher at the US Department of Agriculture Honey Bee Breeding, Genetics, and Physiology Research facility at Baton Rouge facility in Louisiana, came up with this anecdote about Peer[513]:

Some years ago an old acquaintance of mine came out with the idea of requeening a hive with the old queen still there. His name was Don Peer, who lived in what he said was 'bee paradise' in Nipawin, Saskatchewan, Canada. And, for many of you who don't know where that is, it is north of Montana and North Dakota. I have been in the southern part of Saskatchewan, but never that far north.

Figure 2.44   Don Peer and two-queen hives:
(a) with Bill Hamilton [Image ECT-0381]; and
(b) one of Don's many scattered apiaries [Image ECT-0374], Copyright courtesy of the Eva Crane Trust.

A little searching showed that Don had published this finding in 1977 in the *Canadian Beekeeper*[514]. The very notable John Hogg, one of the principle exponents of the Consolidated Brood Nest two-queen hive notes preciently[515]:

> *One seeming exception to this principle is the atypical technique of Peer for requeening without dequeening. In this procedure a ripe queen cell is inserted into a location remote from the broodnest of a queenright colony. The newly hatched (sic emerged) virgin queen will be accepted if she gets to and destroys the unwary queen within that period of time (6-7 hours) when workers, and presumably queens, ignore virgins – possibly because the virgin has no odor to alarm the bees and is as yet unable to secrete stress pheromone.*

Others, such as Jay[516], comparing Peer's achievement with the success of beekeepers who adopted the scheme that were less successful than Peer. Whereas he had managed to replace 80% of 4000 queenright colonies over a three year period in

Saskatchewan in Western Canada – introducing a ripe queen cell to each colony about 2-3 weeks prior to the end of the main honey flow – Skirkevicius[517] in a Lithuanian study had variable success (33 to 100%), while Boch and Avitabile[518] in Eastern Canada only achieved a 15% success rate. Milbrath[519] in reviewing Skirkevicius's study noted that success was only achieved by persistently checking and re-adding queen cells. Of course, Peer was not the first to adopt this supersedure-reliant technique[520] as many commercial beekeepers now routinely requeen colonies this way.

Requeening into a top nucleus colony formed by lifting brood, bees and stores above a nucleus board or a double screen and introducing a queen cell (or a caged queen) has also been widely practiced and is indeed is a very well known queen establishment technique. By then removing the divide and allowing the new well-established queen to supersede the old queen below is widely reported and long preceded Peer's practice. Nevertheless scheme formulated by Peer is time saving, employs cells that are far cheaper and avoids the hassle of locating old queens.

Others contemporaries of Peer, such as Reid[521], add weight to the notion of an old queen being twice as likely to be superseded by a new queen – rather than the other way around – when packages containing new queens were united with overwintered colonies containing old queens.

In 1979, a Frenchman Yves Garezmet met Don Peer subsequently working at Peer's Bee Ranch (Figure 2.45) from May 1983 to October 1989 while he was establishing his own bee business. In his blog[522], Ives records:

> ...*the previous professor of entomology from Madison, Wisconsin, owned and operated 1200 colonies in Nipawin, Saskatchewan. Don offered Yves a job and became his mentor. Just like Don many years before, Yves fell in love with the beautiful hills and prairies surrounding Nipawin. He found an old farm house to rent, and while working for Don during the week, he would manage his own bees in the evenings and week-ends: these were very busy summers for a dad and a mom with little children.*
>
> ...*Upon Don's retirement in 1995, Yves was able to acquire 1000 more hives from Don's outfit, he also bought the quarter of land on which he lived.*

Figure 2.45 Peer's famous bee acreages:
ECT-0371; and
ECT-0372, Copyright courtesy of the Eva Crane Trust.

Peer was also involved in early seasonal studies of infestation of hives by acarine mite (*Acarapis woodi*)[523], an era that preceded the ravages of varroa mite:

> ...package bees infested with A. woodi were installed in two apiaries in early May. Average infestation was initially 27% and it rose to 48% by early June. After decreasing to 30% in early August, it rose to 42% by mid-October. In colonies started with uninfested packages the October level was 7%. In infested nuclei, the infestation increased steadily from 3% in mid-March to 33% in mid-October. As in the 1986 study, brood production was significantly lower in infested colonies than in controls (except on two dates), and this affected the size of the resulting adult population. Preliminary trials with menthol reduced the mite infestation, but the menthol evaporated rather slowly because of low ambient temperatures.

Peer was also an exceptionally keen observer of the changing fortunes of beekeepers[524] in northern Canada:

> Before the 1970s the honey plant we call canola didn't exist. Canola – named for Canada oil – was bred from lesser plants that once yielded a lethal seed. Farmers planted a little of canola's poisonous ancestor – especially during World War II – to make lubricating oil for prairie tractors. Today 20 million acres of canola are planted in Western Canada. The bright yellow flowers attract bees, but the honey from canola crystalizes within days. That's why beekeepers started having problems up in northern Saskatchewan. Beekeepers, of course, learnt to extract early before the honey hardened in their supers. They paid more attention to their bees and honey production soared.

Don Peer, like most exemplary apiarists has almost disappeared from the annals of beekeeping. That said he has taught us a lot about innovative beekeeping practice.

In compiling this retrospective, I've corresponded with that font of Canadian beekeeping practice, Ron Miksha, from whom we learn so much about Don. In his book *Bad Beekeeping*[525] Ron recounts a parallel universe where he ended up leaving a family steeped in beekeeping in Florida to become a Canadian citizen and beekeeper and harvest, like Don Peer, astounding amounts of prairie honey. That was until a mega drought of the 1980s put him out of business.

Ron's recounts a honey packer who kept cards describing the proclivities of the beekeepers he did business with. Don Peer had asked:

Can I see the note card you have on me?

When Don read the card about himself[526], he was surprised to discover it simply said:

Doesn't need money.

Tom Theobald – Where beekeepers fear to tread[527]

Tom Theobald [1942-2021] is the essence of beekeeping folklore (Figure 2.46).

Of a retrospective on Tom, Rachel Gabel writing for *The Fence Post*[528] records Miles McGaughey, an early associate of Tom's, saying that:

> ...he first knew of Theobald after he gave him a stern talking to regarding his speeding while McGaughey was a student at Niwot High School in the late 1970s.

McGaughey, similarly spooked by bees, also noted that each beekeeping book he had borrowed from the Boulder Public Library had Tom Theobald's name on the card as the other borrower. In a serendipitous way Miles also read of his regular column in *The Fence Post* noting that:

> He was the last bee inspector in the state of Colorado and he did it all for free as a service to beekeepers... I always thought he was like the Bee Police, so I kind of kept my head down and avoided him.

Figure 2.46 Tom Theobald, shown on parade with an award-winning float in Boulder Colorado.
Photo: **Miles McGaughey**

Tom Theobald was a great educator and established the Boulder County Beekeeping Association in 1984 to address pesticide issues in the county. He operated a network to warn local beekeepers of the Colorado Department of Agriculture sanctioned aerial spraying plans and took the US Environmental Protection Authority to task in their registration of pesticides that were lethal to bees. Clearly Miles and Tom later got on well, Rachel recording that:

> He [Tom] and McGaughey trucked over 300 packages of bees from California, keeping the bees in Boulder County thriving after the hives were struck with Colony Collapse Disorder.

However it was for Tom Theobald's penchant for operating unconventional two-queen hives that he is best remembered (Figure 2.47). McGaughey records that while most hives in the Boulder district produced a modest 40-80 pounds (18-36 kg) of honey annually, Tom's two-queen hives were producing more like 400 lb (just over 180 kg) each year. Tom, though very much a solitary operator, was a great communicator also publishing in *Bee Culture* and in the *American Bee Journal*. Tom was passionate about bees.

Figure 2.47   Tom Theobald with one of his two-queen tower hives.

In an apiary based 2003 interview[529] Tom talks about how not to set up two-queen hives. He starts by reminding us that the introduction of a caged queen to a queenright colony will always result in the loss of the new queen. He then outlines his method, splitting a hive vertically and introducing a new queen to the upper queenless unit.

Once established, he replaced the division with a double screen, in turn replacing this with a queen excluder to establish a colony with two separate brood nests each with their own honey supers. The technique, replicating that developed by Clayton Farrar, shows how it was assembled detailing the method and pitfalls as well as the superior harvesting performance of the combined workforce compared with the equivalent workforce in two separate colonies. Running two queen hives, up to around 200 was his modus operandi.

He went on to explain that with the arrival of both the tracheal and external mites exacerbated by the accompanying pathogens, he was losing 30% of his colonies every winter curtailing the viability of his operation. Commercial beekeeping had become unviable at least for a beekeeper in Colorado unwilling to migrate his bees.

Kim Flottum, editor of *Bee Culture* first brought to my attention this impressive two-queen Colorado beekeeper. As outlined in a January 2021 *Beekeeping Today* Podcast with Jeff Ott and Kim Flottum[530], we are given a retrospective account of Tom's operation (Figure 2.48):

> *Tom had a small commercial operation in Colorado in the late 70's, running up to 200 or so colonies when he first started using two queen colonies. The biology of running two queens in a colony does make sense if you use the technique Tom perfected, but the better you get, the more lifting you are going to do. His finished production hives had three deeps for brood production and seven mediums for honey storage. At the time, an average to strong colony in his part of Colorado would make about 70 pounds of honey in a season, Tom's two queen colonies averaged between 240 and 270 pounds!*
>
> *The advantages, beside huge honey crops, were that these colonies started with last year's queen, and a new second queen was added in early spring. When the season was over, the new queen almost always prevailed, thus starting the next season with a proven queen. Plus, with a foraging population as large as these colonies produced, there was a vast store house of pollen the next spring for that season's buildup.*

Two-queen honey production is becoming a lost art as it takes a lot of work, timing and an understanding of your area's nectar flows. However, the payoff is big! If you have bees in a location with lots of forage and not too much varroa or pesticide pressure, this approach may work for you.

Figure 2.48   Tom Theobald's fully supered two-queen hives.

In the retrospective of Tom's apiary operations[531] we are told that:

…most hives produce 40 to 80 pounds (9-18 kg) of honey per season, but McGaughey – one of Tom's life long associates – said one of Theobald's two queen hives would produce upwards of 400 pounds (180 kg) per season.

The yields given differ somewhat from those provided in Jeff Ott's and Kim Flottum's interview record with Tom: however the point is that the two-queen hives are very much more productive than equivalent pairs of single queen hives.

What we learn from Theobald's scheme of beekeeping is that, while his honey season was short and sharp, with two queens and hives in optimal condition he could increase his crop severalfold.

Acknowledgements

I would like to thank Dannielle Harden, Jenny and Peter Robinson, John Robinson and Frank Derwent for their co-contributions to Canberra Region *Bee Buzz Box* articles, some published in *The Australasian Beekeeper*, material that formed the basis of some material presented. I also acknowledge the kind permission of the Eva Crane Trust to use their images to illustrate the adventures of Don Peer.

Bibliography

1. *Long live the queen* was published as
   Wade, A. (2017). The honey bee queen Part I: Long live the queen. *The Australasian Beekeeper* **119(6)**:18-24.
   Wade, A. Robinson, J, Robinson, J. and Robinson, P. (2021). The honey bee queen replacement enigma Part I: The colony in transition. *The Australasian Beekeeper* **122(12)**:18-19, 22-24.
   Wade, A. Robinson, J, Robinson, J. and Robinson, P. (2021). The honey bee queen replacement enigma Part II: Requeening practice. *The Australasian Beekeeper* **123(1)**:36-40.
   Wade, A. (2023). Honey bee colony dynamics: – Introduction. *The Australasian Beekeeper* **124(7)**:42-43.
   Wade, A. (2023). Honey bee colony dynamics: Part I Brood nest signatures. *The Australasian Beekeeper* **124(8)**:42-44.
   Wade, A. (2023). Honey bee colony dynamics: Part II Queen cell signatures. *The Australasian Beekeeper* **124(9)**:40-43.

2. Seeley, T.D. (2017a). *Apidologie* **48**:743-754. Life-history traits of wild honey bee colonies living in forests around Ithaca, NY, USA. https://link.springer.com/article/10.1007/s13592-017-0519-1

3. Seeley, T.D. and Morse, R.A. (1976). The nest of the honey bee (*Apis mellifera* L.). *Insectes Sociaux* **23(4)**:495-512. doi:10.1007/BF02223477 http://www.naturalbeekeeping.com.au/Nest%20of%20the%20 Honeybee,%20Seeley%20and%20Morse.pdf https://www.researchgate.net/publication/269996264_The_nest_of_the_honey_bee_Apis_mellifera_L

4. Seeley, T.D. (1989). The honey bee colony as a superorganism.

American Scientist **77(6)**:546-553.
https://www.jstor.org/stable/27856005

5   Heard, T. (2016). *The Australian Native Bee Book*, Chapter 3, Understanding the highly social bees. Sugarbag Bees.

6   Wikipedia (accessed 4 June 2022). Eusociality.
https://en.wikipedia.org/wiki/Eusociality

7   Hughes, W.O.H., Oldroyd, B.P., Beekman, M. and Ratnieks, F.L.W. (2008). Ancestral monogamy shows kin selection is key to the evolution of eusociality. *Science* **320(5880)**:1213-1216. doi:10.1126/science.1156108
https://www.ncbi.nlm.nih.gov/pubmed/18511689

8   Hepburn, R. and Radloff, S.E. (2011). *Honeybees of Asia* pp.57, 374-376, 398-399. Springer, Heidelberg, Dordrecht, London, New York.

9   Hepburn, R. (2011). Chapter 7 Absconding, migration and swarming p.137 *in* Hepburn, R. and Radloff, S.E. (2011). *Honeybees of Asia*.

10 Oliver, R. (2015-2016) series *Understanding colony buildup and decline* provides a detailed account of regulation of bee colony numbers.
See Oliver, R. (2016). *Understanding colony buildup and decline*: Part 9b – The regulation of bee longevity. http://scientificbeekeeping.com/understanding-colony-buildup-and-decline-part-9b/

11 Hepburn, H.R. and Radloff, S.E. (1998). *Honeybees of Africa*. Springer-Verlag. Berlin; New York, 370pp., Chapter 6, Mating and fecundity, pp.133-161 *and* Chapter 7, Absconding, migration and swarming, pp.133-158. http://www.springer.com/gp/book/9783540642213#otherversion=9783642083891
Hepburn, H.R. and Radloff, S.E. (2011). *Honeybees of Asia*. Chapter 5, Swarming, migration and absconding, pp.163-184. https://www.scribd.com/document/334922596/Honeybees-of-Asia

12 Meixner, M.D. and Moritz, R.F.A. (2004). Clique formation of super-sister honeybee workers (*Apis mellifera*) in experimental groups. *Insectes Sociaux* **51(1)**:43-47. doi.org/10.1007/s00040-003-0701-5 https://link.springer.com/article/10.1007/s00040-003-0701-5

13 Free, J.B. (1958). The drifting of honey bees. *The Journal of Agricultural Science* **51(3)**:294-306. doi.org/10.1017/S0021859600035103

14   Collinson, C. (2019). Applying the basics of honey bee biology. UK National Honey Show 14 January 2019. https://www.youtube.com/watch?v=pULjCfKX8_k

15   Möbus, B. (April 1987). The swarm dance and other swarm phenomena Part I. *American Bee Journal* **127(4):**249-251, 253-255.
Möbus, B. (May 1987). The swarm dance and other swarm phenomena Part II. *American Bee Journal* **127(5):**356-362.

16   Farrar, C.L. (1968). Productive management of honey bee colonies. *American Bee Journal* **108(3):**95-97, 141-143.
Farrar, C.L. (1968). Productive management of honey bee colonies. *Apiacta* **4:**22-28. http://www.fiitea.org/cgi-bin/index.cgi
Farrar, C.L. (1968). Productive management of honey-bee colonies. (Vol. 2280). University of Wisconsin--Extension.

17   Butler, C. (1974). *The world of the honeybee,* p.74. Collins New Naturalist, London.

18   Butler, C. (1974) loc. cit. pp.133-134.

19   Hepburn, H.R. and Radloff, S.E. (1998). *Honeybees of Africa*. Chapter 5 Swarming, migration and absconding. 5.3 Supersedure pp.140-141, 147.

20   Butler, C.G. (1957). The process of queen supersedure in colonies of honeybees (*Apis mellifera Linn.*). *Insectes Sociaux* **4(3):**211-223. doi.org/10.1007/BF02222154
https://doi.org/10.1007/BF02222154 https://repository.rothamsted.ac.uk/item/8w2q1/the-process-of-queen-supersedure-in-colonies-of-honeybees-apis-mellifera-linn https://link.springer.com/article/10.1007%2FBF02222154?LI=true
Butler, C.G. (1960). The significance of queen substance in swarming and supersedure in honey bee (*Apis mellifera L.*) colonies. *Proceedings of the Royal Entomological Society of London* **35(7-9):**129-132. doi.org/10.1111/j.1365-3032.1960.tb00681.x http://onlinelibrary.wiley.com/doi/10.1111/j.1365-3032.1960.tb00681.x/full

21   Hepburn, H.R. and Radloff, S.E. (1998). *Honeybees of Africa*. Chapter 5 Swarming , migration and absconding. 5.4.3 Releasing factors in swarming and supersedure. pp.146-147.

22   Butler, C. (1974) loc. cit. p.159.

23   Doolittle, G.M. (1889). Scientific queen-rearing as practically applied being a method by which the best of queen-bees are reared in perfect

accord with nature's ways: for the amateur and veteran in bee-keeping 184pp, p.27. Chicago, Ills. Thomas G. Newman & Son, 923 & 925 West Madison Street. https://archive.org/details/bp_6250849
https://www.biodiversitylibrary.org/bibliography/173039

24 Monty Python Dead parrot sketch (7 December 1969). https://en.wikipedia.org/wiki/Dead_Parrot_sketch

25 Butler, C. (1974) loc. cit. p.164.

26 Hurd, P. (22 August 2018). Epigenetic patterns determine if honeybee larvae become queens or workers. Queen Mary University of London. https://www.qmul.ac.uk/media/news/2018/se/epigenetic-patterns-determine-if-honeybee-larvae-become-queens-or-workers-.html
Paul Hurd presented a lucid presentation on epigenetic queen versus worker development at the Somerset Beekeepers' Association *Honey bees are what they eat* webinar on 3 March 2022.
https://www.eventbrite.co.uk/e/honey-bees-are-what-they-eat-with-dr-paul-hurd-tickets-261796710007

27 Laidlaw, Jr., H.H. (1979). *Contemporary queen rearing,* pp.16-18. Dadant & Sons, Hamilton, Illinois.

28 Winston, M.L. (1987) loc. cit. p.70.

29 Hepburn, H.R. and Radloff, S.E. (1998). *Honeybees of Africa*, p.139.

30 Hogg, J.A. (2006). Colony level honey bee reproduction: The anatomy of reproductive swarming. *American Bee Journal* **146(2):**131-135. http://www.twilightmd.com/Samples/Hogg/Hogg_Halfcomb__Publications/ABJ_2006_February.pdf

31 Demuth, G.S. (1921). Swarm control. *Farmers Bulletin* **1198:**1-45. https://ia601401.us.archive.org/32/items/CAT87202908/farmbul1198.pdf

32 Oliver, R. (2015). *Understanding colony buildup and decline* – Part 7b. Minimizing swarming. http://scientificbeekeeping.com/understanding-colony-buildup-and-decline-part-7b/

33 Simpson, J. (1958a) loc. cit.
Simpson, J. (1959a) loc. cit.
Oliver, R (August 2007). Fat bees – Part I. http://scientificbeekeeping.com/fat-bees-part-1/

34 Winston, M.L. (1987) loc. cit. p.66.

35 Butler (1974) loc. cit.
    Mackenson, O. (1943). The occurrence of parthenogenic females in some strains of honey bees. *Journal of Economic Entomology* **36(3):**465-467. https://academic.oup.com/jee/article-abstract/36/3/465/2202979/The-Occurrence-of-Parthenogenetic-Females-in-Some?redirectedFrom=fulltext doi.org/10.1093/jee/36.3.465 https://academic.oup.com/jee/article-abstract/36/3/465/2202979/The-Occurrence-of-Parthenogenetic-Females-in-Some?redirectedFrom=PDF http://repository.up.ac.za/bitstream/handle/2263/19767/Pirk_Reproductive%282012%29.pdf?sequence=1&isAllowed=y
    Chapman, N.C., Beekman, M., Allsopp, M.H., Rinderer, T.E., Lim, J., Oxley, P.R. and Oldroyd, B.P. (2015). Inheritance of thelytoky in the honey bee *Apis mellifera capensis*. *Heredity* **114:**584-592. http://www.nature.com/hdy/journal/v114/n6/full/hdy2014127a.html

36 Hepburn, H.R. (2001). The enigmatic Cape Honey Bee, *Apis mellifera capensis*. *Bee World* **82(4):**181-191. doi:10.1080/0005772X.2001.11099525

37 Barry, S., Cook, D., Duthie, R., Clifford, D. and Anderson, D. (July 2010). *Future surveillance needs for honeybee biosecurity*. Rural Industries Research and Development Corporation [RIRDC Publication No. 10/107, RIRDC Project No. PRJ-003317] https://www.agrifutures.com.au/wp-content/uploads/publications/10-107.pdf

38 Hepburn, H.R. (2001) loc. cit. p.188.

39 Schneider, S. Deeby, T., Gilley, D. and DeGrandi-Hoffman, G. (2004). Seasonal nest usurpation of European colonies by African swarms in Arizona, USA. *Insectes Sociaux* **51(4):**359-364. doi.org/10.1007/s00040-004-0753-1
    Danka, R.G., Hellmich, R.L. and Rinderer, T.E. (1992). Nest usurpation, supersedure and colony failure contribute to Africanization of commercially managed European honey bees in Venezuela. *Journal of Apicultural Research* **31(3-4):**119-123. doi.org/10.1080/00218839.1992.11101272 https://www.tandfonline.com/doi/abs/10.1080/00218839.1992.11101272

40 Hepburn, H.R. (2001) loc. cit.

41 Winston, M.L. (1987). *The biology of the honey bee*, 281pp. Harvard University Press.
    Winston, M.L. (1992). The biology and management of africanized

honey bees. *Annual Review of Entomology* **37**:173-93. http://www.annualreviews.org/doi/abs/10.1146/annurev.en.37.010192.001133?journalCode=ento https://www.researchgate.net/publication/234150136_The_Biology_and_Management_of_Africanized_Honey_Bees

Winston, M.L., Otis, G.W. and Taylor, Jr., O.R. (1979). Absconding behaviour of the africanized honey bee in South America. *Journal of Apicultural Research* **18(2)**:85-94. doi.org/10.1080/00218839.1979.11099951 https://www.researchgate.net/profile/Nismah_Nukmal/publication/273126221_Discovery_of_Successful_Absconding_in_the_Stingless_Bee_Trigona_Tetragonula_Laeviceps/links/571f04f208aefa648899aa3f/Discovery-of-Successful-Absconding-in-the-Stingless-Bee-Trigona-Tetragonula-Laeviceps.pdf?origin=publication_detail

Winston, M.L. (2014). *Bee time: Lessons from the hive*, Chapter 3, Killer Bees, pp.40-56. Harvard University Press. Cambridge, Massachusetts; London, England.

42 Hepburn, H.R. and Radloff, S.E. (1998). *Honeybees of Africa*. Chapter 4, Introgression and hybridisation in natural populations, pp.103-132.

43 Möbus, B. (1998a). Brood rearing in the winter cluster, Part I. *American Bee Journal* **138(7)**:511-514. https://scientificbeekeeping.com/scibeeimages/Mobus-1998-brood-rearing-in-winter-cluster.pdf.

Möbus, B. (1998b). Rethinking our ideas about the winter cluster, Part II. *American Bee Journal* **138(8)**:587-591. https://scientificbeekeeping.com/scibeeimages/Mobus-1998-winter-cluster-part-2.pdf

Möbus, B. (1998c). The sink: Damp and condensation, Part III: *American Bee Journal* **138(9)**. Published as text at https://polyhive.info/recourses/mobus-bernard-his-work-on-swarming-and-wintering/damp-condensation-and-ventilation/

Möbus, B. (1991). Damp, condensation and ventilation, Part IV. *The Beekeeper's Annual* pp. 90-112. https://www.cabdirect.org/cabdirect/abstract/19910230484

Möbus, B. (1979). Brood rearing in the winter cluster: A new winter cluster model. Apimonda report, Athens pp.244-260. Cited by Mary Ed Fisher (1990). *The Beekeepers Annual* 1991 Honey and Bee. https://books.google.com.au/books/about/The_Beekeepers_Annual.html?id=IUhMAAAAYAAJ&hl=en&redir_esc=yQuiney.

Quiney Honey and Bee [Adrian Quiney] (February 25, 2021). Looking at winter ventilation through the writings of Bernhard Möbus. https://www.youtube.com/watch?v=cx0Rp3gaqwU

44 Hogg, J.A. (2006) loc. cit.

45 Brichter, G. (1921). Two queens in a hive. *British Bee Journal and Bee-Keeper's Adviser* **49**:384. https://ia800203.us.archive.org/17/items/britishbeejourna1921lond/britishbeejourna1921lond.pdf

46 Wade, A. (2018). The honey bee queen Part II: The heiress to the throne. *The Australasian Beekeeper* **119(7)**:24-29.

47 Butler, C.G. (1959a). Bee Department Report of Rothamsted Experimental Station, pp.144-150. https://w3.avignon.inra.fr/dspace/bitstream/2174/402/1/bures%20B12.pdf
Simpson, J. (1957a) loc. cit.

48 Hepburn, H.R. and Radloff, S.E. (1998). *Honeybees of Africa*. Chapter 5, Swarming, migration and absconding. 5.4.1 Frequency distributions for swarming and supersedure, pp.140-141.

49 Theobald, T. (31 July 2003). MROHP [Maria Rogers Oral History Program (Colorado)] Interviews: Tom Theobald (OH1157) Boulder Public Library. https://www.youtube.com/watch?v=q0uwZitB3_o

50 Wade, A. (2017). Bee Buzz Box October 2017. Establishing two queens instead of one queen in a honey bee colony. Part I Principles of introducing and running two-queen colonies. https://actbeekeepers.asn.au/establishing-two-queens-instead-of-one-queen-in-a-honey-bee-colony/

51 Hogg, J.A. (May 1, 1983). Methods for double queening the consolidated broodnest hive: Part 1 The Fundamentals of queen introduction. *American Bee Journal* **123(5)**:383-388. http://www.twilightmd.com/Samples/Hogg/Hogg_Halfcomb___Publications/ABJ_1983_1May.pdf
Hogg, J.A. (June 2, 1983). Methods for double queening the consolidated broodnest hive: The fundamentals of queen introduction. Part 2 Conclusion. *American BeeJournal* **123(6)**:450-454. http://www.twilightmd.com/Samples/Hogg/Hogg_Halfcomb___Publications/ABJ_1983_2June.pdf

52 Oldroyd, B.P. and Wongsiri, S. (2006). *Asian honey bees: Biology, conservation and human interactions*. Harvard University Press. http://trove.nla.gov.au/version/45924294 Australian National Library Dewey Number N 595.799 044

53 Wade, A. (2021). *A history of keeping and managing doubled and two-queen hives,* 177pp. Northern Bee Books, Scout Bottom Farm, Mytholmroyd, West Yorkshire HX7 5JS. https://www.northernbeebooks.co.uk/?s=wade

54 Miller CC. (1890). Queens stinging workers: Two queens in a hive: The young one superseded first. *Gleanings in Bee Culture* **18(24):**875-876. https://www.biodiversitylibrary.org/item/56896#page/767/mode/1up

55 Thompson, J. (27 October 2017). Mysteries in Dr Charles C. Miller's life. *Bee Culture.* https://www.beeculture.com/mysteries-dr-charles-c-millers-life/

56 Peer, D.F. (1977). Requeening with queen cells. *Canadian Beekeeper* **6(8):**89. Cited by Milbrath, M. (September 2020). Beekeeping basics: Working with queen cells, 5pp. https://static1.squarespace.com/static/56818659c21b86470317d96e/t/61101be7f746a1532ac2b768/1628445683539/Milbrath-article_September2020_July24-255PM.pdf

57 Boch, R. and Avitabile, A. (1979). Requeening honeybee colonies without dequeening. *Journal of Apicultural Research* **18(1):**47-51. doi.org/10.1080/00218839.1979.11099943
Forster, I.W. (1972). Requeening honey bee colonies without dequeening. *New Zealand Journal of Agricultural Research* **15(2):**413-419. doi/pdf/10.1080/00288233.1972.10421270
Harris, J.L. (2010). The effect of requeening in late July on honey bee colony development on the Northern Great Plains of North America after removal from an indoor winter storage facility. *Journal of Apicultural Research* **49(2):**159-169. doi:10.3896/IBRA.1.49.2.04 https://www.researchgate.net/publication/272812304_The_effect_of_requeening_in_late_July_on_honey_bee_colony_development_on_the_Northern_Great_Plains_of_North_America_after_removal_from_an_indoor_winter_storage_facility
Invernizzi, C., Harriet, J. and Carvalho, S. (2006). Evaluation of different queen introduction methods in honeybee colonies in Uruguay. *Apiacta* **41:**1-20. https://www.researchgate.net/profile/Jorge-Harriet/publication/238110747_Evaluation_of_different_queen_introduction_methods_in_honeybee_colonies_in_Uruguay/links/63601cc096e83c26eb6ea502/Evaluation-of-different-queen-introduction-methods-in-honeybee-colonies-in-Uruguay.pdf
https://citeseerx.ist.psu.edu/ document?repid=rep1&type=pdf&doi=

Jay, S.C. (1981). Requeening queenright honeybee colonies with queen cells or virgin queens. *Journal of Apicultural Research* **20(2)**:79-83. doi.org/10.1080/00218839.1981.11100476

Reid, G.M. (1979). Requeening honey bee colonies without dequeening using protected queen cells. *New Zealand Beekeeper* **40(3)**:15-17 and XXVI International Beekeeping Congress, Bucharest, pp.249-252.

Reid, G.M. (1977). Requeening honey bee colonies without dequeening using protected queen cells. Apimondia Publishing House. pp.249-253.

58 Farrar, C.L., Miller, L.F., Dunham, W.E., Schaefer, E.A., Holzberlein Jr., J. and Cale, G.H. (1954). Panel for April: How to use two queens for automatic requeening, swarm control, and crop increase. *American Bee Journal* **94(4)**:128-132. https://archive.org/details/sim_american-bee-journal_1954-04_94_4/page/128/mode/1up

Farrar, C.L., Dunham, W.E., Miller, L.F. and Holzberlein, Jr., J.W. (1953). Spotlight: Two-queen management. *American Bee Journal* **93(3)**:107-115, 117. https://archive.org/details/sim_american-bee-journal_1953-03_93_3/page/107/mode/1up

Holzberlein Jr., J.W. (May 1952). Swarm prevention—not swarm control. *American Bee Journal* **92(5)**:195-196. https://archive.org/details/sim_american-bee-journal_1952-05_92_5/page/195/mode/1up

59 Farrar, C.L. (1953). Two-queen colony management. *American Bee Journal* **93(3)**:108-110, 117. Republished as Farrar, C.L. (1953). Two-queen colony management. *Bee World* **34(10)**:189-194. doi.org/10.1080/0005772X.1953.11094821

60 Laidlaw Jr., H.H. (1979) loc. cit.

61 Rhodes, J. and Somerville, D. (May 2003). Introduction and early performance of queen bees: Some factors affecting success. Rural Industries Research and Development Corporation. RIRDC Publication No. 03/049, 44pp. https://www.agrifutures.com.au/wp-content/uploads/publications/03-049.pdf

Rhodes, J. W. (October 2008). Semen production in drone honeybees. Rural Industries Research and Development Corporation. RIRDC Pub. No. 08/130. 95pp. https://www.agrifutures.com.au/wp-content/uploads/publications/08-130.pdf

Brother Adam (2023). Queen rearing: Top quality queens. *The Beekeepers Quarterly* 153:56-57.

62 Sections of *The honey bee swarm* were published as
Wade, A. (2017). Swarming in honey bees Part I: The role of the worker and the queen bee in honey bee swarming. *The Australasian Beekeeper* **119(2)**:30-34.
Wade, A. (2017). Swarming in honey bees Part II: Swarming across the honey bee genus and non-reproductive swarming. *The Australasian Beekeeper* **119(3)**:32-36.
Wade, A. (2017). Swarming in honey bees Part III: Conventional swarm control measures. *The Australasian Beekeeper* **119(4)**:48-51.
Wade, A. (2017). Swarming in honey bees Part IV: Advanced swarm control measures. *The Australasian Beekeeper* **119(5)**:40-42.
Wade, A. (2021). Swarming round-up: Building bees for a bumper honey season. *The Australasian Beekeeper* **123(2)**:20-23. Supplemented by an updated version presented to the Canberra Region Beekeepers Bee Buzz Box in November 2022 as *Requeening with swarm control*.

63 Wheler, G. (1682). *A journey into Greece*, London in Crane, E. (1963). *The archaeology of beekeeping, Chapter 9. Towards movable-frame beekeeping*, pp.196-212. Gerald Duckworth & Co. Ltd, The Old Piano Factory, London.
Wheler, G. (1682). *A journey into Greece, by George Wheler, Esq., in company of Dr Spon of Lyons*. London, 483pp. Printed for William Cademan, Robert Kettlewell, and Awnsham Churchill. *Book VI Athens to Attica* pp.412-413. Detailed by
Crane, E. (1963). *The archaeology of beekeeping,* Chapter 9. Towards movable-frame beekeeping, pp.196-212. Gerald Duckworth & Co. Ltd, The Old Piano Factory, London.
Wheler, G. (1682). The sixth book containing several journeys from Athens to the adjacent places of Attica, Corinth, Bœotia &c, *Athens to Attica: Ordering of bees*. pp.412-413 in *A journey into Greece by George Wheler Esq in company of Dr Spon of Lyons* in six books containing:
  I. A voyage from Greece to Conſtantinople
  II An account of Conſtantinople and adjacent places
  III A voyage through the Leſſer Aſia
  IV A voyage from Zant through ſeveral parts of Greece to Athens
  V. An account of Athens
  VI Several journeys from Athens, into Attica, Corinth, Bœotia &c
With a variety of scuptures.London. Printed for William Cademan, Robert Kettlewell, and Awnſham Churchill at the Paper Head in the New-Exchange, the Hand and Scepter in Fleetſreet, and the Black Swan near Amen Corner. MDCLLXXXII. https://books.google.com.au/books?id=xt1O

AAAAcAAJ&printsec=frontcover&source=gbs_ge_summary_r&cad=0#v=onepage&q&f=false

Wheler, G. (1679). *Voyage d'Italie, de Dalmatie, de Grece, et du Levant, fait és années1675 et 1676-fpar Jacob Spon et George Wheler*, volume 2, 564pp. https://archive.org/details/ned-kbn-all-00003742-001/mode/1up

64 Huber, F. (1806). *New observations on the natural history of bees,* Volume I by François Huber. Letters to M. Bonnet. These letters written between 1787 and 1791 were not published until 1806. http://www.bushfarms.com/huber.htm https://books.google.com.au/books?id=jXdlAAAAMAAJ&printsec=frontcover&dq=huber+leaf+hive+photo&hl=en&sa=X&ved=0ahUKEwiMsvTInvniAhUCfSsKHZ_aDIAQ6AEIMzAC#v=onepage&q&f=false

65 Dzierzon, J., Abbott, C.N., Dieck, H. and Studderd, S. (1882). *Dzierzon's rational bee-keeping or The theory and the practice of Dr. Dzierzon.* Translated by Dieck, H. and Stutterd, S. Edited and revised by Charles Nash Abbott. London: Houlston & Sons, Paternoster Square, Southall: Abbott, Bros. https://ia800201.us.archive.org/17/items/dzierzonsration00stutgoog/dzierzonsration00stutgoog.pdf http://reader.library.cornell.edu/docviewer/digital?id=hivebees5017629#page/5/mode/1up

66 Beesource blog (2016). Jan Dzierżoń (Johann Dzierzon) - True father of beekeeping. Overseas Beekeeper. http://www.beesource.com/forums/showthread.php?328732-Jan-Dzier%26%23380%3Bo%26%23324%3B-(Johann-Dzierzon)-true-father-of-beekeeping

67 Stamp, J. (2013). *The secret to the modern beehive is a one-centimeter air gap.* Smithsonian online at http://www.smithsonianmag.com/arts-culture/the-secret-to-the-modern-beehive-is-a-one-centimeter-air-gap-4427011/

68 Mangum, W.A. (May 1, 2007). Excerpt of *Honey bee biology*: Henry Alley: Pioneer queen producer. *American Bee Journal* **157(5).** https://bluetoad.com/publication/?m=5417&i=397484&view=articleBrowser&article_id=2754069&ver=html5 https://americanbeejournal.com/henry-alley-pioneer-queen-producer/#:~:text=Alley%20used%20an%20empty%20brood,and%20put%20into%20mating%20nucs

Alley, H. (1883). The beekeeper's handy book, or, twenty-two years' experience in queen-rearing, 236pp. https://babel.hathitrust.org/cgi/

pt?id=ncs1.ark:/13960/t9n30dt4v&view=1up&seq=11
Alley, H. (1903). Improved queen-rearing, or, how to rear large, prolific, long-lived queen bees, 67pp. Chas. A. King, Beverly Massachusetts https://archive.org/details/improvedqueenrea00alle

69   Doolittle, G.M. (1889) loc. cit.

70   Miller, C.C. (1911). *Fifty years among the bees.* The A.I. Root Company, Medina, Ohio. https://apiardeal.ro/biblioteca/carti/Straine/EN_-_50_years_among_the_bees_-_dr.C.C.Miller_1911_-_457_pag.pdf http://soilandhealth.org/wp-content/uploads/0302hsted/030208miller/030408miller.PDF

71   Winston, M.L. (1987) loc. cit.

72   Cale, G.H., Banker, R. and Powers, J. (1975). Management for honey production in *The hive and the honey bee.* Dadant & Sons, Hamilton, Illinois, pp.380-384.

73   Winston, M.L., Higo, H.A., Colley, S.J., Pankiw, T. and Slessor, K.N. (1991). The role of queen mandibular pheromone and colony congestion in honey bee (*Apis mellifera* L.) reproductive swarming (Hymenoptera: Apidae). *Journal of Insect Behavior* **4(5):**649-660. doi.org/10.1007/BF01048076 https://link.springer.com/article/10.1007/BF01048076

74   Butler, C. (1974) loc. cit.

75   Butler, C.G. (1959b). Queen substance. *Bee World* **40(11):**269-275. http://tandfonline.com/doi/abs/10.1080/0005772X.1959.11096745

76   Bortolotti, L. and Costa, C. (2014). Chapter 5 Chemical communication in the honey bee society in *Neurobiology of Chemical Communication.* Mucignat-Caretta C. (ed). Boca Raton (F.L.): CRC Press/Taylor & Francis. ncbi.nlm.nih.gov/books/NBK200983/#!po=9.26396 https://www.ncbi.nlm.nih.gov/books/NBK200983/

77   Simpson, J. (1957a). The incidence of swarming among colonies of honeybees in England. *Journal of Agricultural Science* **49(4):**387-393. doi.org/10.1017/S0021859600038375 https://www.cambridge.org/core/journals/journal-of-agricultural-science/article/the-incidence-of-swarming-among-colonies-of-honey-bees-in-england/B4FF7A75BCA3C3E8731D517235A05373
Simpson, J. (1957b). Observations on colonies of honey-bees subjected

to treatments designed to induce swarming. *Proceedings of the Royal Entomological Society London Series A General Entomology* **32(10-12)**:185-192. doi.org/10.1111/j.1365-3032.1957.tb00369.x

Simpson, J. (1958a). The factors which cause *Apis mellifera* to swarm. *Insectes Sociaux* **5(1)**:77-95. https://link.springer.com/article/10.1007%2FBF02222430?LI=true doi.org/10.1007/BF02222430

Simpson, J. (1958b). The problem of swarming in beekeeping practice. *Bee World* **39(8)**:193-202. doi.org/10.1080/0005772X.1958.11095063

Simpson, J. (1959a). Variation in the incidence of swarming amongst colonies of *Apis mellifera* throughout summer. *Insectes Sociaux* **5(1)**:77-95. https://link.springer.com/article/10.1007/BF02223793

Simpson, J. (1959b). The amounts of hive-space needed by colonies of European *Apis mellifera*. *Journal of Apicultural Research* **8(1)**:3-8. doi.org/10.1080/00218839.1969.11100210

Simpson, J. (1962). Work at Rothamsted on honeybee swarming. Rothamsted report for 1962. pp.254-259. http://www.era.rothamsted.ac.uk/eradoc/OCR/ResReport1962-236-259.

Simpson, J. and Riedel, I.B.M. (1963). The factor that causes swarming in honeybee colonies in small hives. *Journal of Apicultural Research* **2(1)**:50-54. doi.org/10.1080/00218839.1963.11100056

Simpson, J. and Riedel, I.B.M. (1964). The emergence of swarms from *Apis mellifera* colonies. *Behaviour* **23(1-2)**:140-148. https://www.jstor.org/stable/4533084

Simpson, J. and Moxley, E. (1971). The swarming behaviour of honeybee colonies in small hives and allowed to outgrow them. *Journal of Apicultural Research* **10(3)**:109-113. doi.org/10.1080/00218839.1971.11099681

Simpson, J. (1973). Influence of hive space-restriction on tendency of honeybee colonies to rear queens. *Journal of Apicultural Research* **12(3)**:183-186. doi.org/10.1080/00218839.1973.11099747

78 Willlams, I.H., Ball, B.V., Tomkins, P.W. and Carreck. (1993). Rothamsted: Cradle of agricultural and apicultural research. *Bee World* **74(2)**:61-74. doi:10.1080/0005772X.1993.11099160

79 Simpson, J. (1957a) loc. cit.
Simpson, J. (1962) loc. cit. pp.254-259.

80 Drake, S.A. (1901). Telling the bees. *New England Legends and Folk Lore*, pp.314-315. Boston: Little Brown and Co. ISBN 978-1-58218-443-2.

https://archive.org/details/bookofnewengland00dra/page/314/mode/1up

81　Bortolotti, L. and Costa, C. (2014) loc. cit.

82　Hogg, J.A. (2006) loc. cit.

83　Hogg, J.A. (1997). Comb honey III: The swarm. *American Bee Journal* **137(12)**:875-883. http://www.twilightmd.com/Samples/Hogg/Hogg_Halfcomb__Publications/ABJ_1997_3December.pdf

84　Möbus, B. (April 1987) loc. cit.
　　Möbus, B. (May 1987) loc.cit.

85　Wikipedia (accessed 1 June 2020). Telling the Bees. Charles Napier Hemy's painting *The widow* (1895). https://en.wikipedia.org/wiki/Telling_the_bees

86　Butler, C.G. (1959c). Extraction and purification of queen substance from queen bees. *Nature* **184**:1871.

87　Wheeler, W.M. (1908). The polymorphism of ants. *Annals of the Entomological Society of America* **1(1)**:38-69. doi:10.1093/aesa/1.1.39. https://archive.org/stream/ants_10548#page/n1/mode/1up

88　Demuth, G.S. (1921) loc. cit.

89　Root, A.I. (1980). *The ABC and XYZ of Bee Culture*, p.507. A.I. Root Company, Medina, Ohio. Huber, F. (1806) loc. cit.

90　Killon, E.E. (1981). *Honey in the comb*. First edition. Dadant & Sons, Hamilton, Illinois.

91　Hepburn, H.R. and Radloff, S.E. (1998). *Honeybees of Africa*. Chapter 5, Swarming, migration and absconding, pp.138-139.

92　Oliver, R. (January 2016). Understanding colony buildup and decline: The regulation of bee longevity. Parts 9a. http://scientificbeekeeping.com/understanding-colony-buildup-and-decline-part-9a/
　　Oliver, R. (February 2016). Understanding colony buildup and decline: The regulation of bee longevity. Parts 9b. https://scientificbeekeeping.com/understanding-colony-buildup-and-decline-part-9b/

93　Oliver, R. (2017). Fat bees - Part 1 http://scientificbeekeeping.com/fat-bees-part-1/

94　Dzierzon, J., Abbott, C.N., Dieck, H. and Studderd, S. (1882) loc. cit. p.7.

95 Brother Adam (1983). In search of the best strains of bees and the results of the crosses and races, 208pp. Northern Bee Books, Hebden Bridge, West Yorkshire, UK.

96 Winston, M.L. (1987) loc. cit.

97 Oliver, R. (March 2014). What's happening to the bees? – Part 2 http://scientificbeekeeping.com/whats-happening-to-the-bees-part-2/

98 Mangum, W.A. (May 1, 2009). Usurpation: A colony taken over by a foreign swarm. *American Bee Journal* **149(5)**. https://americanbeejournal.com/usurpation-a-colony-taken-over-by-a-foreign-swarm/
Mangum, W.A. (2010). Honey bee biology: The usurpation (takeover) of established colonies by summer swarms in Virginia. *American Bee Journal* **150(12)**:1139-1144. Cited by Oliver, R. (2014). What's happening to the bees? – Part 2. http://scientificbeekeeping.com/whats-happening-to-thebees-part-2/
Mangum, W.A. (2012). Colony takeovers (usurpations) by summer swarms: They chose poorly. *American Bee Journal* **153(1)**:73-75. Cited by Oliver, R. (2014) loc. cit.
Mangum, W.A. (2013) Summer swarms with queen balling. *American Bee Journal* **153(2)**:163-165. Cited by Oliver, R. (2014) loc. cit.
Mangum, W.A. (January 1, 2018). More usurpation: Colony take over by summer swarms. *American Bee Journal* **161(1).** https://americanbeejournal.com/usurpation-colony-take-summer-swarms/
Mangum, W.A. (May 1, 2019). Usurpation: A colony taken over by a foreign swarm. *American Bee Journal* **159(5)**:537-541. https://americanbeejournal.com/usurpation-a-colony-taken-over-by-a-foreign-swarm/ https://bluetoad.com/publication/?i=580808&article_id=3356173&view=articleBrowser
Mangum, W. (December 1, 2020). Summer swarms: A prelude to usurpation (colony takeover). *American bee Journal* **160(12)**:1321-1326. https://abj-fe8.kxcdn.com/wp-content/uploads/2021/01/abj_december_2020_final-bt-sm.pdf
https://americanbeejournal.com/summer-swarms-a-prelude-to-usurpation-colony-takeover/
Mangum, W.A. (December 1, 2021). Colony usurpation (takeover): The fourth method of queen replacement – Part 1. *American Bee Journal* **161(12)** https://americanbeejournal.com/colony-usurpation-takeover-the-fourth-method-of-queen-replacement-part-1/
Mangum, W.A. (January 1, 2022). Colony usurpation (takeover):

Situations that stall the takeover process – Part 2. *American Bee Journal* **161(1)** https://americanbeejournal.com/colony-usurpation-takeover-situations-that-stall-the-takeover-process-part-2/

99 Oldroyd, B.P. and Wongsiri, S. (2006) loc. cit.

100 Hepburn, H.R. and Radloff, S.E. (1998). *Honey bees of Africa*, p.133.

101 Winston, M.L. (1987) loc. cit. pp.132-138.

102 Hepburn, H.R. (2011). Absconding, migration and swarming, Chapter 7, pp.113-158. *in* Hepburn, R. and Radloff, S.E. (2011). *Honeybees of Asia*.

103 Oldroyd, B.P. and Wongsiri, S. (2006) loc. cit.

104 Hepburn, H.R. and Radloff, S.E. (1998). *Honeybees of Africa.* Chapter 5, Swarming, migration and absconding. pp.131-161.

105 Robinson, W.S. (2012). Migrating giant honey bees (*Apis dorsata*) congregate annually at stopover site in Thailand. *PLOS One* **7(9)**:e44976. doi:10.1371/journal.pone.0044976 https://www.ncbi.nlm.nih.gov/pmc/articles/PMC3446981/ http://journals.plos.org/plosone/article/file?id=10.1371/journal.pone.0044976&type=printable

106 Underwood, B.A. (1990a). Seasonal nesting cycle and migration patterns of the Himalayan honeybee *Apis laboriosa*. *National Geographic Research* **6(3)**:276-290. https://www.researchgate.net/publication/279756773_Seasonal_nesting_cycle_and_migration_patterns_of_the_Himalayan_honey_bee_Apis_laboriosa
Underwood, B.A. (1992). Impacts of human activities on the Himalayan honey bee, *Apis laboriosa, Honey Bee Resources in Verna*, L.R.(ed). Honeybees in mountain agriculture. Oxford & IBH, New Delhi, Chapter 4, 51-57. http://lib.icimod.org/record/25252/files/c_attachment_411_5173.pdf
Woyke, J., Wilde, J. and Wilde, M. (2001). A scientific note on *Apis laboriosa* winter nesting and brood rearing in the warm zone of the Himalayas. *Apidologie* **32(6)**:601-602. doi:10.1051/apido:2001104 https://www.researchgate.net/publication/239521809_A_scientific_note_on_Apis_laboriosa_winter_nesting_and_brood_rearing_in_the_warm_zone_of_Himalayas
https://www.academia.edu/3397843/A_scientific_note_on_Apis_laboriosa_winter_nesting_and_brood_rearing_in_the_warm_zone_of_Himalayas
https://hal.archives-ouvertes.fr/hal-00891911/document

107 Woyke, J. and Wilde, M. (2003). Periodic mass flights of *Apis laboriosa* in Nepal. *Apidologie* **34(2):**121-127. doi:10.1051/apido:2003002 https://hal.archives-ouvertes.fr/hal-00891762/document
Woyke, J., Wilde, J. and Wilde, M. (2012). Swarming and migration of *Apis dorsata* and *Apis laboriosa* honey bees in India, Nepal and Bhutan. *Journal of Apicultural Science* **56(1):**81-91. doi:10.2478/v10289-012-0009-7
https://sciendo.com/article/10.2478/v10289-012-0009-7 https://sciendo.com/downloadpdf/journals/jas/56/1/article-p81.xml

108 Treza, R. (2013). *Hallucinogen honey hunters - Hunting mad honey*, YouTube, 26 September 2013, 26:39. https://youtu.be/Y_b2i_FvYPw

109 Koeniger, N., Koeniger, G. and Smith, D. (2011). Phylogeny of the genus Apis, Chapter 2, pp.23-50 *in* Hepburn, R. and Radloff, S.E. (2011). *Honeybees of Asia*.

110 Hepburn, H.R. (2011). Absconding, migration and swarming, Chapter 5, pp.133-161 *in* Hepburn, R. and Radloff, S.E. (2011). *Honeybees of Asia*. http://www.springer.com/gp/book/9783540642213#otherversion=9783642083891

111 Mutsaers, M. (2010). Seasonal absconding of honeybees (*Apis mellifera*) in tropical Africa. *Proceedings of the Netherlands Entomological Society Meeting* **21:**55-60. https://www.nev.nl/pages/publicaties/proceedings/nummers/21/55-60.pdf

112 Schneider, S.S. and McNally, L.C. (1992). Factors influencing seasonal absconding in colonies of the African honey bee, *Apis mellifera scutellata*. *Insectes Sociaux* **39(4):**403-423. doi:10.1007/BF01240624
https://www.researchgate.net/publication/227214335_Factors_influencing_seasonal_absconding_in_colonies_of_the_African_honey_bee_Apis_mellifera_scutellata

113 Koetz, A. (2013). The Asian honey bee (*Apis cerana*) and its strains - with special focus on *Apis cerana* Java genotype – Literature review, Department of Agriculture, Fisheries and Forestry, State of Queensland, 58pp.
https://www.planthealthaustralia.com.au/wp-content/uploads/2018/10/Asian-Honey-Bee-Literature-Review.pdf

114 Robinson, W.S. (2011). Observations on the behaviour of a repeatedly absconding *Apis andreniformis* colony in northern Thailand. *Journal of Apicultural Research* **50(4):**292-298. doi:10.3896/IBRA.1.50.4.06 https://

www.researchgate.net/publication/260057687_Observations_on_the_behaviour_of_a_repeatedly_absconding_Apis_andreniformis_colony_in_northern_Thailand

115 Wongsiri, S., Lekprayoon, C., Thapa, R., Thirakupt, K., Rinderer, T.E., Slyvester, H.A., Oldroyd, B.P. and Booncham, U. (1996). Comparative biology of *Apis andreniformis* and *Apis florea* in Thailand. *Bee World* **77(4)**:23-35.
https://www.ars.usda.gov/ARSUserFiles/64133000/PDFFiles/301-400/336-Wongsiri--Comparative%20Biology%20of%20Apis.pdf

116 Hepburn, H.R., Radloff, S.E., Otis, G.W., Fuchs, S., Verma, L.R., Ken, T., Chaiyawong, T., Tahmasebi, G., Ebadi, R. and Wongsiri, S. (2005). *Apis florea:* morphometrics, classification and biogeography. *Apidologie* **36(3)**:359-376. doi:10.1051/apido:2005023 https://hal.archives-ouvertes.fr/hal-00892146/document

117 Demaree, G. (1892). How to prevent swarming. *American Bee Journal* **29(17)**:545-546. https://archive.org/details/sim_american-bee-journal_1892-04-21_29_17/page/545/mode/1up
http://reader.library.cornell.edu/docviewer/digital?id=hivebees6366245_6511_017#page/16/mode/1up

118 Demaree, G.W. (1884). Controlling increase, etc. *American Bee Journal* **20(39)**:619-620. https://www.biodiversitylibrary.org/item/73043#page/625/mode/1up
http://reader.library.cornell.edu/docviewer/digital?id=hivebees6366245_6497_039#page/9/mode/1up
https://ia903406.us.archive.org/4/items/sim_american-bee-journal_1884-09-24_20_39/sim_american-bee-journal_1884-09-24_20_39.pdf
http://bees.library.cornell.edu/cgi/t/text/pageviewer-idx?c=bees;cc=bees;rgn=full%20text;idno=6366245_6497_039;didno=6366245_6497_039;view=image;seq=9;node=6366245_6497_039%3A2.2;page=root;size=s;frm=frameset

119 Dyce, E.J. (1927). A study of the swarm control of bees. MSc Thesis, 213pp. Cornell http://reader.library.cornell.edu/docviewer/digital?id=hivebees7186390#page/131/mode/1up
Downloadable at https://www.worldcat.org/title/957343729

120 Snelgrove, L.E. (1935). *Swarming, its control and prevention.*

Snelgrove and Smith. Pleasant View, Bleadon Hill, Weston-Super-Mare, Avon BS24 9JT.
Snelgrove, L.E. (1963). *Swarming its control and prevention.* 11th Edn, April 1963. Published by Snelgrove, I. Bleadon, Weston-Super-Mare.
Möbus, B. (1979, 1998a, 1998b, 1998c, 1991) loc. cit.

121 Snelgrove, L.E. (1963) loc. cit.

122 Wedmore, E.B. (1945). *A Manual of Beekeeping for English-speaking Beekeepers.* 2nd Revised edition, pp.304-312 republished in 1979 by Bee Books New & Old, Steventon, Hampshire.

123 Hogg, J.A. (2005). The Juniper Hill plan for comb honey production, improved two-queen system. *American Bee Journal* **145(2):**138-141. http://www.twilightmd.com/Samples/Hogg/Hogg_Halfcomb___Publications/ABJ_2005_February.pdf
Hogg, J.A. (2006) loc. cit.

124 Hogg, J.A. (2005) loc. cit.

125 Snelgrove, L.E. (1935) loc. cit.

126 Cannon, D. (2017). In the apiary: Laying workers. *The Australasian Beekeeper* **118(10):**411-412.

127 Hornitzky M., Oldroyd B.P. and Somerville D. (1996). *Bacillus larvae* carrier status of swarms and feral colonies of honeybees (*Apis mellifera*) in Australia. *Australian Veterinary Journal* **73:**116-117. https://eurekamag.com/research/002/762/002762274.php

128 Bailey, L. (1953). The transmission of nosema disease. *Bee World* **34(9):**171-172. doi.org/10.1080/0005772X.1953.11094815
Bailey, L. (1954). The control of nosema disease.
Bee World **35(6):**111-113. doi:10.1080/0005772x.1954.11096676
Cushman, D. (accessed 15 August 2022). Bailey comb change, renewing all combs at the same time. http://www.dave-cushman.net/bee/baileychange.html
Willlams, I.H., Ball, B.V., Tomkins, P.W. and Carreck, N. (1993) loc.cit.

129 Butler, C.G. (1959a) loc. cit.
Simpson, J. (1957a) loc. cit.
Collison, C.H. (12 April 2017). Swarm management. Extension Service of Mississippi State University, cooperating with US Department of Agriculture, Publication P1817, 4pp. https://wilsoncountybeekeepers.org/2017/04/12/swarm-management-publication-by-dr-clarence-collison-

posted/
https://wcbeekeepers.files.wordpress.com/2017/01/swarm_management_collison_msu-p1817.pdf
Collison, C. (2005). Do you know? Swarming and nectar? *Bee Culture* **133(4)**:63-65. https://archive.org/details/sim_bee-culture_2005-04_133_4/page/63/mode/1up
Collison, C.H. (12 April 2017). Swarm management. Extension Service of Mississippi State University, cooperating with US Department of Agriculture, Publication P1817, 4pp. https://wilsoncountybeekeepers.org/2017/04/12/swarm-management-publication-by-dr-clarence-collison-posted/
https://wcbeekeepers.files.wordpress.com/2017/01/swarm_management_collison_msu-p1817.pdf
Collison, C.H. (2018). Swarming behavior: A closer look. *Bee Culture*. https://www.beeculture.com/a-closer-look-20/

130 *Direct requeening* was published as a two-part series
Wade, A. (2021). Requeening for beginners and old hands Part I: A guide for beginners. *The Australasian Beekeeper* **123(4)**:32-34.
Wade, A. (2021). Requeening for beginners and old hands Part II: A guide for old hands. *The Australasian Beekeeper* **123(5)**:18-20.

131 Demaree, G. (1892) loc. cit.

132 Haydak, M.H., Schaefer, H.A., Miller, E.S., Killion, C.E. and Killion, E., Lyle, N.I., Holzberlein, J.W., Engle, C.S. and Cale, G.H. (1952). Swarming round-up. *American Bee Journal* **92(5)**:181, 189-198. https://archive.org/details/sim_american-bee-journal_1952-05_92_5/page/189/mode/1up
Farrar, C.L., Miller, L.F., Dunham, W.E., Schaefer, E.A., Holzberlein, Jr., J. and Cale, G.H. (1954) loc. cit.

133 Farrar, C.L., Dunham, W.E., Miller, L.F. and Holzberlein, Jr., J.W. (1953) loc. cit.

134 Haydak, M.M. (1952). The causes of swarming. *American Bee Journal* **92(5)**:189-190. https://archive.org/details/sim_american-bee-journal_1952-05_92_5/page/189/mode/1up

135 Möbus, B. (April 1987) loc. cit.
Möbus, B. (May 1987) loc. cit.

136 Rahmlow, H.J. (ed.) (September 1952/August 1953). Schaefer new beekeepers federation president. *Wisconsin Horticulture* **43(5)**:119.

https://search.library.wisc.edu/digital/AJ6L5XONZV6VLQ85/search?
https://search.library.wisc.edu/digital/AGQK7M5AI7PHYN8K/full/AR73X5UBJSUYI586

137 Schaefer, H.A. (1952). Swarm prevention. *American Bee Journal* **92(5):**190-191. https://archive.org/details/sim_american-bee-journal_1952-05_92_5/page/190/mode/1up
Schaefer, H.A. (1954). The Shaefer two-queen swarm control method. *American Bee Journal* **94(4):**131. https://ia804500.us.archive.org/23/items/sim_american-bee-journal_1954-04_94_4/sim_american-bee-journal_1954-04_94_4.pdf
https://archive.org/details/sim_american-bee-journal_1954-04_94_4/page/131/mode/1up

138 Killion, E.E. (1981) loc. cit.

139 Morrison, W.K. (1907). Gravenhorst's system in Germany: The peculiar form of his hive and frame. *Gleanings in Bee Culture* **35(1):**30-31. https://babel.hathitrust.org/cgi/pt?id=umn.31951d00953180r;view=1up;seq=24

140 Miller, C.C. (1911). Rearing queens in hive with laying queen, pp.310-312 *in* Miller, C.C. (1911) loc. cit.

141 Killion, C.E. and Killion, E. (1952). Swarm control and queen rearing in comb honey. *American Bee Journal* **92(5):**190-191. https://archive.org/details/sim_american-bee-journal_1952-05_92_5/page/192/mode/1up

142 Demuth, G.S. (1919). Commercial comb honey production. US Department of Agriculture Farmers' Bulletin 1039, 41pp. https://ia804509.us.archive.org/26/items/CAT87202962/farmbul1039.pdf
Demuth, G.S. (1917). Comb honey. US Department of Agriculture Farmers' Bulletin 503. Washington Government Printing Office. https://ia802209.us.archive.org/21/items/CAT87201971/farmbul0503.pdf

143 Killion, E.E. (1981) loc.cit. pp.79-80, 96.

144 How flow works (accessed 2 June 2022). https://www.honeyflow.com.au/pages/how-flow-works#:~:text=Plastic%20foundation%20in%20beehives%20is,tube%20and%20into%20your%20jar

145 Lyle, N.I. (1952). Swarm control with the nucleus system. *American Bee Journal* **92(5):**194-195. https://archive.org/details/sim_american-bee-journal_1952-05_92_5/mode/1up

146 Holzberlein Jr., J. (1954). Another way to start two-queen colonies. *American Bee Journal* **94(4)**:131-132. https://archive.org/details/sim_american-bee-journal_1954-04_94_4/page/131/mode/1up

147 Jay, S.C. (1981) loc. cit.

148 Holzberlein Jr., J.W. (May 1952) loc. cit.

149 Dunham, W.E. (1943). The modified two-queen system. *American Bee Journal* **83(5)**:192-194, 203. https://archive.org/details/sim_american-bee-journal_1943-05_83_5/page/192/mode/1up
Dunham, W.E. (1947). Modified two-queen system for honey production. *Bulletin of the Agricultural Extension Service, the Ohio State University* **281**:1-16. Downloadable at https://www.google.com/search?q=Dunham%2C+W.E.+(1947).+Modified+two-queen+system+for+honey+production.+Bulletin+of+the+Agricultural+Extension+Service%2C+the+Ohio+State+University&oq=Dunham%2C+W.E.+(1947).++Modified+two-queen+system+for+honey+production.++Bulletin+of+the+Agricultural+Extension+Service%2C+the+Ohio+State+University&aqs=chrome..69i57.1180j0j7&sourceid=chrome&ie=UTF-8
Dunham, W.E. (1948a). Modified two-queen system for honey production. [Part of Bulletin No. 281, issued March 1947 by Agricultural Extension Service, The Ohio State University, Columbus, Ohio. https://library.osu.edu/collections/rg.22.o]
Dunham, W.E. (1948b). Modified two-queen system for honey production. *Gleanings in Bee Culture* **76(5)**:277-281. https://archive.org/details/sim_gleanings-in-bee-culture_1948-05_76_5/page/277/mode/1up
Dunham, W.E. (1951). The Ohio modified two-queen system. *Gleanings in Bee Culture* **79(4)**:212-214. https://archive.org/details/sim_gleanings-in-bee-culture_1951-04_79_4/page/212/mode/1up
Dunham, W.E. (1953). The modified two-queen system for honey production. *American Bee Journal* **93(3)**:111-112. https://archive.org/details/sim_american-bee-journal_1953-03_93_3/page/111/mode/1up
Dunham, W.E. (1954). Dunham's modified two-queen plan. *American Bee Journal* **94(4)**:130-131. https://archive.org/details/sim_american-bee-journal_1954-04_94_4/page/130/mode/1up

150 Demaree, G. (1892) loc. cit.

151 Engle, C.S. (1952). Automatic Demareeing. *American Bee Journal*

92(5):197. https://archive.org/details/sim_american-bee-journal_1952-05_92_5/page/197/mode/1up

152. Cale, G.H. (1952). Emergency swarm control. *American Bee Journal* **92(5)**:198. https://archive.org/details/sim_american-bee-journal_1952-05_92_5/page/198/mode/1up

153. Cale Sr., G.H., Banker, R. and Powers, J. (1979). Management for honey production, Chapter XII, pp.382-383 *in The hive and the honey bee*. (1979). Extensively revised edition, fifth printing. Dadant & Sons, Hamilton, Illinois.
Cale, G.H. (1943). Relocation as a means of swarm control. *American Bee Journal* **83(5)**:190-191. https://archive.org/details/sim_american-bee-journal_1943-05_83_5/page/190/mode/1up

154. Cale, G.H. (1954). Reversal, separation and reunion. *American Bee Journal* **94(4)**:128-132. https://archive.org/details/sim_american-bee-journal_1954-04_94_4/page/132/mode/1up

155. Miller, L.F. (1954). Two queens to reclaim the weak colonies. *American Bee Journal* **94(4)**:130. https://archive.org/details/sim_american-bee-journal_1954-04_94_4/page/130/mode/1up
Victors, S. (2001). Two queen hive system from package bees. Alaska Wildflower Honey, 8pp. http://sababeekeepers.com/files/Two_Queen_Hive_System_From_Package_Bees.pdf
Winston, M. and Mitchell, M. (1986). Timing of package honey bee (Hymenoptera: Apidae) production and use of two-queen management in southwestern British Columbia, Canada. *Journal of Economic Entomology* **79(4)**:952-956. doi:10.1093/JEE/79.4.952

156. Miller, L.F. (1953). Crop insurance with two queens. *American Bee Journal* **93(3)**:113, 117. https://archive.org/details/sim_american-bee-journal_1953-03_93_3/page/113/mode/1up

157. Shook, S. (May 13, 2017). Honey industry of Porter County referencing *The Country Gentleman* (November 23, 1918) **865**:29. http://www.porterhistory.org/2017/05/honey-industry-of-porter-county.html

158. Miller, E.S. (1952). Swarm control with extracted honey. *American Bee Journal* **92(5)**:192-193. https://archive.org/details/sim_american-bee-journal_1952-05_92_5/page/192/mode/1up

159. Miller, E.S. (May 1943). Swarm prevention. *American Bee Journal*

83(5):194, 203. https://archive.org/details/sim_american-bee-journal_1943-05_83_5/page/194/mode/1up

160 Miller, E.S. (1932). Mating queen from a top story. *American Bee Journal* **72(10)**:403. https://archive.org/details/sim_american-bee-journal_1932-10_72_10/page/403/mode/1up

161 *Art thou a skeppist* was published as a two-part series as
Wade, A. and Derwent, F. (2019). Art thou a skeppist? Part I: The origins and practice of skep beekeeping.
The Australasian Beekeeper **121(2)**:34-39.
Derwent, F. and Wade, A. (2020). Art thou a skeppist? Part II: From the skep to the modern frame hive.
The Australasian Beekeeper **122(1)**:30-34.

162 Neighbour, A. (1866). *The apiary or Bees, bee-hives and bee-culture being a familiar account of the habits of bees and the most improved methods of management, with full directions, adapted for the cottager, farmer, or scientific apiarian.* Kent and Co., Paternoster Row; Geo. Neighbour and Sons, 149 Regent Street, and 127, High Holborn, London. https://victoriancollections.net.au/media/collectors/51d110e42162ef12e06aa06b/items/5341f0202162ef0a84dbc1c6/item-media/5341f0c52162ef0a84dbc6e3/original.pdfbooks?id=T0QDAAAAQAAJ&pg=PA129&dq=alfred+neighbour&hl=en&sa=X&ved=0ahUKEwir277v1PLiAhWHaCsKHecGDSMQ6AEIKDAA#v=onepage&q=alfred%20neighbour&f=false

163 Wikipedia (accessed 10 November 2022). *Beehive.* https://en.wikipedia.org/wiki/Beehive#Skeps

164 Kritsky, G. (2017). Beekeeping from antiquity through the middle ages. *Annual Review of Entomology* **62**:249-264. doi.org/10.1146/annurev-ento-031616-035115

165 Levett, J. (1634). *The ordering of bees: or, the history of managing them:* From time to time, with their hony and waxe, shewing their nature and breed: As also what trees, plants, and hearbes are good for them, and namely what are hurtfull: together with the extraordinary profit arising from them. Set forth on a dialogue resolving all doubts whatsoever. Thomas Harper, for John Harison. Accessible online through the National Library of Australia at http://gateway.proquest.com/openurl?ctx_ver=Z39.88-2003&res_id=xri:eebo&rft_val_fmt=&rft_

id=xri:eebo:image:203420

166 Dams, M. and Dams, L. (1977). Spanish rock art depicting honey gathering during the Mesolithic. *Nature* **268(5617):**228-230. doi:10.1038/268228a0 *See also*
Wikipedia (accessed 24 November 2022). Beekeeping. https://en.wikipedia.org/wiki/Beekeeping

167 Harley MS 3244 (1236-c1250). Bestiary. Rubric: 'Incipit liber de natura bestiarum', incipit: 'Leo fortissimus bestiarum'.Decoration: Bees f. 57v. http://www.bl.uk/manuscripts/Viewer.aspx?ref=harley_ms_3244_f057v

168 University of Aberdeen special collections and museums (accessed 10 November 2022). *The Aberdeen Bestiary* - ms 24 (c. 1200).
The Aberdeen Bestiary. Creation of the animals, Folio 63r - De apibus: Of bees. https://www.abdn.ac.uk/bestiary/ms24/f63r

169 Poets.org (accessed 9 July 2022). How doth the busy little bee. https://poets.org/poem/how-doth-little-busy-bee

170 Bowen, A.H. (1923). The old straw skep.
American Bee Journal **63(3):**126-127.
https://archive.org/details/sim_american-bee-journal_1923-03_63_3/page/126/mode/1up

171 Morrison, W.K. (1907) loc. cit.

172 Hands (accessed 27 November 2022). Hands: Of bees & bee skeps. https://www.youtube.com/watch?v=99MBkslFhGU&feature=youtu.be

173 Butler, C. (1609). *The feminine monarchie*: or a treatise concerning bees, or a hive ordering of them: wherein, the truth found out by experience and diligent observation, discovereth the idle and sound concepts which may have written about this subject. Printed by Joseph Barnes, Oxford. Also titled and reprinted as Butler, C. (1623). The historie of bees. Shewing their admirable nature, and properties, their generation, and colonies, their government, loyaltie, art, industrie, enemies, warres, magnamimitie, &c. together with the right ordering of them from time to time: and the sweet profit arising thereof. Printed by John Haviland for Roger Jackson, Fleetstreet London, 217pp. Chapter 3, Of the hives, and the dressing of them, pp.62-79. https://books.google.com.au/books/about/The_Feminine_Monarchie_Or_the_Historie_o.html?id=f5tbAAAAMAAJ&redir_esc=y

174 Southerne, E. (1593). *Bee Culture*: A treatise concerning the right use and ordering of bees: newlie made and set forth according to the authors owne experience: (which by any heretofore hath not been done). Orwin for Thomas Woodcocke, 18pp.
http://find.galegroup.com.rp.nla.gov.au/mome/quickSearch.do?now=1558780323606&inPS=true&prodId=MOME&userGroupName=nla
Available online at National Library of Australia http://gateway.proquest.com/openurl?ctx_ver=Z39.88-2003&res_id=xri:eebo&rft_val_fmt=&rft_id=xri:eebo:image:18224

175 Seeley, T.D. and Morse, R.A. (1976) loc. cit.
Loftus, J.C., Smith, M.L. and Seeley, T.D. (2016). How honey bee colonies survive in the wild: testing the importance of small nests and frequent swarming. *PLOS One* **11(3)**:e0150362. https://journals.plos.org/plosone/article?id=10.1371/journal.pone.0150362
Seeley, T.D. (2017a) loc. cit.

176 Tusser, T. (1557a). *A Hundreth Good Pointes of Husbandrie*.
http://www.archive.org/stream/fivehundredpoint08tussuoft/fivehundredpoint08tussuoft_djvu.txt

177 Tusser, T. (1557b). *A Hundreth Good Pointes of Husbandrie*.
http://www.luminarium.org/renascence-editions/tusser1.html

178 Richardson, H.D. (1849). *The hive and the honey-bee*: With plain directions for obtaining a considerable annual income from this branch of rural economy. To which is added, an account of the diseases of bees, with their remedies; Remarks as to their enemies, and the best mode of protecting the bees from their attacks. https://ia800200.us.archive.org/26/items/hivehoneybeewith00rich/hivehoneybeewith00rich.pdf

179 Winston, M.L. (2014) loc. cit.

180 Galle, H.-K. (Ed): Fuchs, P. and Nemes, Z. (Co-Eds) E2962. IWF 9 (1986). *Skep beekeeping: Work during the cast swarm period/Summer work during the heather bloom* (1979 and 1983). [Mitteleuropa Nördliches Niedersachsen - Arbeiten zue Zeit der Nachschwärme in einer Korgimkerei] Encyclopaedia Cinematographer iwf.de
https://av.tib.eu/media/14375

181 Neighbour, A. (1866) loc. cit. p.139.

182 J.A. (1683). Directions for the making of colonies for bees, and a new

invented model of hive, to improve them, whereby without killing, may be enjoyed the fruit of their labour. Tuesday 16 June 1683, Letter Number VI. pp.374-381. A collection for improvement of husbandry and trade *in* Houghton, J. (1684). *Husbandry and trade improved: Being a collection of many valuable materials relating to corn, cattle, coals, hops, wool, &c. with a complete catalogue of the several sorts of earths, and their proper product; the best sorts of manure for each; with the art of draining and flooding of lands as also full and exact histories of trades, as malting, brewing, &c. the description and structure of instruments for husbandry, and carriages, with the manner of their improvement. An account of the rivers of England, &c. and how far they maybe made navigable; of weights and measures, of woods, cordage, and metals; of building and stowage, the vegetation of plants, &c. with many other useful particulars, communicated by several eminent members of the Royal Society, to the Collector,* Vol. IV, 414pp. Woodman and Lyon, Russel Street, Covent Garden, London. M,DCC,XXVII. https://books.google.com.au/books?id=vL0PAAAAQAAJ&pg=PA374&lpg=PA374&dq=Directions+for+the+making+of+colonies+for+bees,+and+a+new+invented+model+of+hive,+to+improve+them,+whereby+without+killing,+may+be+enjoyed+the+fruit+of+their+labour.&source=bl&ots=QgGYcqa-9m&sig=ACfU3U1ojgkFSNoHpI__5SVl1JPQewpjxg&hl=en&sa=X&ved=2ahUKEwjHo6fYlcXiAhXg63MBHZLbDdUQ6AEwAXoECAgQAQ#v=onepage&q=Directions%20for%20the%20making%20of%20colonies%20for%20bees%2C%20and%20a%20new%20invented%20model%20of%20hive%2C%20to%20improve%20them%2C%20whereby%20without%20killing%2C%20may%20be%20enjoyed%20the%20fruit%20of%20their%20labour.&f=false

183 Bevan, E. (1827). The honey-bee: Its natural history, physiology and management. Baldwin, Cradock and Joy. https://www.gutenberg.org/ebooks/67107
Bevan, M.D. and van Voorst, E. (1838). The honey-bee, its natural history, physiology, and management. *Annals and Magazine of Natural History* [*Journal of Natural History*] **2(10):**293-294. doi:10.1080/00222933809496677

184 Hartlib, S.E. (1655). The Hartlid papers: The reformed commonwealth of bees', Part 2, p.49. https://www.dhi.ac.uk/hartlib/view?docset=pamphlets&docname=pam_55

185 Huber, F. (1806) loc. cit.

186 Dzierzon, J., Abbott, C.N., Dieck, H. and Studderd, S. (1882) loc. cit.

187 Langstroth, L.L. (1853). *Langstroth on the hive and the honey-bee: A beekeeper's manual.* Northampton: Hopkins, Bridgman & Company. https://www.biodiversitylibrary.org/item/115026#page/9/mode/1up
http://www.thebeeyard.org/wp-content/uploads/2014/02/The.Hive_.And_.The_.Honey_.Bee_.Langstroth.1853.pdf
Langstroth, L.L. (1857). *Langstroth on the hive and the honey-bee: A beekeeper's manual* Second Edition, enlarged, and illustrated with numerous engravings. C.M. Saxton & Co., Agricultural Booksellers – 40 Fulton Street, New York in Chapter VII *On the advantages which ought to be found in an improved hive,* pp.98-113. http://www.southeastalabamabeekeepers.com/files/A_Practical_Treatise_on_the_Hive_and_Honey_Bee.pdf
Langstroth, L.L. (1863). *Langstroth on The Hive and The Honey-Bee: A bee keeper's manual.* Third Edition, revised, and illustrated with seventy-seven engravings. C.M. Saxton & Co., 25 Park Row, New York in Chapter VII *Requisites of a complete hive,* pp.95-108. https://www.biodiversitylibrary.org/item/120282#page/13/mode/1up
also republished as
Langstroth, L.L. (1873). *A practical treatise on the hive and honey-bee [Langstroth on the honey bee].* Third Edition, revised, and illustrated with seventy-seven engravings, Philadelphia:J.B. Lippincott & Co. https://archive.org/details/practicaltreati00lange
Dadant, C (1913). Langstroth on the hive & honey bee. Revised by Dadant. Twentieth Century Edition. Dadant & Sons Hamilton, Hancock County, III, USA. https://ia600204.us.archive.org/28/items/cu31924003227679/cu31924003227679.pdf

188 Huber, F. (1814). *New observations upon bees* translated from the French by Dadant, C.P. Editor of the *American Bee Journal,* Hamilton, Illinois, 1926. [Also entitled *Huber's observations on bees* and *New observations upon bees.*] https://victoriancollections.net.au/media/collectors/51d110e42162ef12e06aa06b/items/537c1a572162ef06a03aafc4/item-media/537c1b7a2162ef06a03ab6cd/original.pdf

189 United States Patent Office (1912). Lorenzo L. Langstroth, of Philadelphia, Pennsylvania. Beehive. Specification forming part of Letters Patent No. 9,800, dated October 5, 1852; Reissued May 26, 1863, No. 1,484, 10pp. https://patents.google.com/patent/US9300?Oq=beehive+honey+langstroth
https://docs.google.com/viewer?url=patentimages.storage.googleapis.com/pdfs/US9300.pdf
Langstroth, L.L. (1853). Advertisement for L.L. Langstroth's movable comb hive. Patented October 5, 1852, pp.xv-xvi. *in Langstroth on the hive and the honey-bee: A beekeepers manual.* Hopkins, Bridgman & Company, Northampton. https://archive.org/details/langstrothonhiv00lang/page/n24/mode/1up

190 Sieling, P. (October 1996). Build the original Langstroth hive. *Bee Culture* http://www.beeculture.com/build-original-langstroth-hive/

191 Miksha, R. (December 25, 2021). *Bad Beekeeping Blog.* Langstroth's Christmas present. https://badbeekeepingblog.com/2021/12/25/langstroths-christmas-present-2/

192 Johansson, T.S.K. and Johansson, M.P. (1967). Lorenzo L. Langstroth and the bee space. *Bee World* **48(4):**133-143. doi:10.1080/0005772X.1967.11097170

193 Root, A.I. (1980) loc. cit. pp.75, 333-343.

194 Eckert, J.E. and Shaw, F.R. (1960). *Beekeeping.* New York: Macmillan. Chapter 3 Beekeeping equipment: *The modern frame hive,* pp.37-38.

195 Neighbour, A. (1866) loc. cit.

196 Neighbour, G. (May 1887). George Neighbour and Sons' bee-hives and apparatus. *The British Bee Journal and Bee-Keepers' Adviser* 5(40):**45.** https://books.google.com.au/books?id=afpHAAAAYAAJ&pg=PA43-IA3&dq=alfred+neighbour+beekeeper&hl=en&sa=X&ved=0ahUKEwiDqo2diPLiAhWYfH0KHYnMA6gQ6AEINDAC#v=onepage&q=alfred%20neighbour%20beekeeper&f=false

197 Crane, E. and Walker, P. (1999). Early English beekeeping: the evidence from local records up to the end of the Norman period. Reprinted from *The Local Historian* **29(3):** August 1999. https://www.evacranetrust.org/uploads/document/400fe90f27902b22bd24552c495914998fc978c5.pdf

198 Möbus, B. (1998a, 1998b, 1998c, 1991) loc. cit.

199 Mitchell, D. (2016). Ratios of colony mass to thermal conductance of tree and man-made nest enclosures of *Apis mellifera*: Implications for survival, clustering, humidity regulation and *Varroa destructor*. *International Journal of Biometereology* **60(5)**:629-638. doi:10.1007/s00484-015-1057-z https://pubmed.ncbi.nlm.nih.gov/26335295/
Mitchell D. (2017). Honey bee engineering: top ventilation and top entrances. *American Bee Journal* **157(8)**:887-889. https://eprints.whiterose.ac.uk/141140/5/honeybee_engineering.pdf
Mitchell, D. (2019). Nectar, humidity, honey bees (*Apis mellifera*) and varroa in summer: a theoretical thermofluid analysis of the fate of water vapour from honey ripening and its implications on the control of *Varroa destructor. Journal of The Royal Society Interface* **16(156)**:20190048. doi:10.1098/rsif.2019.0048 https://royalsocietypublishing.org/doi/10.1098/rsif.2019.0048
Mitchell, D. (2019). Thermal efficiency extends distance and variety for honey bee foragers: analysis of the energetics of nectar collection and desiccation by *Apis mellifera*. *Journal of the Royal Society Interface* **16(150)**:20180879. doi:10.1098/rsif.2018.0879 https://royalsocietypublishing.org/doi/10.1098/rsif.2018.0879 http://eprints.whiterose.ac.uk/141143/3/rsoc23122018-manuscriptD.pdf
Mitchell, D. (2019). To save honey bees we need to design them new hives. *The Conversation* 12 September 2019. https://theconversation.com/to-save-honey-bees-we-need-to-design-them-new-hives-121792
Mitchell, D. (February 2019). Profile of Derek Mitchell PhD Candidate in Mechanical Engineering, University of Leeds. *The Conversation*. https://theconversation.com/profiles/derek-mitchell-680930
LIFD Early Career Researcher Spotlight: Derek Mitchell (24 September 2021). Leeds Institute for fluid dynamics. https://fluids.leeds.ac.uk/2021/09/24/lifd-early-career-researcher-spotlight-derek-mitchell/

200 September in the Rohan Hours (BNF Latin 9471, fol. 13r), (c.1430-1435). Heures. (1401-1500). Bibliothèque Nationale de France, Département des Manuscrits. https://gallica.bnf.fr/ark:/12148/btv1b10515749d/f35.item

201 Rutilius Taurus Aemilianus Palladius (c. 350). *Palladius on Husbandrie. From the unique ms. of about 1420 A.D. in Colchester Castle.* Baeton Lodge, B. (Ed). with a ryme index edited by Sidney J.H. Herrtage, London: Published for The Early English Text Society by N. Trübner & Co., 57 & 59, Ludgate Hill, p.37. [Palladius' Latin text on agriculture *De Re Rustica, Opus agriculturae* translated into Middle English in *Of Husbandry*] https://ia800901.us.archive.org/27/items/palladiusonhusbo00palluoft/

palladiusonhusbo00palluoft.pdf Medieval Sourcebook: Palladius: *On Husbandry*, c. 350, p.37 lines 1009-1011. *De opium castris. 145*. https://sourcebooks.fordham.edu/source/350palladius-husbandry.asp

202 *Bee space* was presented in the Canberra Region Beekeepers *Bee Buzz Box* by Alan Wade and Frank Derwent in July 2020.

203 Möbus, B. (1998c) loc. cit.

204 Seeley, T.D. (2019). The lives of bees: The untold story of the honey bee in the wild. Princeton University Press, Chapter 5, The nest, pp.99-139.

205 Iodice, I. (John Grubb, pers. comm.). The *Beekeeping Naturally* Kenyan Top Bar Hive.

206 Cushman, D.A. (2011). Spacing of frames in bee hives. http://www.dave-cushman.net/bee/framespacing.html
Beekeeping.IsGood (25 May 2015). Variability in natural cell size and comb spacing. https://www.beekeeping.isgood.ca/behaviour/comb-building/variability-in-natural-cell-size-and-comb-spacing

207 Huber, F. (1814) loc. cit.

208 Stamp, J. (2013) loc. cit.

209 Beespoke Info (3 February 2015). Which frames. http://beespoke.info/?s=manley+frame

210 Flow Hive (2021). What are the dimensions of the flow frames? https://www.honeyflow.com/faqs/all/what-are-the-dimensions-of-the-flow-frames/p/70#:~:text=The%20Flow%20Frames%20are%20designed,ends%20is%20set%20to%2051mm

211 Cushman, D.A. (accessed 5 November 2022). Queen excluder types used in bee hives. http://www.dave-cushman.net/bee/excludertypes.html.

212 Laffan, J., Somerville, D., Annand, N. and Rhodes, J. (2016). NSW Department of Primary Industries (2016). Bee Agskills: A practical guide to farm skills, 114pp. https://books.google.com.au/books/about/Bee_AgSkills.html?id=W3fQBgAAQBAJ&redir_esc=y

213 Australian Honey Bee Industry Council (9 June 2011). Beekeeping basics - Certificate II Participants Learning Guide RTE2217A, Construct and repair beehives, 29pp. https://fliphtml5.com/flqg/pfix/basic

214 Warré, É (1948). *L'Apiculture Pour Tous* (12th edition). Translated as *Beekeeping for All* by Patricia and David Heaf, January 2007, Lightening

Source, UK. http://annemariemaes.net/wp-content/uploads/2013/10/beekeeping_for_all.pdf

215 Ministry of Agriculture, Fisheries and Food (1960). The British National Hive advisory leaflet 367.
http://www.peak-hives.co.uk/wp-content/uploads/2009/11/leaflet-367b.pdf

216 odfrank (2 January 2010). Jumbo dimensions. https://www.beesource.com/forums/archive/index.php/t-236688.html

217 Erickson E.H., Lusby, D.A., Hoffman, G.D. and Lusby E.W. (1990). On the size of cells. Part I. Speculations on foundation as a colony management tool. *Gleanings in Bee Culture* **118(2)**:98-101. https://archive.org/details/sim_gleanings-in-bee-culture_1990-02_118_2/page/98/mode/1up
Baudoux, U. (1933). The influence of cell size. *Bee World* **14(4)**:37-41. doi.org/10.1080/0005772X.1933.11093211

218 Math.net. Area of a hexagon. https://www.math.net/area-of-a-hexagon

219 *Honey bee hive architecture* was presented in the Canberra Region Beekeepers *Bee Buzz Box* by Frank Derwent and Alan Wade in March 2020 and revised in July 2022.

220 Seeley, T.D. and Morse, R.A. (1976) loc. cit.
Morse, R.A., Layne, J.N., Visscher, P.K. and Ratnieks, F. (1993). Selection of nest cavity volume and entrance size by honey bees in Florida. *Florida Scientist* **56(3)**:163-167. https://www.jstor.org/stable/24320554
Seeley, T.D. (1977). Measurement of nest cavity volume by honeybees (*Apis mellifera*). *Behavioral Ecology and Sociobiology* **2(2)**:201-227. doi:10.1007/BF00361902

221 Wikiwand (accessed 13 July 2022). Horizontal top-bar hive. https://www.wikiwand.com/en/Horizontal_top-bar_hive

222 Crane, E. (1999). The world history of beekeeping and honey hunting. Routledge, New York.

223 Bush, M. (28 July 2005). *Beesource*. https://www.beesource.com/threads/popular-hives-dadant-v-langstroth.190779/#:~:text=A%20Dadant%20hive%20is%2011,8%22%20and%20of%20various%20depths

224 Seeley, T.D. (2019) loc. cit. Chapter 11, Darwinian beekeeping, pp.277-292.
Seeley, T. (2017). Darwinian beekeeping: An evolutionary approach to

apiculture. *American Bee Journal* **157(3)**:277-282. https://www.researchgate.net/publication/318784675_Darwinian_beekeeping_An_evolutionary_approach_to_apiculture Reprinted at Natural Beekeeping Trust https://www.naturalbeekeepingtrust.org/darwinian-beekeeping

225 Horr, B.Z. (1998). My intensive two-queen management system means bigger crops. *American Bee Journal* **138(7)**:507-510.

226 Dugat, M. (1948). The skyscraper hive. Authorized translation of La Ruche Gratteciel à Plusieurs Reines by Norman C. Reeves. Faber and Faber, 24 Russell Square, London. https://www.abebooks.com/servlet/SearchResults?an=dugat+father&bi=h&ds=5&n=100121503&sortby=1&tn=skyscraper+hive&cm_sp=mbc-_-ats-_-filter
Butler, C.G. (17 July 1948). The skyscraper hive. *Nature* **162(4107)**:87.

227 Warré, É. (1948) loc. cit.

228 United States Patent Office (1912) loc. cit.
Langstroth, L.L. (1853) loc. cit.

229 Huber, F. (1806) loc. cit.

230 Dzierzon, J., Abbott, C.N., Dieck, H. and Studderd, S. (1882) loc. cit.

231 Langstroth, L.L. (1857) loc.cit.

232 Simpson, J. (1958b) loc. cit.

233 Wade, A. (2021) loc. cit.

234 British Bee-Keepers' Association Quarterly Converzatione (April 6, 1893). *British Bee Journal, Bee-Keepers' Record and Adviser* **21(563)**:126, 132-134. https://www.biodiversitylibrary.org/item/83769#page/142/mode/1up

235 Farrar, C.L. (October 1936). Two-queen vs. single-queen colony management. *Gleanings in Bee Culture* **64(10)**:593-596. https://archive.org/details/sim_gleanings-in-bee-culture_1936-10_64_10/page/593/mode/1up

236 Eckert, J.E. (1937). The duo or two-queen hive. *Gleanings in Bee Culture* **65(3)**:137. https://archive.org/details/sim_gleanings-in-bee-culture_1937-03_65_3/page/137/mode/1up

237 Gouget, C.W. (July 1953). The three-queen pyramid. *Gleanings in Bee Culture* **81(7)**:395-398. https://archive.org/details/sim_gleanings-in-bee-

culture_1953-07_81_7/page/395/mode/1up

238 Crane, E. (1980). Multiple queen hives and hyper hives in *Perspectives in world agriculture: Apiculture*, Chapter 10. pp.261-294. Farnham Royal, UK: Commonwealth Agricultural Bureaux.
https://www.evacranetrust.org/uploads/document/b00fb1cb4f88f874217e28dcf04930fb5639e1bd.pdf

239 Moeller, F.E. (April 1976). Two queen system of honeybee colony management. Production Research Report 161, Agricultural Research Service, United States Department of Agriculture, 15pp., Washington DC 20402.
http://naldc.nal.usda.gov/download/CAT87210713/PDF http://mesindus.ee/files/52221134-2-queen-management.pdf

240 Hesbach, W. (2015). Two queens, one hive=lots of honey. *Honey Bee Suite.* https://honeybeesuite.com/two-queens-one-hivelots-of-honey/
Hesbach, W. (2016). The horizontal two-queen system. *Bee Culture* **144(3)**:63-66. http://www.beeculture.com/the-horizontal-two-queen-system/
Nabors, R. (1 July 2015). Beekeeping topics: My two-queen hive experiment – Part I. *American Bee Journal.* https://americanbeejournal.com/my-two-queen-hive-experiment/
Nabors, R. (1 August 2015). Beekeeping topics: My two-queen hive experiment – Part II. *American Bee Journal.* https://americanbeejournal.com/my-two-queen-hive-experiment-part-ii/

241 Kleinhenz, M., Bujok, B., Fuchs, S. and Tautz, J. (2003). Hot bees in empty broodnest cells: Heating from within. *Journal of Experimental Biology* **206(Pt 23)**:4217-4231. https://www.researchgate.net/publication/9037048_Hot_bees_in_empty_broodnest_cells_heating_from_within

242 Kastberger, G., Waddoup, D., Weihmann, F. and Hoetz, T. (2016). Evidence for ventilation through collective respiratory movements in giant honeybee (*Apis dorsata*) nests. *PLOS One* **11(8)**:e0157882. doi:10.1371/journal.pone.0157882

243 Tan, K., Yang, S., Wang, Z-W., Radloff, S. and Oldroyd, B. (2012). Differences in foraging and broodnest temperature in the honey bees *Apis cerana* and *A. mellifera*. *Apidologie* **43(6)**:618-623.
https://hal.science/hal-01003658/file/hal-01003658.pdf https://www.tib.eu/en/search/id/BLSE%3ARN321836196/Differences-in-foraging-

and-broodnest-temperature/
Jones J., Helliwell, P., Beekman, M., Maleszka, R.J. and Oldroyd, B.P. (2005). The effects of rearing temperature on developmental stability and learning and memory in the honey bee, *Apis mellifera*. *Journal of Comparative Physiology A* **191**:1121-1129. https://www.researchgate.net/profile/Benjamin-Oldroyd/publication/7695427_The_effects_of_rearing_temperature_on_developmental_stability_and_learning_and_memory_in_the_honey_bee_Apis_mellifera/links/0fcfd50adb8aa98895000000/The-effects-of-rearing-temperature-on-developmental-stability-and-learning-and-memory-in-the-honey-bee-Apis-mellifera.pdf

244 Mitchell, D. (2016) loc. cit.
Mitchell, D. (2019) loc. cit.
Stabentheiner, A., Kovac, H. and Brodschneider, R. (2010). Honeybee colony thermoregulation: Regulatory mechanisms and contribution of individuals in dependence on age, location and thermal stress. *PLOS One* **5(1)**:e8967. https://www.ncbi.nlm.nih.gov/pmc/articles/PMC2813292/ https://journals.plos.org/plosone/article?id=10.1371/journal.pone.0008967

245 Somerville, D. and Collins, D. (2014). Screened bottom boards. Rural Industries Research and Development Corporation. RIRDC Publication No 14/061, 47pp. RIRDC Project No PRJ-006390. https://www.agrifutures.com.au/wp-content/uploads/publications/14-061.pdf

246 *Going overboard* was published as
Wade, A. and Derwent, F. (2022). Going overboard: Part I – Hive dividers. *The Australasian Beekeeper* **123(8)**:10-13.
Wade, A and Derwent, F. (2022). Going overboard: Part II – Hive lids, bottom boards and feeders. *The Australasian Beekeeper* **123(9)**:10-14.

247 British Bee Journal Eds (1886). Count Zorzi's uncapping machine. *British Bee Journal and Bee-Keepers' Adviser* **14(199)**:166. https://www.biodiversitylibrary.org/item/83067#page/180/mode/1up

248 Snelgrove, L.E. (1981). Swarming: Its control and prevention. Thirteenth Edition, Snelgrove & Smith, Pleasant View, Bleadon Hill, Weston-Super-Mare, Avon BS24 9JT.

249 British Bee Journal Eds (1886). Mr Corneil's super. *British Bee Journal and Bee-Keepers' Adviser* **14(230)**:531-532. https://www.biodiversitylibrary.org/item/83067#page/535/mode/1up

250 British Bee Journal Eds (1891). Queens passing through excluder zinc.

British Bee Journal, Bee-Keepers' Record and Adviser **19(450)**:176. https://www.biodiversitylibrary.org/item/83767#page/186/mode/1up

251 Crane, E. (1999) loc. cit.

252 Warré, E. (1948) loc. cit.

253 Seeley, T.D. (2010). Honeybee Democracy. Princeton University Press, Princeton, NJ, Chapter 5, pp.99-117.

254 de Guzman, L.I., Williams, G.R., Khongphinitbunjong, K. and Chantawannakul, P. (2017). Ecology, life history, and management of *Tropilaelaps* mites. *Journal of Economic Entomology* **110(2)**:319-332. https://doi.org/10.1093/jee/tow304

255 Sommerville, D. and Collins, D. (2014) loc. cit.

256 Seeley, T.D. and Morse, R.A. (1976) loc. cit.

257 *Let us spray* was presented as a two part series in the Canberra Region Beekeepers *Bee Buzz Box* by Alan Wade in March and April 2022.

258 Bortolotti, L. and Costa, C. (2014) loc. cit.

259 Butler, C.G., Callow, R.K. and Johnston, N.C. (1962). The isolation and synthesis of queen substance, 9-oxodec-trans-2-enoic acid, a honeybee pheromone. *Proceedings of the Royal Society B: Biological Sciences* **155(960)**:417-432. doi:10.1098/rspb.1962.0009

260 Free, J.B. (1987). Pheromones of social bees. Comstock Publishing Associates, Ithaca, NY, 218pp.

261 Slessor, K.N., Kaminski, L.A., King, G.G.S, Borden, J.H. and Winston, M.L. (1988). Semiochemical basis for retinue response to queen honey bees. *Nature* **332(6162)**:354-356. doi:10.1038/332354a0

262 Bortolotti and Costa (2014) loc. cit.

263 Maisonnasse, A., Alaux, C., Beslay, D., Crauser, D., Gines, C., Plettner, E. and Le Conte, Y. (2010). New insights into honey bee (*Apis mellifera*) pheromone communication. Is the queen mandibular pheromone alone in colony regulation? *Frontiers in Zoology* **7(1)**:18. doi:10.1186/1742-9994-7-18 https://frontiersinzoology.biomedcentral.com/articles/10.1186/1742-9994-7-18

264 Keeling, C.I., Slessor, K.N., Higo, H.A. and Winston, M.L. (2003). New components of the honey bee (*Apis mellifera* L.) queen retinue pheromone. *Proceedings of the National Academy of Sciences* **100(8)**:4486-

4491. doi:10.1073/pnas.0836984100 https://www.ncbi.nlm.nih.gov/pmc/articles/PMC153582/

265 Bortolotti and Costa (2014) loc. cit.

266 Peso, M. and Barron, A.B. (2014). The effects of brood ester pheromone on foraging behaviour and colony growth in apicultural settings. *Apidologie* **45(5):**529-536. doi:10.1007/s13592-014-0270-9 https://link.springer.com/article/10.1007/s13592-014-0270-9
Arnold, G., Le Conte, Y., Trouiller, J., Hervet, H., Chappe, B. and Masson, C. (1994). Inhibition of worker honeybee ovaries development by a mixture of fatty acid esters from larvae. *Comptes Rendus de l'Academie des Sciences, Serie III: Sciences de la Vie* **317(6):**511-515. https://hal.inrae.fr/hal-02713615
Oldroyd, B.P., Wossler, T.C. and Ratnieks, F.L.W. (Sep 2001). Regulation of ovary activation in worker honey-bees (*Apis mellifera*): Larval signal production and adult response thresholds differ between anarchistic and wild-type bees. *Behavioral Ecology and Sociobiology* **50(4):**366-370. doi.org/10.1007/s002650100369 https://www.jstor.org/stable/4601977

267 Maisonnasse, A., Lenoir, J.-C., Beslay, D., Crauser, D. and Le Conte, Y. (2010). E-β-ocimene, a volatile brood pheromone involved in social regulation in the honey bee colony (*Apis mellifera*). *PLOS One* **5(10):**e13531. https://www.ncbi.nlm.nih.gov/pmc/articles/PMC2958837/

268 Collison, C. (23 October 2015). A closer look: Tarsal glands/footprint pheromone. *Bee Culture* https://www.beeculture.com/a-closer-look-tarsal-glands-footprint-pheromone/

269 Lensky, Y. and Seifert, H. (1980). The effect of volume, ventilation and overheating of bee colonies on the construction of swarming queen cups and cells. Physiology Part A: *Comparative Biochemistry and Physiology* **67(1):**97-101. doi:10.1016/0300-9629(80)90413-2
Lensky, Y. and Slabezki, Y. (1981). The inhibiting effect of the queen bee (*Apis mellifera* L.) foot-print pheromone on the construction of swarming queen cups. *Journal of Insect Physiology* **27(5):**313-323. doi.org/10.1016/0022-1910(81)90077-9
Lensky, Y., Cassier, P., Finkel, A., Teeshbee, A., Schlesinger, R., Delorme-Joulie, C. and Levinsohn, M. (1984). The tarsal glands of honeybee queens, workers and drones. II. Biological effect. [Les glandes tarsales de l'abeille mellifique (*Apis mellifera* L.) reins, ouvières et faux-bourdons

(Hymenoptera, Apidae). II Rôle biologique.] *Annales des Sciences Naturelles, Zoologie et Biologie Animale*, Paris **6(13)**:1675-175. https://eurekamag.com/research/001/717/001717071.php
Lensky, Y., Finkel, A., Cassier, P., Teeshbee, A. and Schlesinger, R. (1987). The tarsal glands of the honeybee *Apis mellifera* L., queens, workers and drones — chemical characterization of footprint secretions. *Honeybee Science* **8(3)**:97-102. Eureka Magazine Accession: 001717070. https://eurekamag.com/research/001/717/001717070.php
Collison, C. (23 October 2015) loc. cit.

270 Seeley, T.D. (2010) loc. cit.

271 Wikipedia lists twenty nine principal components of lavender oil.
Wikipedia (accessed 9 November 2022). Lavender oil. https://en.wikipedia.org/wiki/Lavender_oil
One other group reports some forty individual compounds.
Dong, G., Bai, X., Aimila, A., Aisa, A. and Maiwulanjiang, M. (2020). Study on lavender essential oil chemical compositions by GC-MS and improved pGC. *Molecules* **25(140)**:3166. https://www.ncbi.nlm.nih.gov/pmc/articles/PMC7397202/ doi:10.3390/molecules25143166
Yet another group reports some seventy eight compounds.
Smigielski, K., Raj, A.Krosowiak, K. and Gruska, R. (2009). Chemical composition of the essential oil of *Lavandula angustifolia* cultivated in Poland. *Journal of Essential Oil Bearing Plants* **12(3)**:338-347. doi:10.1080/0972060x.2009.10643729

272 Horridge, A. (2019). *The discovery of a visual system: The honeybee*, 256pp. CABI, Wallingford, Oxfordshire.

273 Wikipedia (accessed 16 November 2022). Nasonov pheromone. https://en.wikipedia.org/wiki/Nasonov_pheromone Image at
Wikimedia Commons (accessed 16 November 2022). File:Nikolai Nasonov 3.jpg https://commons.wikimedia.org/w/index.php?curid=45861527

274 Pickett, J.A., Williams, I.H., Martin, A.P. and Smith M.C. (1980). The Nasonov pheromone of the honey bee, *Apis mellifera* L. (Hymenoptera, Apidae), Part I. Chemical characterisation. *Journal of Chemical Ecology* **6(2)**:425-434. doi.org/10.1007/BF01402919
Williams, I.H., Pickett, J.A. and Martin, A.P. (1981). Nasonov pheromone of the honeybee, *Apis mellifera* L. (Hymenoptera, Apidae), Part II. Bioassay of chemical components with foragers. *Journal of Chemical Ecology* **7(2)**:225-237. doi:10.1007/BF00995745

Pickett, J.A., Williams, I.H., Smith, M.C. and Martin, A.P. (1981). Nasonov pheromone of the honey bee, *Apis mellifera* L. (Hymenoptera, Apidae), Part III. Regulation of pheromone composition and production. *Journal of Chemical Ecology* **7(3)**:543-554. doi:10.1007/bf00987702

275 Burlew, R. (2017). Honey bee pheromones: Common scents. https://www.honeybeesuite.com/honey-bee-pheromones-common-scents/

276 Koetz, A. (2013) loc. cit.

277 Naik, D.G., Gadre, R.V., Kapadi, A.H., Singh, M.K., Suryanarayana, M.C. and Kshirsagar, K.K. (1988). Nasonov gland pheromone of the Indian honeybee, *Apis cerana indica*. *Journal of Apicultural Research* **27(4)**:205-206.
doi/abs/10.1080/00218839.1988.11100803

278 Matsuyama, S., Suzuki, H. and Sasagawa, H. (2000). Chemical analysis of worker pheromone component of Asian honeybees. Proceedings of the 4th Asian apiculture association international conference, Kathmandu. Matsuyama S., Suzuki T. and Sasagawa H. (1997). Semiochemicals in the Japanese honeybee, *Apis cerana japonica* Rad. *Zoological Science* **14**:49.

279 Butler, C.G., Fletcher, D.J.C. and Watler, D. (1969). Nest-entrance marking with pheromones by the honeybee *Apis mellifera L.* and by a wasp, *Vespula vulgaris* L. *Animal Behaviour* **17**:142-147. doi:10.1016/0003-3472(69)90122-5

280 Kastberger, G., Raspotnig, G., Biswas, S. and Winder, O. (2010). Evidence of Nasonov scenting in colony defence of the giant honeybee *Apis dorsata*. *Ethology* **104(1)**:27-37. doi:10.1111/j.1439-0310.1998.tb00027.x

281 Winston, M.L. (1987) loc. cit. pp.131-132.

282 Pickett, J.A., Williams, I.H. and Martin, A.P. (1982). (Z)-11-Eicosen-1-ol, an important new pheromonal component from the sting of the honey bee, *Apis mellifera* L. (Hymenoptera, Apidae.). *Journal of Chemical Ecology* **8(1)**:163-175. doi.org/10.1007/BF00984013

283 Cassier, P., Tel-Zur, D. and Lensky, Y. (1994). The sting sheaths of honey bee workers (*Apis mellifera* L.). Structure and alarm pheromone secretion. *Journal of Insect Physiology* **40(1)**:23-32. doi:10.1016/0022-1910(94)90108-2

284 Camargos, A.F., Cossolin, J.F.S., Martínez, L.C., Gonçalves, W.G., dos Santos, M.H., Zanuncio, J.C. and Serrão, J.E. (2020). Morphology

and chemical composition of the Koschewnikow gland of the honey bee *Apis mellifera* (Hymenoptera: Apidae) workers engaged in different tasks. *Journal of Apicultural Research* **59(5)**:1037-1048. doi:10.1080/00218839.2020.1736 https://www.tandfonline.com/doi/full/10.1080/00218839.2020.1736781

Lensky, Y., Cassier, P., Rosa, S. and Grandperrin, D. (1991). Induction of balling in worker honeybees (*Apis mellifera* L.) by stress pheromone from Koschevnikov glands of queen bees: Behavioural, structural and chemical study. *Comparative Biochemistry and Physiology* – Part A: *Physiology* **100(3)**:585-594. doi.org/10.1016/0300-9629(91)90374-L

285 Collison, C. (2022). A closer look: Colony defence — Guarding and stinging. *The Australasian Beekeeper* **123(8)**:40-42.
Collison, C. (July 2021). A closer look: Colony defense. *Bee Culture* **149(7)**:35-38. https://www.beeculture.com/wp-content/uploads/2021/06/July2021Combinedv1.pdf

286 Villar, G., Wolfson, M.D., Hefetz, A. and Grozinger, C.M. (2017). Evaluating the role of drone-produced chemical signals in mediating social interactions in honey bees (*Apis mellifera*). *Journal of Chemical Ecology* **44(1)**:1-8. doi:10.1007/s10886-017-0912-2

287 Wanner, K.W., Nichols, A.S., Walden, K.K.O., Brockmann, A., Luetje, C.W. and Robertson, H.M. (2007). A honey bee odorant receptor for the queen substance 9-oxo-2-decenoic acid. *Proceedings of the National Academy of Sciences* **104(36)**:14383-14388. doi.org/10.1073/pnas.0705459104 http://www.pnas.org/content/104/36/14383.fulpdf

288 Leoncini, I., Le Conte, Y., Costagliola, G., Plettner, E., Toth, A.L., Wang, M., Huang, Z., Bécard, J.-M., Crauser, D., Slessor, K.N. and Robinson, G.E. (2004). Regulation of behavioral maturation by a primer pheromone produced by adult worker honey bees. *Proceedings of the National Academy of Sciences* **101(50)**:17559-17564. doi:10.1073/pnas.0407652101

289 *Do you carry a spare tyre?* was was presented in the Canberra Region Beekeepers *Bee Buzz Box* by Alan Wade in May 2019.

290 Wikipedia (accessed 11 July 2022). Edmond Hoyle. https://en.wikipedia.org/wiki/Edmond_Hoyle

291 *Should I feed my dog?* was presented in the Canberra Region Beekeepers *Bee Buzz Box* by Alan Wade in April 2019.

292 RSPCA South Australia (June 2018). Port Noarlunga South man convicted for failing to feed three dogs. https://www.rspcasa.org.au/prosecutions-2017-18/#:~:text=Port%20Noarlunga%20South%20man%20convicted,%2C%20Cooper%2C%20Ralph%20and%20Ruby

293 Seeley, T.D. (2017) loc. cit.

294 Hepburn, H.R. and Radloff, S.E. (1998). *Honeybees of Africa.* Chapter 1 Biogeographical perspective. 1.4 Phenology of flowering. pp.13-16.

295 ACT Government Territory and Municipal Services *Code of practice for beekeeping in residential areas of the ACT* (2016), 7pp. https://www.environment.act.gov.au/__data/assets/pdf_file/0003/901983/Code-of-Practice-for-Beekeeping-in-Residential-Areas-of-the-ACT-2016.pdf

296 New South Wales Department of Primary Industry (2022). NSW DPI endorses the Australian Honey Bee Industry *Biosecurity Code of Practice* https://www.dpi.nsw.gov.au/animals-and-livestock/bees/biosecurity-code-of-practice#:~:text=NSW%20DPI%20endorses%20the%20Australian,industry%20from%20pests%20and%20diseases.

297 *Rivers of honey: Keeping bees in doubled hives and two-queen colonies* was prepared as a support document for a webinar presented by Alan Wade and Dannielle Harden to the Somerset Beekeepers' Association on the 6th of January 2022. It was reproduced in the Canberra Region Beekeepers *Bee Buzz Box* in February 2022.

298 Wikipedia (accessed 11 July 2022). Hoyle, E. (1741). Edmund Hoyle. https://en.wikipedia.org/wiki/Edmond_Hoyle

299 Natural History Museum of Los Angeles (2015). Invertebrate palaeontology. *Apis henshawi* Cockerell, 1907 (Henshaw's honey bee). https://research.nhm.org/ip/Apis-henshawi/

300 Seeley, T.D. and Morse, R.A. (1976) loc. cit.
Seeley, T.D. (2010) loc. cit.

301 Demaree, G. (1892) loc. cit.

302 Farrar, C.L. (1968) loc. cit.

303 Wade, A. (2021). *A history of keeping and managing doubled and two-queen hives*, 177pp. Northern Bee Books, West Yorkshire.

304 British Bee-Keepers' Association Quarterly Conversazione (April 7, 1892). Report of meeting of 31 March 1892. *British Bee Journal, Bee-Keepers' Record and Adviser* **20(511)**:132-133. https://www.biodiversitylibrary.

org/item/83768#page/147/mode/1up
British Bee-Keepers' Association Quarterly Conversazione (April 6, 1893). *British Bee Journal, Bee-Keepers' Record and Adviser* **21(563):**126, 132-134. https://www.biodiversitylibrary.org/item/83769#page/142/mode/1up
Editorial notices &c. (November 12, 1896). British Bee-Keepers' Association Conversazione.
British Bee Journal, Bee-Keepers' Record and Adviser **24(752):**451-453. https://www.biodiversitylibrary.org/item/83772#page/460/mode/1up

305 Alexander, E.W. (September 1, 1907). A plurality of queens without perforated zinc: How the queens are introduced: The advantages of the plural-queen system. *Gleanings in Bee Culture* **35(17):**1136-1138. https://babel.hathitrust.org/cgi/pt?id=umn.31951d00953180r&view=1up&seq=660

306 Cruadh (December 15, 1907). Plurality of queens: Another plan for introduction: Two or more queens to a colony. *Gleanings in Bee Culture* **35(24):**1592-1593. https://www.biodiversitylibrary.org/item/74679#page/1608/mode/1up

307 Medicus (February 10, February 17 and February 24, 1910). A two-queen system: Some remarks on the adaptability for a heather district. *British Bee Journal and Bee-Keepers' Adviser* **38(1442-1444):**55-56, 64-66, 75-77. https://www.biodiversitylibrary.org/item/83077#page/67/mode/1up
https://ia800301.us.archive.org/1/items/britishbeejourna1910lond/britishbeejourna1910lond.pdf

308 Banker, R. (1968). A two-queen method used in commercial operations. *American Bee Journal* **108(5):**180-182. Republished as Banker R. (1968). A two-queen method used in commercial operations. *Apiacta* **2:**1-4. http://www.fiitea.org/cgi-bin/index.cgi?sid=&zone=cms&action=search&categ_id=53&search_ordine=descriere
Banker, R. (1979). Part B. Two-queen colony management in *The Hive and the Honey Bee*, Dadant & Sons, Hamilton, Illinois, Chapter XII, pp.404-410, 412.

309 Cale, G.H. (June 1952). The effect of the two-queen system on the harvest: An interview with Dr C.L. Farrar. *American Bee Journal* **92(6):**236-237. https://archive.org/details/sim_american-bee-journal_1952-06_92_6/page/236/mode/1up

310 *Social organisation of ants, bees and wasps* was presented in the Canberra Region Beekeepers *Bee Buzz Box* by Alan Wade in a two part series in October and November 2018 as Social organisation of ants, bees and wasps Part I – The origins of eusociality and Part II – The eusocial hymenoptera fauna.

311 Wilson, E.O. and Hölldobler, B. (2005). Eusociality: origin and consequences. *Proceedings of the National Academy of Sciences of the United States of America* **102(38):**3367-13371. http://www.pnas.org/content/pnas/102/38/13367.full.pdf
Nowak, M.A., Tarnita, C.E. and Wilson, E.O. (2010). The evolution of eusociality. *Nature* **466(7310):**1057-1062. doi:10.1038/nature09205 https://www.ncbi.nlm.nih.gov/pmc/articles/PMC3279739/

312 Look and Learn History Picture Archive (accessed 26 November 2022). A wedding assailed by a swarm of bees. https://www.lookandlearn.com/history-images/M512371/A-wedding-assailed-by-a-swarm-of-bees?t=2&q=interruption&n=9

313 Hughes, W.O.H., Oldroyd, B.P., Beekman, M. and Ratnieks, F.L.W. (2008) loc. cit.
Cardinal, S. and Danforth, B.N. (2011). The antiquity and evolutionary history of social behavior in bees. *PLOS One* **6(6):**e21086 http://journals.plos.org/plosone/article?id=10.1371/journal.pone.0021086

314 Wikipedia (accessed 15 July 2022). Banded sugar ant. https://en.wikipedia.org/wiki/Banded_sugar_ant
Wikipedia (accessed 15 July 2022). Meat ant. https://en.wikipedia.org/wiki/Meat_ant

315 Wikipedia (accessed 15 July 2022). *Ogyris genoveva*. https://en.wikipedia.org/wiki/Ogyris_genoveva
Wikipedia (accessed 15 July 2022). *Jalmenus evagoras*. https://en.wikipedia.org/wiki/Jalmenus_evagoras

316 The Australian Museum (28 February 2022). Paper wasps. https://australian.museum/learn/animals/insects/paper-wasps/

317 Heard, T. (2016) loc. cit.

318 *Global distribution of honey bees* was published as
Wade, A. (2017). A review of the global distribution of honey bees. *The Australasian Beekeeper* **118(10):**447-452; **118(11):**485.

319 Sheppard, W. and Meixner, M. (2003). *Apis mellifera pomonella*, a new

honey bee subspecies from Central Asia. *Apidologie* **34(4)**:367-375. https://hal.archives-ouvertes.fr/hal-00891797/document

320 Engel, M.S., Hinojosa-Díaz, I.A.and Rasnitsyn, A.P. (2009). A honey bee from the Miocene of Nevada and the biogeography of *Apis* (Hymenoptera: Apidae: Apini). *Proceedings of the California Academy of Sciences*, Series 4, **60(3)**:23-38. http://susquehannabeekeepers.com/pdfs/A_honey_bee_from_the_Miocene_of_Nevada_and_the_bio.pdf

321 Hepburn, H.R. and Radloff, S.E. (1998). *Honeybees of Africa,* p.52.

322 Wikipedia (accessed 25 July 2022). Honey bee. https://commons.wikimedia.org/wiki/File:Apis_distribution_map.svg

323 Cervancia, C.R. (2002). Biodiversity of *Apis cerana* Fabricius in the Philippines. http://agris.fao.org/agris-search/search.do?recordID=PH2002000835

324 Arias M.C. and Sheppard, W.S. (2005). Phylogenetic relationships of honey bees (Hymenoptera:Apinae:Apini) inferred from nuclear and mitochondrial DNA sequence data. *Molecular Phylogenetics and Evolution* **37**:25-35. doi:10.1016/j.ympev.2005.02.017 https://www.ncbi.nlm.nih.gov/pubmed/16182149

325 Lo, N., Gloag, R.S., Anderson, D.L. and Oldroyd, B.P. (2010). A molecular phylogeny of the genus *Apis* suggests that the giant honey bee of the Philippines, *A. breviligula* Maa, and the Plains Honey Bee of southern India, *A. indica* Fabricius, are valid species. *Systematic Entomology* **35(2)**:226-233. The Royal Entomological Society. doi:10.1111/j.1365-3113.2009.00504.x

326 Wikimedia Commons (accessed 2 April 2023). *Apis florea* distribution map. https://commons.wikimedia.org/wiki/File:Apis_florea_distribution_map.svg

327 Wikimedia Commons (accessed 2 April 2023). *Apis andreniformis* distribution map. https://commons.wikimedia.org/wiki/File:Apis_andreniformis_distribution_map.svg

328 Koetz, A. (2013) loc. cit.

329 Engel, M.S. (1999). The taxonomy of recent and fossil honey bees (Hymenoptera: Apidae: *Apis*). *Journal of Hymenoptera Research* **8(2)**:165-196. https://kuscholarworks.ku.edu/bitstream/handle/1808/16476/Engel_JoHR_8%282%29165.

pdf?sequence=1&isAllowed=y

330 Anderson, D.L. (2002). *Varroa*-bee relationships – what they tell us about controlling *Varroa* mites on the European honey bee. *Apiacta* **3**:9.
http://www.fiitea.org/cgi-bin/index.cgi?sid=&zone=cms&action=search&categ_id=53&search_ordine=descriere
http://www.fiitea.org/foundation/files/2002/D.L.%20ANDERSON.pdf

331 Wikimedia Commons (accessed 2 April 2023). *Apis andreniformis* distribution map.
https://commons.wikimedia.org/wiki/File:Apis_cerana_distribution_map.svg
Radloff, S.E., Hepburn, C., Hepburn, H.R., Fuchs, S., Hadisoesilo, S., Tan, K., Engel, M.S. and Kuznetsov, V. (2010). Population structure and classification of *Apis cerana*. *Apidologie* **41(6)**:589-601. http://www.apidologie.org/articles/apido/full_html/2010/06/m08176/m08176.html
https://hal.archives-ouvertes.fr/hal-00892035v1
A very similar map showing eastward expanded distribution of *Apis cerana javanica* was reproduced recently by Robert Owen.
Owen, R. (2022). *Apis cerana* genetics. *The Australasian Beekeeper* **124(6)**:34-35.

332 Google Maps (accessed 2 April 2023). See also Wikimedia Commons (accessed 2 April 2023). *Apis nuluensis* distribution map. https://commons.wikimedia.org/wiki/File:Borneo_map_with_borders.png

333 Wikimedia Commons (accessed 2 April 2023). *Apis nigrocincta* distribution map.
https://commons.wikimedia.org/wiki/File:Apis_nigrocincta_distribution_map.svg
see also Tanaka, H., Roubik, D.W., Kato, M., Liew, F. and Gunsalam, G. (2001). Phylogenetic position of *Apis nuluensis* of northern Borneo and phylogeography of *A. cerana* as inferred from mitochondrial DNA sequences. *Insectes Sociaux* **48(1)**:44–51. doi:10.1007/pl00001744

334 Anderson, D.L. Halliday, R.B. and Otis G.W. (1997). The occurrence of *Varroa underwoodi* (Acarina: Varroidae) in Papua New Guinea and Indonesia. *Apidologie* **28(3-4)**:143-147. doi.org/10.1051/apido:19970305
https://hal.archives-ouvertes.fr/hal-00891413/document

335 Wikipedia (accessed 2 April 2023). *Apis koschevnikovi* distribution map.
https://commons.wikimedia.org/wiki/File:Apis_koschevnikovi_distribution_map.svg

336 Whitfield, C.W., Behura, S.K., Berlocher, S.H., Clark, A.G. Johnston, J.S., Sheppard, W.S., Smith, D.R., Suarez, A.V., Weaver, D. and Tsutsui, N.D. (2006). Thrice out of Africa: ancient and recent expansions of the honey bee, *Apis mellifera*. *Science.* **314(5799)**:642-645. http://www.life.illinois.edu/suarez/publications/Whitfield_etal2006Science.pdf
https://sib.illinois.edu/suarez/local/suarez/uploads/2020/01/Whitfield_etal2006Science.pdf

337 Cridland, J.M., Tsutsui, N.D. and Ramírez, S.R. (2017). The complex demographic history and evolutionary origin of the western honey bee, *Apis mellifera*. *Genome Biology and Evolution* **9(2)**:457-472. https://academic.oup.com/gbe/article/9/2/457/2970293/The-Complex-Demographic-History-and-Evolutionary?searchresult=1 doi:10.1093/gbe/evx009

338 Brother Adam (1983) loc. cit.

339 Ruttner, F., Tassencourt, L. and Louveaux, J. (1978). Biometrical statistical analysis of the geographic variability of *Apis mellifera* L. I Material and methods. *Apidologie* **9(4)**:363-381. https://hal.archives-ouvertes.fr/hal-00890475

340 Winston, M.L. (1992) loc. cit.

341 Engel, M.S., Hinojosa-Díaz, I.A. and Rasnitsyn, A.P. (2009). A honey bee from the Miocene of Nevada and the biogeography of *Apis* (Hymenoptera: Apidae: Apini). *Proceedings of the California Academy of Sciences*, Series 4, **60(3)**:23-38. http://susquehannabeekeepers.com/pdfs/A_honey_bee_from_the_Miocene_of_Nevada_and_the_bio.pdf
Hepburn, H.R. and Radloff, S.E. (eds) (2011). Honeybees of Asia. Chapter 2, Classification of honey bees. 2.8 Worker size and fossil record. pp.46-47.
Google Books (accessed 16 November 2022). https://books.google.com.au/books?id=zNuwxBP8MTUC&pg=PA47&lpg=PA47&dq=apis+miocene&source=bl&ots=u66tgaJsGU&sig=WZZOGV7gIks7XlsERAYKj_aAOT4&hl=en&sa=X&ved=0ahUKEwi6mr6566XRAhWFGJQKHZgTCOoQ6AEIFjAD#v=onepage&q=apis%20miocene&f=false

342 *The giant honey bees* was published as
Wade, A. (2022a). The giant honey bees: Part I – The sting is in the tail. *The Australasian Beekeeper* **124(4)**:30-35.
Wade, A. (2022b). The giant honey bees: Part II – The tale is in the sting (miss-titled The sting is in the tail). *The Australasian Beekeeper*

124(5):30-34.

343 Engle, M.S. (2006). A giant honey bee from the Middle Miocene of Japan (Hymenoptera: Apidae). *American Museum Novitates* 3504(1):1-12. doi:10.1206/0003-0082(2006)504[0001:AGHBFT]2.0.CO;2

344 Michener, C.D. (2007). *The Bees of the World*, second edition. The Johns Hopkins University Press, Baltimore, Maryland.

345 Fuchs, S., Koeniger, N. and Tingek, S. (1996). The morphometric position of *Apis nuluensis* Tingek, Koeniger and Koeniger, 1996 within cavity-nesting honey bees. *Apidologie* 27(5):397-405. doi:10.1051/apido:19960507 https://www.researchgate.net/publication/248855256_The_morphometric_position_of_Apis_nuluensis_Tingek_Koeniger_and_Koeniger_1996_within_cavity-nesting_honey_bees
Arias, M.C,. Tingek, S., Kelitu, A.and Sheppard, W.S. (1996). *Apis nuluensis* Tingek, Koeniger and Koeniger, 1996 and its genetic relationship with sympatric species inferred from DNA sequences. *Apidologie* 27(5):415-422. https://hal.archives-ouvertes.fr/hal-00891385/document
Koeniger N., Koeniger G., Gries M., Tingek S. and Kelitu A. (1996). Reproductive isolation of *A nuluensis,* Tingek, Koeniger and Koeniger 1996, by species specific mating time. *Apidologie* 27(5):353-360. doi:10.1051/apido:2000125
https://hal.archives-ouvertes.fr/hal-00891379/document
Damus, M.S. and Otis, G.W. (1997). A morphometric analysis of *Apis cerana* F and *Apis nigrocincta* Smith populations from Southeast Asia. *Apidologie* 28(5):309-323. doi.org/10.1051/apido:19970507 http://www.apidologie.org/articles/apido/abs/1997/04/Apidologie_0044-8435_1997_28_5_ART0007/Apidologie_0044-8435_1997_28_5_ART0007.html

346 Wikimedia Commons (accessed 2 April 2023). *Apis dorsata* distribution map.
https://commons.wikimedia.org/wiki/File:Apis_dorsata_distribution_map.svg

347 Lo, N., Gloag, R.S., Anderson, D.L. and Oldroyd, B.P. (2010) loc. cit.

348 Plant Health Australia (accessed 15 April 2022). Giant honey bees. https://beeaware.org.au/archive-pest/giant-honey-bees/#ad-image-0

349 Gregory, M., and Jack, C. (February 2022). Himalayan giant honey bee, cliff honey bee *Apis laboriosa* Smith (Insecta: Hymenoptera: Apidae). University of Florida. https://edis.ifas.ufl.edu/publication/IN1348

350 New South Wales Department of Primary Industries (accessed 16 April 2022). European honey bee. https://www.dpi.nsw.gov.au/biosecurity/plant/bees-and-wasps/european-honeybee

351 Kekeçoğlu, M., Bouga, M., Soysal, M.I. and Harizanis, P. (2007). Morphometrics as a tool for the study of genetic variability of honey bees. *Tekirdağ Ziraat Fakültesi Dergisi* (*Journal of Tekirdağ Agricultural Faculty*) **4(1)**:7-15. https://www.researchgate.net/publication/38107507_Morphometrics_as_a_Tool_for_the_Study_of_Genetic_Variability_of_Honey_Bees
Ruttner, F. (1988). *Biogeography and taxonomy of honeybees*. Springer Verlag, Berlin.

352 Bee Aware: Plant Health Australia (accessed 16 April 2022). Asian honey bee. https://beeaware.org.au/archive-pest/asian-honey-bee/#ad-image-0

353 Bee Aware: Plant Health Australia (accessed 16 April 2022). Dwarf honey bees. https://beeaware.org.au/archive-pest/dwarf-honey-bees/#ad-image-0

354 Cao, L.-F., Zheng, H.-Q., Chen, X., Niu, D.-F., Hu, F.-L. and Hepburn, H.R. (2012). Multivariate morphometric analyses of the giant honey bees, *Apis dorsata* F. and *Apis laboriosa* F. in China. *Journal of Apicultural Research* **51(3)**:245-251. doi:10.3896/IBRA.1.51.3.05

355 Hepburn, R. and Radloff, S.E. (2011) loc. cit.
Cushman, D.A. (accessed 19 July 2007). Morphometry: Called morphology in some parts of the world. http://www.dave-cushman.net/bee/morphometry.html

356 Chuttong, B., Buawangpong, N. and Burgett, M.D. (2019). Drone production by the giant honey bee *Apis dorsata* F. (Hymenoptera: Apidae). *Sociobiology* **66(3)**:475. https://www.researchgate.net/publication/337437506_Drone_Production_by_the_Giant_Honey_Bee_Apis_dorsata_F_Hymenoptera_Apidae

357 Morse, R.A. (1969). The biology of *Apis dorsata* in the Philippines. *International Union for the Study of Social Insects Proceedings* pp.185-187. https://cataglyphis.fr/Actes-SF-UIEIS/IUSSI-Bern-1969/IUSSI-Bern-1969-Morse.pdf
Morse, R.A. and Laigo, F.M. (1969). *Apis dorsata* in the Philippines (including an annotated bibliography) Laguna. *Monograph of Philippine Association of Entomologists* **1**:1-96. https://www.worldcat.org/title/apis-dorsata-in-the-philippines-including-an-annotated-bibliography/

oclc/63374047#similar
https://www.amazon.com/Apis-dorsata-Philippines-bibliography-Entomologists/dp/B0007JS7TK
Additional notes and correspondence are posited with The National Library of Wales [Llyfrgell Genedlaethol Cymru] as GB 0210 BEEIBRA International Bee Research Association (IBRA) Records: Reference code: CP1/29. Vtls006326885 and physically located at ARCH/MSS (GB0210) Morse R.A. and Laigo, F.M. (1968). The biology of *Apis dorsata* parts iii, v, vii and ix,] https://archiveshub.jisc.ac.uk/search/archives/cd8e66d5-b792-306a-9073-7d6b54d3b8ce?component=c82b8c51-4ac6-38cc-b020-809d542ff657
Seeley, T.D., Seeley, R.H. and Akratanakul, P. (1982). Colony defense strategies of the honeybee in Thailand. *Ecological Monographs* **52(1)**:43-63. doi.org/10.2307/2937344 https://www.jstor.org/stable/2937344
Seeley, T.D. (2010) loc. cit.
Moritz, R.F.A., Kraus, F.B., Kryger, P. and Crewe, R.M. (2007). The size of wild honeybee populations (*Apis mellifera*) and its implications for the conservation of honeybees. *Journal of Insect Conservation* **11(4)**:391-397. doi:10.1007/s10841-006-9054-5
Hepburn, R., Duangphakdee, O., Phiancharoen, M. and Radloff, S. (2010). Comb wax salvage by the Red Dwarf Honeybee, *Apis florea* F. *Journal of Insect Behavior* **23(2)**:159-164. doi:10.1007/s10905-010-9205-0
Karlsson, T. (1990). The natural nest of the Asian hive bee (*Apis cerana*) in Bangladesh: A minor field study. Working paper, International Rural Development Centre, Swedish University of Agricultural Sciences No. 134, 35pp. https://www.cabi.org/ISC/abstract/19910229922

358 Akratanakul, P. (1986) p.43. *Beekeeping in Asia*, 121pp. Food and Agriculture Services Bulletin 68/4, Rome. https://victoriancollections.net.au/items/5313c4ab2162ef22accd9360

359 Wade, A. (2022c) loc.cit.

360 Sakagami, S.F., Matsumura, T. and Ito, K. (1980). *Apis laboriosa* in Himalaya, the little known world largest honeybee (Hymenoptera, Apidae). *Insecta Matsumurana* **19**:47-78. https://eprints.lib.hokudai.ac.jp/dspace/bitstream/2115/9801/1/19_p47-77.pdf

361 Anderson, D.L. and Morgan, M.J. (2007). Genetic and morphological variation of bee-parasitic *Tropilaelaps* mites (Acari: Laelapidae): New and re-defined species. *Experimental and Applied Acarology* **43(1)**:1-24. https://openresearch-repository.anu.

edu.au/bitstream/1885/51418/2/01_Anderson_Genetic_and_morphological_2007.pdf
https://pubmed.ncbi.nlm.nih.gov/17828576/ doi:10.1007/s10493-007-9103-0

362 Shearer, D.A., Boch, R., Morse, R.A. and Laigo, F.M. (1970). Occurrence of 9-oxodec-trans-2-enoic acid in queens of *Apis dorsata, Apis cerana* and *Apis mellifera. Journal of Insect Physiology* **16(7)**:1437-1441. doi.org/10.1016/0022-1910(70)90142-3 https://www.sciencedirect.com/science/article/abs/pii/0022191070901423

363 Woyke, J., Wilde, J. and Reddy, C. (February 2004). The gentleness of *Apis dorsata* verified while investigating brood-cross fostering and hygienic behavior. Proceedings of the seventh Asian Apicultural Association Conference and tenth BEENET Symposium and Technofora, pp.87-91. https://www.researchgate.net/publication/251570611_2004_The_gentleness_of_Apis_dorsata_verified_while_investigating_brood-cross_fostering_and_hygienic_behavior

364 Roepke, W. (1930). Beobachtungen an Indischen Honigbienen, insbesondere an *Apis dorsata* F. Meded. LandbHoogesh, 26pp. H. Veenman & Zonen — Wageningen. https://library.wur.nl/WebQuery/wurpubs/fulltext/293809

365 Lindauer, M. (1956). Über die Verständigung bei indischen Bienen. *Zeitschrift für vergleichende Physiologie* **8(6)**:521-557. doi.org/10.1007/BF00341108 https://link.springer.com/article/10.1007/BF00341108

366 Fuchs, S. and Tautz, J. (2011). Colony defence and natural enemies, Chapter 17, pp.369-395 *in* Hepburn, H.R. and Radloff, S.E. (eds) (2011). *Honeybees of Asia.*

367 Kastberger, G., Weihmann, F., Zierler, M. and Hötzl, T. (2014). Giant honeybees (*Apis dorsata*) mob wasps away from the nest by directed visual patterns. *Naturwissenschaften* **101(11)**:861-873. doi:10.1007/s00114-014-1220-0
https://www.ncbi.nlm.nih.gov/pmc/articles/PMC4209238/#MOESM2
https://link.springer.com/article/10.1007/s00114-014-1220-0
Kastberger, G., Weihmann, F. and Höetzl, T. (2013).
Social waves in giant honeybees (*Apis dorsata*) elicit nest vibrations. *Naturwissenschaften* **100(7)**:595-609.
doi:10.1007/s00114-013-1056-z
Kastberger, G., Hötzl, T., Maurer, M., Kranner, I., Weiss, S. and Weihmann,

F. (2014). Speeding up social waves: Propagation mechanisms of shimmering in giant honeybees. *PLOS One* **9(1)**:e86315. doi:10.1371/journal.pone.0086315

Seeley, T.D., Seeley, R.H. and Akratanakul, P. (1982) loc. cit.

Oldroyd, B.P. and Wongsiri, S. (2006) loc. cit.

Woyke, J., Wilde, J., Wilde, M. and Cervancia, C. (2006). Abdomen flipping of *Apis dorsata breviligula* worker bees correlated with temperature of nest curtain surface. *Apidologie* **37(4)**:501-505. doi: 10.1051/apido:2006032 https://www.apidologie.org/articles/apido/pdf/2006/04/m6040.pdf

368 See video clips at
Kastberger, G., Hötzl, T., Maurer, M., Kranner, I., Weiss, S., and Weihmann, F. (2014) loc. cit. https://www.ncbi.nlm.nih.gov/pmc/articles/PMC4209238/#MOESM2

369 Ramya, J. and Rajagopal, D. (2008). Morphology of the sting and its associated glands in four different honey bee species. *Journal of Apicultural Research* **47(1)**:46-52. doi:10.1080/00218839.2008.11101422
Hepburn and Radloff (2011) loc. cit.

370 Kipling, R. (29 July 1895). *The Second Jungle Book,* Red Dog. Macmillan Publishers, London. The 1906 edition, New York, The Century Co. https://ia902704.us.archive.org/35/items/junglebooksecond00kipl/junglebooksecond00kipl.pdf

371 Wikipedia (accessed 14 May 2022). Dhole. https://en.wikipedia.org/wiki/Dhole

372 Radloff, S.E., Hepburn, H.R. and Engel, M.S. (2011). Chapter 1, The Asian species of *Apis*, 1.4 The giant honey bees, pp.13-16 *in* Hepburn, R. and Radloff, S.E. (2011). *Honeybees of Asia.*

373 Oldroyd, B.P. and Wongsiri, S. (2006) loc. cit.

374 Paar, J., Oldroyd, B.P., Huettinger, E. and Kasterberger, G. (2004). Genetic structure of an *Apis dorsata* population: The significance of migration and colony aggregation. *Journal of Heredity* **95(2)**:119-126. doi:10.1093/jhered/esh026

375 Hepburn, H.R. and Radloff, S.E. (2011). Chapter 3 Biogeography, pp.31-68 and Hepburn, H.R. (2011). Chapter 7, Absconding, migration and swarming, pp.133-158 *in* Hepburn, H.R. and Radloff, S.E. *Honeybees of*

Asia.

376 Woyke, J., Wilde, J. and Wilde, M. (2012) loc. cit.

377 Robinson, W.S. (2012) loc. cit.

378 Underwood, B.A. (1990a) loc. cit.
Underwood, B.A. (1992) loc. cit.

379 Woyke, J. and Wilde, M. (2003) loc. cit.
Woyke, J., Wilde, J. and Wilde, M. (2012) loc. cit.

380 Tan, N.Q. and Ha, D.T. (2002). Socio-economic factors in traditional rafter beekeeping with *Apis dorsata* in Vietnam. *Bee World* **83(4)**:165-170. doi.org/10.1080/0005772X.2002.11099559 https://www.tandfonline.com/doi/abs/10.1080/0005772X.2002.11099559

381 Kitnya, N., Prabhudev, M.V., Bhatta, C.P., Pham, T.H., Nidup, T., Megu, K., Chakravorty, J., Brockmann, A. and Otis, G.W. (2020). Geographical distribution of the giant honey bee *Apis laboriosa* Smith, 1871 (Hymenoptera, Apidae). *ZooKeys* **951**:67-81. https://zookeys.pensoft.net/article/49855/

382 Woyke, J., Wilde, J. and Wilde, M. (2001) loc. cit.

383 Underwood, B.A. (1990a) loc. cit.
Underwood, B.A. (1990b). The behaviour and energetics of high-altitude survival by the Himalayan honeybee, *Apis laboriosa*. PhD thesis, Cornell University, Ithaca, USA. 144pp. https://www.cabdirect.org/cabdirect/abstract/19910230590
Dyer, F.C. and Seeley, T.D. (1986). Interspecific comparisons of endothermy in honey-bees (*Apis*): Deviations from the expected side-related patterns. *Journal of Experimental Biology* **127(1)**:1-26. https://citeseerx.ist.psu.edu/document?repid=rep1&type=pdf&doi=fdffbcd28d23b1c5a3b6cee86a5407baeb2bd00c

384 Roubik, D.W., Sakagami, S.F. and Kudo, I. (1985). A note on distribution and nesting of the Himalayan honey bee *Apis laboriosa* Smith (Hymenoptera: Apidae). *Journal of the Kansas Entomological Society* **58(4)**:746-749. https://www.jstor.org/stable/25084723

385 Gregory, M. and Jack, C. (2022) loc. cit.

386 Woyke, J., Wilde, J. and Wilde, M. (2001) loc. cit.

387 Joshi, S.R., Ahmad, F. and Gurung, M.B. (2004). Status of Apis laboriosa populations in Kaski district, western Nepal. *Journal of*

Apicultural Research **43(4):**176-180. https://www.researchgate.net/publication/292620321_Status_of_Apis_laboriosa_populations_in_Kaski_district_western_Nepal

Ahmad, F., Gurung, M.B. and Joshi, S.R. (2003). The Himalayan cliff bee *Apis laboriosa* & honey hunters of Kaski, 71pp. https://shop.beesfordevelopment.org/products/the-himalayan-cliff-bee-apis-laboriosa-honey-hunters-of-kaski (accessed 20 May 2022 at The Himalayan Cliff Bee *Apis laboriosa* – ICIMOD https://lib.icimod.org › files › attachment_124PDF)

388   Batra, S.W.T. (1996). Biology of *Apis laboriosa* Smith, a pollinator of apples at high altitude in the Great Himalaya Range of Garhwal, India, (Hymenoptera: Apidae). *Journal of the Kansas Entomological Society* **69(2):**177-181. https://www.jstor.org/stable/25085665

389   University of Florida. (accessed 17 May 2022). Featured creatures: *Apis dorsata* Fabricius (Insecta: Hymenoptera: Apidae) https://entnemdept.ufl.edu/creatures/MISC/BEES/Apis_dorsata.htm

390   Lo, N., Gloag, R.S., Anderson, D.L. and Oldroyd, B.P. (2010) loc. cit.

391   Lewes, G.H. (before 1885). The life and works of Goethe, Volume 1, p.359. Dent: London. Everyman's Library Dutton: New York. https://ia803002.us.archive.org/27/items/lifeworksofgoeth0000lewe/lifeworksofgoeth0000lewe.pdf

392   Starr, C.K., Schmidt, P.J. and Schmidt, J.O. (1987). Nest-site preferences of the giant honey bee, *Apis dorsata* (Hymenoptera: Apidae), in Borneo. *Pan-Pacific Entomologist* **63(1):**37-42. http://www.ckstarr.net/cks/1987-MEGAPIS.pdf

393   Maa, T.C. (1953). An inquiry into the systematics of the tribus Apidini or honeybees (Hym.). *Treubia* **21(3):**525-640. https://e-journal.biologi.lipi.go.id/index.php/treubia/article/view/2669/2299

394   Sakagami, S.F., Matsumura, T. and Ito, K. (1980) loc. cit.
Darlington, Jr., P. (1957). *Zoogeography: The geographical distribution of animals*, 694pp, John Wiley, New York. Available at the Australian National Library. https://catalogue.nla.gov.au/Record/2516011

395   Nagir, M.T., Atmowidi, T. and Kahono, S. (2016). The distribution and nest-site preference of *Apis dorsata binghami* at Maros Forest, South Sulawesi, Indonesia. *Journal of Insect Biodiversity* **4(23):**1-4. doi:10.12976/jib/2016.4.23 https://www.researchgate.net/

publication/311972586_The_distribution_and_nest-site_preference_of_Apis_dorsata_binghami_at_Maros_Forest_South_Sulawesi_Indonesia

396 Lo, N., Gloag, R.S., Anderson, D.L. and Oldroyd, B.P. (2010) loc. cit.

397 Morse, R.A. and Laigo, F.M. (1969) loc. cit.

398 Woyke, J., Wilde, J. and Wilde, M. (2012) loc. cit.

399 Anderson, D.L. and Morgan, M.J. (2007) loc. cit.

400 de Guzman, L.I., Williams, G.R., Khongphinitbunjong, K. and Chantawannakul, P. (2017) loc. cit.

401 Anderson, D.L. and Morgan, M.J. (2007) loc. cit.

402 de Guzman, L.I., Williams, G.R., Khongphinitbunjong, K. and Chantawannakul, P. (2017) loc. cit.

403 Laigo, F.M. and Morse, R.A. (1968). The mite *Tropilaelaps clareae* in *Apis dorsata* colonies in the Philippines. *Bee World* **49(3)**:116-118. doi.org/10.1080/0005772X.1968.11097211

404 Morse, R.A. and Laigo, F.M. (1969) loc. cit.

405 Franco, S. (2018). OIE Terrestrial Manual, Chapter 3.2.6. Infestation of honey bees with *Tropilaelaps* spp. pp.765-776. OIE World Organisation for Animal Health. https://www.oie.int/fileadmin/Home/fr/Health_standards/tahm/3.02.06_TROPILAELAPS.pdf

406 *Phoretic honey bee mites* was published as
Wade, A. (2022c). Phoretic honey bee mites.
The Australasian Beekeeper **123(7)**:14-18.

407 Warrit, N. and Lekprayoon, C. (2011). Chapter 16 Asian honeybee mites, pp.347-368 in Hepburn, R. and Radloff, S.E. [eds] (2011). *Honeybees of Asia*. https://www.bijenhouders.nl/files/Bijengezondheid/Cornelissen/Warrit_Lekprayoon_2011_Asian%20honey%20bee%20mites%20(1).pdf

408 Refaei, G.S., Zeid, W.R.A. and Roshdy, O.M. (2018). Incidence of parasitic and non-parasitic mites of honeybee, *Apis mellifera* (Linnaeus). *Journal of Plant Protection and Pathology* **9(12)**:873-875. doi:10.21608/jppp.2018.44098

409 Walter, D.E., Beard, J.J., Walker, K.L. and Sparks, K. (2002). Of mites and bees: A review of mite–bee associations in Australia and a revision of Raymentia Womersley (Acari: Mesostigmata: Laelapidae), with the description of two new species of mites from *Lasioglossum*

(Parasphecodes) spp. (Hymenoptera: Halictidae). *Australian Journal of Entomology* **41(2)**:128-148. https://doi.org/10.1046/j.1440-6055.2002.00280.x

410   The poems of Jonathon Swift, Volume 1, 1733. On poetry – A rhapsody. p.264. https://archive.org/details/poemsofjonathans01swifuoft/page/274/mode/1up

411   Walter, D.E., Krantz, G. and Lindquist, E. (1996).
Acari: The mites. http://tolweb.org/acari
Linquist, E.E. (1984). Current theories on the evolution of major groups of Acari and on their relationships with other groups of Arachnida, with consequent implications for their classification. Proceedings of the 6th International Congress of Acarology, Edinburgh, Scotland, Sept. 1982. Vol. 1. Chichester, England, Ellis Horwood Ltd, pp.28-62. http://museum.wa.gov.au/catalogues-beta/bibliography/current-theories-evolution-major-groups-acari-and-their-relationships-other-groups

412   Delfinado-Baker, M. and Baker, E.W. (1982). Notes on honey bee mites of the genus *Acarapis* hirst (Acari: Tarsonemidae). *International Journal of Acarology* **8(4)**:211-226. doi.org/10.1080/01647958208683299 http://www.tandfonline.com/doi/abs/10.1080/01647958208683299?journalCode=taca20

413   Delfinado, M.D. and Baker, E.W. (1974). Varroidae, a new family of mites on honey bees (Mesostigmata: Acarina). *Journal of the Washington Academy of Sciences* **64(1)**:4-10. https://www.jstor.org/stable/24535743

414   de Guzman, L.I., Williams, G.R., Khongphinitbunjong, K. and Chantawannakul, P. (2017) loc. cit.

415   Chantawannakul, P., de Guzman, L.I., Li, J. and Williams, G.R. (2016). Parasites, pathogens, and pests of honeybees in Asia. *Apidologie* **47(3)**:301-324. https://hal.archives-ouvertes.fr/hal-01532338/document

416   Warrit and Lekprayoon (2011) loc. cit.

417   Sammataro, D., Gerson, U. and Needham, G. (2000). Parasitic mites of honey bees: Life history, implications, and impact. *Annual Review of Entomology* **45**:519-548. https://www.ars.usda.gov/ARSUserFiles/31186/ann.rev.samm.pdf

418   Seeley, T.D. (2019) loc. cit.

419   Lee, T. (Fri 16 July 2021). Dutch honey bees resistant to varroa mite

imported to Australia to help guard against the pest. ABC Landline. https://www.abc.net.au/news/2021-07-16/bee-imports-to-protect-against-varroa-mite/100289356

420 Lo, N., Gloag, R.S., Anderson, D.L. and Oldroyd, B.P. (2010) loc. cit.

421 de Guzman, L.I., Williams, G.R., Khongphinitbunjong, K. and Chantawannakul, P. (2017) loc. cit.

422 Anderson, D.L. and Morgan, M.J. (2007) loc. cit.

423 Ramsey, S. (4 November 2021). *Tropilaelaps* mites: A fate worse than *Varroa* with Dr Samuel Ramsey: Tell us what you think! Somerset Beekeepers Association webinar Thursday 4 November 2021 at GMT 1900.
Very ominously, Stuart Robert reports the finding of *Tropilaelaps* mite in Uzbekistan signalling that it may well soon spread to and devastate beekeeping in Russia and Europe. Robert, S. (March 2023). Newsround: Yet another pest. *The Beekeepers Quarterly* **151**:8.

424 New South Wales Department of Primary Industries (29 June 2022). Stakeholder update - *Varroa* mite biosecurity emergency: email to registered NSW beekeepers.

425 *Beetle mania* was published as
Wade, A., Robinson, J. and Robinson, P. (2022). Beetle mania. *The Australasian Beekeeper* **123(12)**:42-45.

426 Sgt. Pepper's Lonely Hearts Club Band (1967). https://en.wikipedia.org/wiki/Sgt._Pepper%27s_Lonely_Hearts_Club_Band

427 Neumann, P. and Elzen, P.J. (2004). The biology of the small hive beetle (*Aethina tumida*, Coleoptera: Nitidulidae): Gaps in our knowledge of an invasive species. *Apidologie* **35(3)**:229-247. doi:10.1051/apido:2004010
Müerrie, T. and Neumann, P. (2004). Mass production of small hive beetle (*Aethina tumida*, Coleoptera: Nitidulidae). *Journal of Apicultural Research* **43(3)**:144-145. doi:10.1080/00218839.2004.11101125

428 Annand, N. (March 2008). Small hive beetle management options. New South Wales Department of Primary Industry *PrimeFacts*, 7pp. https://www.dpi.nsw.gov.au/__data/assets/pdf_file/0010/220240/small-hive-beetle-management-options.pdf

429 Bowman, P. (24 April 2014). Catching small hive beetle:
How to prepare and deploy lantern traps. https://www.youtube.com/

watch?v=YHUmK5SlzXU

430 Levot, G. (2012). Commercialisation of the small hive beetle harbourage device. Rural Industries Research and Development Corporation publication no. 11/122, 46pp. https://www.agrifutures.com.au/wp-content/uploads/publications/11-122.pdf

431 Ritchie, N. (2022). Deep hive bases and trays for controlling small hive beetle. *The Australasian Beekeeper* **121(10):**24-25.

432 Burke, R. (2022). Cleaning up small hive beetle slime-outs. *The Australasian Beekeeper* **123(8):**45-47.

433 Ellis, J.D. and Ellis, A. (2019). Small hive beetle, *Aethina tumida* Murray (Insecta: Coleoptera: Nitidulidae), 5pp. University of Florida IFAS Extension Publication #EENY-474. https://edis.ifas.ufl.edu/pdf/IN/IN85400.pdf https://entnemdept.ufl.edu/creatures/misc/bees/small_hive_beetle.htm

434 Neumann, P. and Elzen, P.J. (2004) loc. cit.

435 Oldroyd B.P. and Allsopp M.H. (2017). Risk assessment for large African hive beetle. Research report, Rural Research and Development Corporation, Canberra publication no.16/054. https://agrifutures.com.au/wp-content/uploads/publications/16-054.pdf
Oldroyd, B.P. and Allsopp, M.H. (2017). Risk assessment for large African hive beetles (*Oplostomus* spp.) – a review. *Apidologie* **48(4):**495-503. doi:10.1007/s13592-017-0493-7
Diedrick, W. and Jack, C. (2021). Large African hive beetle *Oplostomus fuligineus* (Olivier) (Insecta: Coleoptera: Scarabaeidae). University of Florida, Institute of Food and Agricultural Sciences, 4pp. https://edis.ifas.ufl.edu/publication/IN1309 https://edis.ifas.ufl.edu/pdf/IN/IN130900.pdf
This publication updates an earlier review entitled Oldroyd, B. and Allsopp, M. (2006) Risk assessment for large African hive beetle. Rural Industries R&D Corporation, 8pp. https://www.agrifutures.com.au/wp-content/uploads/publications/16-054.pdf

436 BeeAware at Plant Health Australia. Large African hive beetle now high priority pest. https://beeaware.org.au/archive-news/large-african-hive-beetle-now-high-priority-pest/

437 Halcroft, M., Spooner-Hart, R. and Neumann, P. (2011). Behavioral defense strategies of the stingless bee, *Austroplebeia australis*, against the

small hive beetle, *Aethina tumida*. *Insectes Sociaux* **58(2)**:245-253. doi.org/10.1007/s00040-010-0142-x

438   *Where rust and moth doth corrupt* was presented to Canberra Region Beekeepers *Bee Buzz Box* by Alan Wade in January 2019.

439   Kwadha, C.A., Ong'amo, G.O., Ndegwa, P.N., Raina, S.K. and Fombong, A.T. (2017). The biology and control of the Greater Wax Moth, *Galleria mellonella*. *Insects* **8(2)**:61-77. doi:10.3390/insects8020061

440   Somerville, D. (August 2007). Wax moth. PrimeFacts No. 658, NSW Wales Department of Primary Industries.
https://www.dpi.nsw.gov.au/__data/assets/pdf_file/0010/176284/wax-moth.pdf

441   Wikipedia (accessed 28 May 2022). *Aphomia sociella*. https://en.wikipedia.org/wiki/Aphomia_sociella

442   Wikipedia (accessed 28 May 2022). *Vitula edmandsii*. https://en.wikipedia.org/wiki/Vitula_edmandsii

443   Wikipedia (accessed 28 May 2022). Mediterranean flour moth. https://en.wikipedia.org/wiki/Mediterranean_flour_moth

444   Wikipedia (accessed 28 May 2022). Indian meal moth. https://en.wikipedia.org/wiki/Indianmeal_moth

445   Agriculture Victoria (accessed 3 November 2022). Wax moth a beekeeping pest. https://agriculture.vic.gov.au/biosecurity/pest-insects-and-mites/priority-pest-insects-and-mites/wax-moth-a-beekeeping-pest

446   *Ascosphaera apis* also effects bumblebees and at least one species of carpenter bee. The genus including *Ascosphaera callicarpa* affect a range of bee species.
Hornitzky, M (2014). Literature review of chalkbrood: a fungal disease of honeybees NSW Agriculture RIRDC Publication No 01/150 https://www.hgsc.bcm.edu/sites/default/files/images/review_Chalkb.pdf

447   *Lomaria passim* is reported to cause colony losses possibly in association with Nosema. Cannon, D. Ed. (September 2017). Readers Letters: Reply to Question 3 *Lomaria passim? The Australasian Beekeeper* (2017) **119(5)**:8-9.

448   Affects some species of *Bombus* and has been occasionally detected in *Apis mellifera*. Maharramov, J., Meeus, I., Maebe, K., Arbetman, M., Morales, C., Graystock, P., Hughes, W.O.H., Plischuk, S., Lange, C.E., de Graaf, D.C.,

Zapata, N., de la Rosa, J.J.P., Murray, T.E., Brown, M.J.F., Smagghe, G. and Wicker-Thomas, C. (2013). Genetic variability of the neogregarine *Apicystis bombi*, an etiological agent of an emergent bumblebee disease. *PLOS One* **8(12)**:e81475. doi:10.1371/journal.pone.0081475
https://journals.plos.org/plosone/article?id=10.1371/journal.pone.0081475

Plischuk, S., Meeus, I., Smagghe, G. and Lange, C. E. (2011). *Apicystis bombi* (Apicomplexa: Neogregarinorida) parasitizing *Apis mellifera* and *Bombus terrestris* (Hymenoptera: Apidae) in Argentina. *Environmental Microbiology Reports* **3(5)**:565-568. doi.org/10.1111/j.1758-2229.2011.00261.x

449   This bacterium is probably transmitted by *Varroa destructor*.
Burritt N.L., Foss, N.J., Neeno-Eckwall, E.C., Church, J.O., Hilger, A.M., Hildebrand, J.A., Warshauer, D.M., Perna, N.T. and Burritt, J.B. (2016). Sepsis and hemocyte loss in honey bees (*Apis mellifera*) infected with *Serratia marcescens* strain *sicaria*. *PLOS One* **11(12)**:e0167752. https://journals.plos.org/plosone/article?id=10.1371/journal.pone.0167752

450   This Australian beetle has not been reported to affect stored beekeeping material here in Australia but is causing serious damage in California.
Cannon, D. (Editorial Notes (2017)). *The Australasian Beekeeper* **118(8)**:341.

451   Small hive beetle may be controlled by *Kodamea ohmeri*, a fungus symbiotic with SHB, is also a likely pathogen for immunocompromised people. Other fungi, *Metarhizium anisopliae* and *Beauveria bassiana* show potential for control of larvae of SHB.
Leemon, D. (August 2012). In-hive fungal biocontrol of small hive beetle. RIRDC Publication No. 12/012. http://era.daf.qld.gov.au/id/eprint/5646/1/12-012.pdf

452   Kwadha, C.A., Ong'amo, G.O., Ndegwa, P.N., Raina, S.K. and Fombong, A.T. (2017) loc.cit.

453   Charrière, J.-D. and Imdorf, A. (1999). Protection of honey combs from wax moth damage. *American Bee Journal* **139**:627-630. https://www.researchgate.net/publication/294472170_Protection_of_honey_combs_from_wax_moth_damage

454   *Varroa* species are natural parasites of the Asian honey bee, *Apis cerana* and its close relatives *A. nululensis* (or *Apis cerana nuluensis*), *Apis nigrocincta* and *Apis koschevnikovi*.

Anderson, D.L. and Trueman, J.W.H. (2000). *Varroa jacobsoni* (Acari: Varroidae) is more than one species. *Experimental and Applied Acarology* **24(3)**:165-189. doi.org/10.1023/A:1006456720416

The key viruses associated with Varroa destructor are DWV, KBV, ABPV, SBPV and IAPV. The paralysis viruses are more virulent than more common DWV some forms of which appear to be avirulent.

Martin, S.J. (2001). The role of *Varroa* and viral pathogens in the collapse of honeybee colonies: a modelling approach. *Journal of Applied Ecology* **38**:1082-1093. doi.org/10.1046/j.1365-2664.2001.00662.x http://onlinelibrary.wiley.com/doi/10.1046/j.1365-2664.2001.00662.x/full

455 Fourteen RNA bee viruses are recorded by Bailey and Ball (they review a menagerie of other bee diseases) although an assessment by Muñoz and coworkers indicates there may be twenty four species.

Bailey, L. and Ball, B.V. (1991). *Honey Bee Pathology* 2nd edn. Academic Press, London, UK. https://www.sciencedirect.com/book/9780120734818/honey-bee-pathology

Muñoz, I., Cepero, A., Pinto, M.A., Martín-Hernández, R., Higes, M. and De la Rúa, P. (2014). Presence of *Nosema ceranae* associated with honeybee queen introductions. Infection, *Genetics and Evolution* **23**:161-168. doi:10.1016/j.meegid.2014.02.008 https://coloss.org/articles/1/1

456 DWV type A pathogenic and B are non pathogenic variants. Mordecai et al note that their discovery of a new master variant of DWV has important implications for the positive identification of the true pathogen within global honey bee populations.

Mordecai, G.J., Wilfert, L., Martin, S.J., Jones, I.M. and Schroeder, D.C. (2015). Diversity in a honey bee pathogen: first report of a third master variant of the *Deformed Wing Virus* quasispecies. *International Society for Microbial Ecology Journal* **10(5)**:1264-1273. doi.org/10.1038/ismej.2015.178

Deformed Wing Virus also has very closely related viruses: *Kakugo Virus*, *Varroa destructor Virus*, and asymptomatic Egypt Bee Virus variously affecting bee temperament and variously associated with and without *Varroa destructor*.

Zioni, N., Soroker, V. and Chejanovsky, N. (2011). Replication of *Varroa destructor Virus* 1 (VDV-1) and a *Varroa destructor virus* 1-deformed wing virus recombinant (VDV-1-DWV) in the head of the honey bee. *Virology* **417(1)**:106-112. doi:10.1016/j.virol.2011.05.009 https://www.ncbi.nlm.nih.gov/pubmed/21652054

457 There are two distinct strains of SBPV.
de Miranda, J.R., Dainat, B., Locke, B., Cordoni, G., Berthoud, H., Gauthier, L., Neumann, P., Budge, G.E., Ball, B.V. and Stoltz, D.B. (2010). Genetic characterization of slow bee paralysis virus of the honeybee (*Apis mellifera* L.). *Journal of General Virology* **91**:2524-2530. doi:10.1099/vir.0.022434-0 www.nationalbeeunit.com/downloadDocument.cfm?id=954
Slow Bee Paralysis Virus is closely related to *Moku Virus*.
Mordecai, G.J., Brettell, L.E., Pachori, P., Villaobos, E.M., Martin, Jones, I.M. and Schroeder, D.C. (2016). *Moku Virus;* a new Iflavirus found in wasps, honey bees and *Varroa*. *Scientific Reports* **6**:34983. doi.org/10.1038/srep34983 http://www.nature.com/articles/srep34983

458 There are a novel range of bee RNA viruses including ARV-1 and ARV-2. Remnant, E.J., Shi, M., Buchmann, G., Blacquière, T., Holmes, E.C. Madeleine Beekman, M. and Ashe, A. (2017). A diverse range of novel RNA viruses in geographically distinct honey bee populations. *Journal of Virology* **91(16)**:e00158-17.
https://www.ncbi.nlm.nih.gov/pmc/articles/PMC5533899/

459 Many viruses are vectored by *Varroa*.
Levin, S., Galbraith, D., Sela, N., Erez, T., Grozinger, C.M. and Chejanovsky, N. Presence of Apis Rhabdovirus-1 in populations of pollinators and their aprasites from two continents. *Frontiers in Microbiology* **8**:2482.
https://www.ncbi.nlm.nih.gov/pmc/articles/PMC5732965/

460 Heard, T. (2016) loc. cit. pp.205-206.

461 *E.W. Alexander – The North American inventor of the two-queen hive* is unpublished but complements an extensive outine of Alexanders contribution to two-queen beekeeping made in author's 2022 *A history of keeping and managing doubled and two-queen hives.* loc. cit.

462 Root, H.H. (ed) (1910). *Alexander's writings on practical bee culture,* third edition, 124pp. A.I. Root Company, Medina, Ohio. https://archive.org/details/cu31924003065244
The second edition can be found at
https://www.biodiversitylibrary.org/item/115381#page/7/mode/1up

463 Alexander, E.W. (April 1, 1907). Laying queens: Is it practical to have two or more in one colony during the summer season? *Gleanings in Bee Culture* **35(17)**:473-474. https://babel.hathitrust.org/cgi/pt?id=umn.319

51d00953180r&view=1up&seq=269
Alexander, E.W. (May 15, 1907). Queen-rearing: Some questions answered concerning the age of drones, the two-queen system and other matters. *Gleanings in Bee Culture* **35(10)**:694. https://babel.hathitrust.org/cgi/pt?id=umn.31951d00953180r&view=1up&seq=398
Alexander, E.W. (September 1, 1907) loc. cit.
Alexander, E.W. (December 1, 1907). The plural queen system: All queens but one disappear at the end of the season. *Gleanings in Bee Culture* **35(23)**:1496. https://babel.hathitrust.org/cgi/pt?id=hvd.32044106180268&view=1up&seq=566

464 Wade, A. and Derwent, F. (2020). *Bee Buzz Box* – April 2020. Early forays into two-queen hive management Part I – A plurality of queens. Canberra Region Beekeepers.
Robinson, T.P. (December 15, 1907). Plurality of queens in one hive: What are the advantages. *Gleanings in Bee Culture* **35(24)**:1598. https://www.biodiversitylibrary.org/item/74679#page/1614/mode/1up

465 Hopkins, I. (1907). Bee-Culture (2nd Ed.), Bulletin No. 5. New Zealand Department of Agriculture, 37pp. John Mackay, Government Printer, Wellington. https://en.wikisource.org/wiki/Bee-Culture_(2nd_ed.)
Protherue, J. (1921). Fitting the Alexander feeder. *American Bee Journal* **61(4)**:144. https://ia802702.us.archive.org/19/items/americanbeejourn6061hami/americanbeejourn6061hami.pdf

466 Morse, R.A. (November 1967). A short history of the Empire State Honey Producers' Association, 32pp. Cornell University Ithaca, NY. https://static1.squarespace.com/static/54d903ace4b0979f4046c07c/t/54ef6668e4b00716705cebc2/1424975464417/history+of+the+empire+state+honey+producers.pdf

467 Egbert W. Alexander. Findagrave.com https://www.findagrave.com/memorial/109883800/egbert-w_-alexander

468 *George Wells – The British inventor of the doubled hive* was presented by Alan Wade to the Canberra Region Beekeepers *Bee Buzz Box* in May 2021 as *The Discovery of the doubled hive*.

469 Wade, A. (2021) loc. cit.

470 Moeller, F.E. (April 1976) loc. cit.
Hogg, J.A. (1981). The consolidated two-queen brood nest and queen behavior: Queens co-exist in contact through an excluder. *American Bee*

Journal **121(1)**:36-42. http://www.twilightmd.com/Samples/Hogg/Hogg_Halfcomb___Publications/ABJ_1981_January.pdf

471 British Bee-Keepers' Association Quarterly Converzatione (April 7, 1892). Report of meeting of 31 March 1892. *British Bee Journal, Bee-Keepers' Record and Adviser* **20(511)**:132-133. https://www.biodiversitylibrary.org/item/83768#page/146/mode/1up

472 Eds The Bee-Keepers' Record and Adviser (April 1892). Working two queens in each hive. *The Bee-Keepers' Record and Adviser* **3(27)**:40. https://babel.hathitrust.org/cgi/pt?id=coo.31924065068953&view=1up&seq=56 https://books.google.com.au/books?id=tDwoAQAAIAAJ&printsec=frontcover&source=gbs_ge_summary_r&cad=0#v=onepage&q&f=false
Eds The Bee-Keepers' Record and Adviser (May 1892). Working two queens in each hive. *The Bee-Keepers' Record and Adviser* **3(28)**:55-56. https://babel.hathitrust.org/cgi/pt?id=coo.31924065068953&view=1up&seq=71 https://babel.hathitrust.org/cgi/pt?id=uc1.b3339275&view=1up&seq=31&q1=wells https://books.google.com.au/books?id=mbVJAAAAYAAJ&printsec=frontcover&source=gbs_ge_summary_r&cad=0#v=onepage&q&f=false.
Eds The Bee-Keepers' Record and Adviser (February 1893). Two queens in each hive: The Wells system. *The Bee-Keepers' Record and Adviser* **4(37)**:9-10. https://babel.hathitrust.org/cgi/pt?id=coo.31924065068870&view=1up&seq=27

473 Mr Wells on the two-queen system (February 1893). *The Bee-Keepers's Record and Adviser* **4(37)**:18-19. https://babel.hathitrust.org/cgi/pt?id=coo.31924065068870&view=1up&seq=36 https://books.google.com.au/books?id=vLVJAAAAYAAJ&printsec=frontcover&source=gbs_ge_summary_r&cad=0#v=onepage&q=Wells&f=false

474 Eds Bee-Keepers' Record and Adviser (April 1893). Is the Wells or double-queen system to be a success? *The Bee-Keepers' Record and Adviser* **4(39)**:41-42. https://babel.hathitrust.org/cgi/pt?id=coo.31924065068870&view=1up&seq=59 https://books.google.com.au/books?id=vLVJAAAAYAAJ&printsec=frontcover&source=gbs_ge_summary_r&cad=0#v=onepage&q&f=false

475 J.N.A., Lancashire (1896). Foreign bees: Keeping pure races of Italians apart. *The Bee-Keepers' Record* **14(72)**:11. https://babel.hathitrust.org/cgi/pt?id=coo.31924065068862&view=1up&seq=21

https://www.google.com.au/books/edition/Bee_keepers_Record/-LJJAAAAYAAJ?hl=en&gbpv=1&dq=Bee-keepers%27+Record+1897&pg=RA2-PA172&printsec=frontcover

476 E.R. (1897). Bees in trees. *The Bee-Keepers' Record* **15(89)**:86. https://babel.hathitrust.org/cgi/pt?id=coo.31924065068862&view=1up&seq=310
Pugh, H. (1897). Stocking a Wells hive. *The Bee-Keepers' Record* **15(88)**:73-74. https://babel.hathitrust.org/cgi/pt?id=coo.31924065068862&view=1up&seq=297 https://books.google.com.au/books?id=-LJJAAAAYAAJ&printsec=frontcover&source=gbs_ge_summary_r&cad=0#v=onepage&q=bees%20in%20trees&f=false

477 Austin, L. (1974). Autobiography of Leonard (Len) Austin. Unpublished, courtesy of Sarah Austin, great great granddaughter of George Wells (received 15 July 2022).

478 British Bee-Keepers' Association Quarterly Conversazione (October 18, 1894). Editorial, Notices, &c. British Bee Journal, *Bee-Keepers' Record and Adviser* **22(643)**:411-413. https://www.biodiversitylibrary.org/item/83770#page/421/mode/1up

479 Editors British Bee Journal (April 20, 1893). The double-queen system: A visit to Mr Wells' apiary. *British Bee Journal, Bee-Keepers' Record and Adviser* **21(565)**:151-153. https://www.biodiversitylibrary.org/item/83769#page/161/mode/1up

480 Mr Wells's apiary (July 6, 1899). Homes of the honey bee: The apiaries of our readers. *British Bee Journal, Bee-Keepers' Record and Adviser* **27(889)**:262-263. https://www.biodiversitylibrary.org/item/83775#page/274/mode/1up
Mr Wells's apiary (October, 1899). Homes of the honey bee: The apiaries of our readers. *The Bee-Keepers' Record* **17**:252-254.

481 Woodley, W. (March 23, 1893). Notes by the Way [Letter 1376]. *British Bee Journal, Bee-Keepers' Record and Adviser* **21(561)**:115-116. https://www.biodiversitylibrary.org/item/83769#page/125/mode/1up

482 Humble 833 (May 2019). William Woodley – An introduction. https://www.beehiveyourself.co.uk/william-woodley-beekeeper-1845-1923-from-beedon-berks-england/
Homes of the honey bee. The apiaries of our readers – No. IV. *The Bee-Keepers' Record* **15(85)**:22-24. https://babel.hathitrust.org/cgi/pt?id=

coo.31924065068862&view=1up&seq=245&size=125 https://books.google.com.au/books?id=-LJJAAAAYAAJ&printsec=frontcover&source=gbs_ge_summary_r&cad=0#v=onepage&q=wells&f=false

Heap, C.H. (February 1911). William Woodley at home. *The Bee-Keepers' Record* **29**:23-24. https://books.google.com.au/books/about/Bee_keepers_Record.html?id=obNJAAAAYAAJ&redir_esc=y

483 Herrod-Hempsall, W. (1930/1936). Bee-keeping new and old described with pen and camera, vol. 1 (1930). British Bee Journal, London. in Cushman, D. (accessed 22 November 2022). Movable frame hives: Re-inventing the wheel, p.6. http://www.dave-cushman.net/bee/moveableframehives.pdf

484 Wade, A. (2020). *Bee Buzz Box* February 2020. Doubling hives – Part II The Wells System. https://actbeekeepers.asn.au/bee-buzz-box-february-2020-doubling-hives-part-ii-the-wells-system/

485 Wells. G. (1894). *Guide Book pamphlet on the two-queen system of bee keeping.* G.F. Gay. Snodland Steam Printing Works, Malling Road, 15pp.

486 Doolittle, G.M. (1889) loc. cit.
Doolittle, G.M. (1891). Contraction and comb honey: The right and wrong kind again, as discussed by Doolittle. *Gleanings in Bee Culture* **19(7)**:269. https://www.biodiversitylibrary.org/item/69193#page/223/mode/1up

487 Dayton, C.W. (1891). Contradiction: The right and the wrong kind. *Gleanings in Bee Culture* **19(5)**:167-168. https://archive.org/details/sim_gleanings-in-bee-culture_1891-03-01_19_5
Dayton, C.W. (1891). Superseding the old queen: Having queens fertilized in full colonies containing a laying queen. *Gleanings in Bee Culture* **19(20)**:815-816. https://archive.org/details/sim_gleanings-in-bee-culture_1891-10-15_19_20

488 Tinker, G.L. (1892). Ramblers hive-hobby riding. *Gleanings in Bee Culture* **20(4)**:119. Use of excluder for section comb production. https://archive.org/details/sim_gleanings-in-bee-culture_1892-02-15_20_4
J.G.K. (1892). Queens fertilised in full colonies with a laying queen. *The Bee-Keepers' Record and Adviser* **3(25)**:18-19. https://babel.hathitrust.org/cgi/pt?id=coo.31924065068953&view=1up&seq=30 https://books.google.com.au/books?id=mbVJAAAAYAAJ&printsec=frontcover&source=gbs_ge_

summary_r&cad=0#v=onepage&q=J.G.K.&f=false
Tinker, G.L. (1892). Queens fertilised in full colonies with a laying queen. *The Bee-Keepers' Record and Adviser* **3(27):**45-46. https://babel.hathitrust.org/cgi/pt?id=coo.31924065068953&view=1up&seq=61 https://books.google.com.au/books?id=mbVJAAAAYAAJ&printsec=frontcover&source=gbs_ge_summary_r&cad=0#v=onepage&q=J.G.K.&f=false
Taythhall, A.B. (1892). Queens fertilised in full colonies with a laying queen. *The Bee-Keepers' Record and Adviser* **3(27):**51-52. https://babel.hathitrust.org/cgi/pt?id=coo.31924065068953&view=1up&seq=68&size=125 https://books.google.com.au/books?id=mbVJAAAAYAAJ&printsec=frontcover&source=gbs_ge_summary_r&cad=0#v=onepage&q=J.G.K.&f=false

489 Miller, C.C. (1911). Rearing queens in hive with laying queen, pp.310-312 in Miller, C.C. (1911) loc. cit.

490 Miller, C.C. (1892). Two queens to one colony. *Gleanings in Bee Culture* **20(11):**416. https://archive.org/details/sim_gleanings-in-bee-culture_1892-06-01_20_11

491 *Stringy Hughston's coffin hives* was published as Wade, A. and Cannon, D. (2021). Stringy Hughston's tripled hives. *The Australasian Beekeeper* **123(6):**33-35.

492 Gouget, C.W. (July 1953) loc. cit.

493 Wells. G. (1894) loc cit.

494 Eva Crane Gallery (1967). Sid Murdoch's super coffin hives (images ECT-1558 - ECT-1560) at Manjimup and Ken Gray's coffin hives (ECT-1561 - ECT-1572) in Wandoo woodland. https://www.evacranetrust.org/gallery/australia

495 Crane, E. (1980) loc. cit.

496 Holzberlein Jr., J.W. (1955). Some whys and hows of two-queen management. *Gleanings in Bee Culture* **83(6):**344-347. https://archive.org/details/sim_gleanings-in-bee-culture_1955-06_83_6/page/344/mode/1up

497 Wade, A. (2021). *A history of keeping and managing doubled and two-queen hives.* loc. cit.

498 *Don Peer – The wisdom of lost beekeeping practice* was presented by Alan Wade to the Canberra Region Beekeepers Bee Buzz Box in June 2022.

499 Miksha, R. (August 2018). How to predict the honey flow. *Bad Beekeeping Blog*. https://badbeekeepingblog.com/2018/08/02/how-to-predict-the-honey-flow/

500 Crane, E. (2003). *Making a Beeline: My journeys in sixty countries 1949-2000.* International Bee Research Association.

501 Miksha, R. (June 2019). Remembering Eva Crane: Beekeeper and physicist. *Bad Beekeeping Blog*. https://badbeekeepingblog.com/2019/06/12/remembering-eva-crane-beekeeper-and-physicist/

502 Miksha, R. (August 2001). Honey combing – Selling your gold. *Bad Beekeeping Blog*. http://www.badbeekeeping.com/chc_04.htm

503 Peer, D.F. (1955). The foraging range of the honey bee. Part 1 PhD. Thesis, University of Wisconsin, Madison, WI. Part I: The foraging range of the honeybee; Part II: The mating-range of the honeybee, 102pp.
Peer, D.F. and Farrar, C.L. (1956). The mating range of the honey bee. *Journal of Economic Entomology* **49(2)**:254-256. doi.org/10.1093/jee/49.2.254 https://academic.oup.com/jee/article-abstract/49/2/254/2205999?login=false
Peer, D. F. (1956). Multiple mating of queen honey bees. *Journal of Economic Entomology* **49(6)**:741-743. doi:10.1093/jee/49.6.741 https://academic.oup.com/jee/article-abstract/49/6/741/2206208?login=false
Peer, D.F. (1957). Further studies on the mating range of the honey bee, *Apis mellifera* L. *The Canadian Entomologist* **89(3)**:108-110. https://www.cambridge.org/core/journals/canadian-entomologist/article/abs/further-studies-on-the-mating-range-of-the-honey-bee-apis-mellifera-l/8B126801D410AEFDBE08C4A457B729AA

504 Tarpy, D.R., Delaney, D.A., Seeley, T.D. and Nieh, J.C. (2015). Mating frequencies of honey bee queens (*Apis mellifera* L.) in a population of feral colonies in the northeastern United States. *PLOS One* **10(3)**:e0118734. doi:10.1371/journal.pone.0118734
Tarpy, D.R., Nielsen, R., and Nielsen, D.I. (2004). A scientific note on the revised estimates of effective paternity frequency in *Apis*. *Insectes Sociaux* **51(2)**:203-204. doi.org/10.1007/s00040-004-0734-4

505 Peer, D.F. and Farrar, C.L. (1956) loc. cit.
Peer, D. F. (1956) loc. cit.
Peer, D.F. (1957) loc. cit.

506 Root, H.H. (ed) (1910) loc. cit.

507 Mikska, R. (June 2013). Fire in the shop. *Bad Beekeeping Blog*. https://badbeekeepingblog.com/2013/06/03/fire-in-the-shop/

508 Mikska, R. (August 2013). Time to say goodbye – or maybe not. *Bad Beekeeping Blog*. https://badbeekeepingblog.com/2013/08/09/time-to-say/

509 Peer, D.F. (1978). A warm method of wintering honey bee colonies outdoors in cold regions. *Canadian Beekeeping* **7(3)**:33, 36. Abstract by Richard Jones (2005). Bibliography of Commonwealth Apiculture (Google Books). https://books.google.com.au/books?id=sOBiT384y3YC&pg=PA57&lpg=PA57&dq=Peer,+D.F.+(1978)

510 Dick, A. (February 2003). Honeybeeworld.com A Beekeeper's Diary. https://www.honeybeeworld.com/diary/2003/diary020103.htm

511 Crane, E. (1966). Canadian bee journey. *Bee World* **41**:55-65, 132-148. https://www.evacranetrust.org/uploads/document/a6e5fe111233c705b26c4a35d3f0022a54d15d9d.pdf

512 Peer, D.F. (1965). Two-queen management with package bees. *Bee-Lines* **21**:3-7. Cited by Crane (1966) loc. cit.
Peer, D.F. (1969). Two-queen management with package colonies (*Apis mellifera*). *American Bee Journal* **109(3)**:88-89. Cited by Valle, A.G.G., Guzmán-Novoa, E., Benítez, A.C. and Rubio, J.A.Z. (2004). The effect of using two bee (*Apis mellifera* L.) queens on colony population, honey production, and profitability in the Mexican high plateau. *Téc Pecu Méx* **42(3)**:361-377. http://cienciaspecuarias.inifap.gob.mx/index.php/Pecuarias/article/viewFile/1404/1399

513 Taber, S. (2002). Requeen your bees without first removing the old queen. *American Bee Journal* **142(4)**:275-276. https://www.researchgate.net/publication/297976630_Requeen_your_bees_without_first_removing_the_old_queen
Szabo, T.I. (1981). Requeening honeybee colonies with queen cells. *Journal of Apicultural Research* **21(4)**:208-211. doi:10.1080/00218839.1982.11100543

514 Peer, D.F. (1977) loc. cit.

515 Hogg, J.A. (May 1, 1983) loc. cit.
Hogg, J.A. (June 2, 1983) loc. cit.

516 Jay, S.C. (1981) loc. cit.

517 Skirkevicius, A. (1963). New method of requeening during the main honey flow. *Lzua Moksliniai Darbai* **3(19)**:25-41.
Skirkevicius, A. (1965). New method of requeening. *Lietuvois Žemdirbystės Mokslas Tyrimo Institutas Darbo* **9**:59-62.

518 Boch, R. and Avitabile, A. (1979) loc. cit.

519 Milbrath, M. (July 2020) loc. cit.
Milbrath, M. (September 2020) loc. cit.

520 Forster, I.W. (1972) loc. cit.
Szabo, T.I. (1982) loc. cit.

521 Reid, G.M. (1977). Letters to the editor. *Canadian Beekeeper* **6(12)**:155. Cited by Jay (1981) loc.cit.
Reid, G.M. (1979). Requeening honey bee colonies without dequeening using protected queen cells. *New Zealand Beekeeper* **40(3)**:15-17 and XXVI International Beekeeping Congress, Bucharest, pp.249-252.

522 Garez, I. (2022). Yves Garez Honey Inc. https://ca.linkedin.com/in/yves-garez-9b1b93107

523 Gruszka, J., Peer, D. and Tremblay, A. (1990). The impact of *Acarapis woodi* on the development of package bee colonies. *Beelines* **89**:9-24. https://www.cabdirect.org/cabdirect/abstract/19910230771
Peer, D., Gruszka, J. and Tremblay, A. (1987). Preliminary observations on the impact of *Acarapis woodi* on the development of package bee colonies. *Saskatchewan Beekeepers Association* **81**:1-8. https://scholar.google.com.au/scholar?hl=en&as_sdt=0%2C5&q=Peer%2C+D.%2C+Gruszka%2C+J.+and+Tremblay

524 Miksha, R. (2016). How to keep honey from granulating before extracting. *American Bee Journal* **156(8)**:899-901. https://bluetoad.com/publication/?i=318073&article_id=2528241&view=articleBrowser&ver=html5
Downloadable as pdf at https://www.researchgate.net/publication/320233431

525 Miksha, R. (2004). Honey Crop, p.70: *Bad Beekeeping*, 307pp. Trafford Publishing, Victoria, B.C., Canada.

526 Miksha, R. (27 April 2017). Billy Bee and Doyon – Canadian honey forever. https://badbeekeepingblog.com/2017/04/27/billy-bee-and-

doyon-canadian-honey-forever/

527 *Tom Theobald – Where beekeepers fear to tread* was presented by Alan Wade to the Canberra Region Beekeepers *Bee Buzz Box* in December 2022.

528 Gabel, R. (13 November 2021). Telling the bees: Beekeeper Tom Theobald's passing. *The Fence Post* https://www.thefencepost.com/news/telling-the-bees-beekeeper-tom-theobalds-passing/

529 Theobald, T. (31 July 2003) loc. cit.

530 Ott, J. and Flottum, K. (4 January 2021). Two queen honey production with Tom Theobald (S3, E32). Beekeeping Today Podcast: *Bee Culture.* https:// podcastaddict.com/episode/117841608

531 Gabel, R. (13 November 2021) loc. cit.

Milton Keynes UK
Ingram Content Group UK Ltd.
UKHW051912291024
450402UK00005B/49